D0498639

ALSO BY BARRY WERTH

*Banquet at Delmonico's*

*31 Days*

*The Scarlet Professor*

*Damages*

*The Billion-Dollar Molecule*

*The Architecture and Design of Man and Woman*
(coauthored with Alexander Tsiaras)

*From Conception to Birth*
(coauthored with Alexander Tsiaras)

Werth, Barry,
The antidote : inside
the world of new pharma
2014.
33305230458097
mi        03/20/14

# THE
# ANTIDOTE

## INSIDE THE WORLD OF NEW PHARMA

## BARRY WERTH

SIMON & SCHUSTER

NEW YORK   LONDON   TORONTO   SYDNEY   NEW DELHI

Simon & Schuster
1230 Avenue of the Americas
New York, NY 10020

Copyright © 2014 by Barry Werth

All rights reserved, including the right to reproduce this book or portions thereof in any form whatsoever. For information, address Simon & Schuster Subsidiary Rights Department, 1230 Avenue of the Americas, New York, NY 10020.

First Simon & Schuster hardcover edition February 2014

SIMON & SCHUSTER and colophon are registered trademarks of Simon & Schuster, Inc.

For information about special discounts for bulk purchases, please contact Simon & Schuster Special Sales at 1-866-506-1949 or business@simonandschuster.com.

The Simon & Schuster Speakers Bureau can bring authors to your live event. For more information or to book an event, contact the Simon & Schuster Speakers Bureau at 1-866-248-3049 or visit our website at www.simonspeakers.com.

Interior design by Kyoko Watanabe
Jacket design by Christopher Lin
Jacket photograph by Lauren Burke/Getty Images

Manufactured in the United States of America

10   9   8   7   6   5   4   3   2   1

Library of Congress Cataloging-in-Publication Data

Werth, Barry.
    The antidote : inside the world of new pharma / Barry Werth.  — First Simon & Schuster hardcover edition.
        p.   cm.   Includes index.
      1.  Vertex Pharmaceuticals Incorporated.    2.  Drug Industry—United States.
3.  Drug Industry—history—United States.    4.  History, 20th Century—United
States.    5.  History, 21st Century—United States.    6.  Technology,
Pharmaceutical—United States.    I.  Title.
    RA401.A3
    338.4'76151—dc23                          2013039646

ISBN 978-1-4516-5566-7
ISBN 978-1-4516-5569-8 (ebook)

*For my remarkable mother, Hilda Werth*

# CONTENTS

# CAST OF CHARACTERS

Job titles indicate significant posts at Vertex unless otherwise specified. Years represent total time with the company.

**John Alam:** former executive vice president for medicines development and chief medical officer (1997–2006).

**Richard Aldrich:** former senior vice president and chief business officer (1989–2000).

**Bob Beall:** CEO and chairman of the Cystic Fibrosis Foundation.

**Joshua Boger:** founder and former CEO and chairman; director (1989–).

**Ken Boger:** former general counsel (2001–2011); brother of the founder.

**John Condon:** senior vice president of pharmaceutical operations and manufacturing (2005–).

**Bo Cumbo:** former vice president of sales; leader of Incivek commercial team (2010–2012).

**Matthew Emmens:** former CEO, president, and chairman (2005–2012).

**Russ Fleischer:** senior clinical analyst at the Food and Drug Administration, division of antiviral products; chief examiner for Incivek.

**Bink Garrison:** former senior vice president and "catalyst"; drove Vertex's values and vision process (2004–2009).

**Trish Hurter:** senior vice president for pharmaceutical development (2004–).

**Keith Johnson:** cystic fibrosis patient; clinical study participant for Kalydeco.

**Robert Kauffman:** chief medical officer and senior vice president (1997–).

**Adam Koppel:** managing director at Brookside Capital; major Vertex investor.

**Ann Kwong:** former vice president, hepatitis C franchise (1997–2012).

**Jeffrey Leiden:** chairman, president, and current CEO (2012–).

**John McHutchison:** outside clinical investigator for Incivek; later, senior vice president for liver disease therapeutics at Gilead.

**Peter Mueller:** chief scientific officer and executive vice president for global R&D (2002–).

**Mark Murcko:** former chief technology officer and former chair of the scientific advisory board (1990–2011).

**Paul Negulescu:** vice president of research; San Diego site head (2001–).

**Eric Olson:** former vice president and cystic fibrosis program leader (2001–2013).

**Michael Partridge:** vice president of investor relations (1997–).

**Geoffrey Porges:** senior analyst for global biotechnology at AllianceBernstein.

**Amit Sachdev:** senior vice president for global government strategy, market access, and value (2007–).

**Charles Sanders:** former chairman (1996–2010).

**Vicki Sato:** former president (1992–2005).

**Ian Smith:** executive vice president and chief financial officer (2001–).

**John Thomson:** vice president of strategic R&D networks (1989–).

**Roger Tung:** former vice president of drug discovery (1989–2004).

**Fred Van Goor:** head of cystic fibrosis biology, cystic fibrosis research program (2001–).

**Jack Weet:** former vice president of regulatory affairs (2009–2011).

**Nancy Wysenski:** former chief commercial officer (2009–2012).

# THE
# ANTIDOTE

Joshua Boger

Courtesy of Vertex Pharmaceuticals, circa 1994

# INTRODUCTION

## Why I Went Back Inside Vertex

Twenty years ago, I wrote a book about a bold and bruising quest. It told the story of a group of entrepreneurial young scientists who left the world's best drug company—the most admired business in America year after year—because they were confident they would be more productive on their own, starting from scratch. They aimed to design better drugs, atom by atom. Most people across the industry thought their project in a refitted construction company garage in Cambridge, Massachusetts— to build an organization that could produce dramatically improved medicines to transform the lives of people with serious diseases—was a pipe dream, a money pit, a consuming act of arrogance, an exhausting feat of hubris, a fool's errand.

"Don't you think this is five years too early?" founding scientist and president Joshua Boger was often asked. "Yes," he would say, "but five years from now it'll be five years too late."

I found their passionate belief in science and in themselves, brimming with high purpose and combative glee, stirring and infectious as I followed them around for a couple of years while they tried to get their cash-starved company, Vertex Pharmaceuticals, off the ground. It was a rocky, exhilarating, eye-opening ride. The chase for new leads was fierce, not just against "Mother Merck" but also top academic labs, including those led by some of their own scientific advisors, who they feared were sharing Vertex's most prized insights with its rivals. When Boger settled

1

for a tie in a race to publication against one of them, a Harvard professor, he told me: "I'll take it. But I want to rub his nose in the dirt and step on his head."

Such was the knife-edge between cooperation and competition in the new biopharmaceutical order. Whatever unease I felt at witnessing up close how ferocious capitalism and scientific rivalries—rather than, say, altruism—drove the search for new lifesaving drugs receded in the wake of Vertex's precocious early success. Boger assembled a team of talented, rampantly motivated biologists, chemists, biophysicists, and computer scientists while he and his chief lieutenant tap-danced their way around the world to raise the money they would need to compete with the pharmaceutical behemoths. Though they were spectacularly outspent and outmanned in every area, he let them organize themselves, rather than try to direct them from above. He let them fail, time and again, until they came up with better approaches. He was a visionary goal setter, an inspirer.

Against all odds, within four years Vertex proved it could compete at the forefront of drug research, against the industry leaders, in several major areas at once. It had gone public and Wall Street considered it a hot stock. What I saw impressed me as a worthy, honest, compelling, even noble effort both to beat and influence the world around it—a world where life-changing new drugs were getting harder and harder to find despite the best efforts of hundreds of companies employing tens of thousands of equally gifted and passionate researchers and spending hundreds of billions of dollars on research and development.

That was the story I told in *The Billion-Dollar Molecule*. I was encouraged by the company's progress; pleased, too, that the book was acclaimed as an insightful look inside the world of commercial medicine. But I understood that the upstart-biotech-looks-promising version of events that I had reported wasn't the full story, or even the main one. Boger had set out to build a drug company, but Vertex hadn't yet produced a drug. Nowhere near it. For him and the other company pioneers, the larger prize wasn't organizing a research group to find better compounds; it was to build a business that could go head-to-head with the world's most profitable drugmakers against the hardest diseases, involving some of competitive capitalism's most complicated science and most cutthroat marketing maneuvers.

I'd described the opening skirmish, not the war.

❖

The modern pharmaceutical industry emerged from one of the great triumphs of twentieth-century science. Before the 1940s, there were medicines and companies that made them, but no one had invented a method for actively finding and developing new drugs. Profits in medicine were disdained as suspect—immoral—and the companies were essentially manufacturers of fine chemical compounds. Since their products could do as much harm as good, integrity was key. Then university laboratories advanced a new approach: microbial screening. Systematically harvesting large numbers of chemicals from "good bugs" and feeding them to "bad bugs," then monitoring and improving their activity, drugmakers produced and brought to patients the first antibacterials that had been actively sought and developed.

The chase was on: for new diseases to treat, testing strategies, business opportunities, scientists, alliances with leading doctors, prestige, and money. As with all things in America, World War II was the great catalyst. Just as the companies were flexing their research and development arms to tackle other diseases, the government enlisted them in the war effort. In 1941 the Germans were rumored to have isolated the chemical secretion of the adrenal cortex, cortisone, and given it to their pilots, amping them up, emboldening them. Battlefield wounds and home-front contagions drove the need for better antibiotics, vaccines, pain relievers, and surgical products. Drugmakers were marshaled to counter the threat of a pharmacologic arms race. By midcentury, US companies had more than matched the government's urgency, and were racing ahead, developing new biological models to screen against. Profits began to pour in. Wall Street stood up and took notice. The companies grew spectacularly.

Merck, where Boger started his career in 1979 after getting a PhD in chemistry at Harvard and doing a postdoctoral stint with future Nobel laureate Jean-Marie Lehn, was their paragon. It best represented the qualities that the industry exalted, a patient-centered, high-science focus combined with unrivaled organizational commitment to R&D. It wasn't always the most profitable drug company—Pfizer and others were better at making money—but its research campuses in New Jersey and outside Philadelphia attracted the most promising scientists. It was where you wanted to be, the top of the pyramid.

In the 1970s and 1980s, with the swift expansion of government-sponsored research spurred on by the "war on cancer," and as the universities and Wall Street simultaneously discovered a bonanza in the life sciences, there was an explosion in medical understanding, and the low-hanging fruit were quickly plucked. Merck's labs launched the first or second significant drugs for cholesterol, hypertension, osteoporosis, and asthma, as well as a class of pain medications known as COX-2 inhibitors. At Merck as elsewhere, scientists burned to do pathbreaking work on new medical frontiers, but increasingly, in management suites and boardrooms across the industry, the consequences of success yielded a conservative strategic consensus: move cautiously rather than struggle to produce breakthroughs; settle for modest "quick-to-market" improvements where treatments already exist, and where the resulting products can be aggressively marketed to doctors and people with chronic diseases.

Gradualism held zero appeal for Boger. "Now, I don't think there's anything wrong with bringing an incremental advance to the marketplace; you're not a bad person," he says. "It's just I don't want to do that; life's too short." Biotechnology companies by now had joined the competition. A few top university professors or government scientists with a tantalizing idea could raise tens of millions of dollars, go out and test it, then go public—*public*—when all they had to sell to investors was a theory and the only certainty in their business model was years and years of progressively more unprofitable darkness. Wall Street blew hot and cold, periodically falling hard for their stories of genetic breakthroughs and miracle cures before returning to its senses. Merck, recognizing Boger's talents (if not buying into his ideas about building better drugs by applying advances from the biotech, software, and computer graphics industries), encouraged him to do his experiment, letting him piece together a team in immunology. But he quickly felt thwarted, impatient. Pent-up.

His frustrations crystallized in the late 1980s, as many things did across the medical world, with the AIDS crisis. Drugmakers at first ignored the epidemic, seeing a small market. Off-the-shelf compounds were ineffective and toxic. When Merck entered the arena, many doctors, public health officials, and even some activists felt that the cavalry had arrived. Boger's closest scientific friend in the company, a brilliant and brash young biologist named Irving Sigal, led Merck's project, and Boger cleared the decks in his group to help. CEO Roy Vagelos announced he was "damn optimistic" about Merck's chances. In late 1988, returning from a meeting in Europe, Sigel was killed when Pan Am Flight 103, carrying 259 people and a terrorist organization's bomb in a cargo container, exploded in a fireball over Lockerbie, Scotland. He was thirty-five. Merck scrambled to recover from its loss.

Within a month, Boger was gone.

❖

So I was there when Vertex set out in its garage to overtake the "bigs." And what I saw were staggering contrasts. The major pharmaceutical companies were lumbering along; mightily equipped, cash-rich, charging higher and higher prices while bringing out fewer and fewer important new drugs, their reputation for putting profit before patients replayed and reinforced in the AIDS epidemic. It was fifteen years into the war on cancer, and cancer was winning in a rout. The biotechs had yet to pay out, and Wall Street was skittish about their high failure rate and the chronic risk and volatility of an industry where horizons were measured in decades. It was into this environment that Boger led his young company.

Now leap ahead to early 2011: the grinding recovery from the worst financial crisis in eighty years, the raging political storm over Obamacare, a drumbeat of lurid press reports about the drug business, revealing an industry in crisis and under siege. Vertex, after twenty-two years and $3.6 billion in losses, was about to launch its first drug under its own name, a major breakthrough against the leading cause of advanced liver disease. It had a second drug nearing regulatory approval that promised to revolutionize the treatment of the most common inherited fatal disease in the United States and Europe. Just as the world around it was shuddering, Vertex was poised to soar. What better vantage point for witnessing the mounting collision of medicine, money, and society?

I went back inside Vertex to learn what it takes—to succeed in science and business, yes, but also in fleshing out and struggling to achieve a radical vision of a better future. Could a group of very bright, very determined people make a difference in a market dominated by profits and Wall Street? Could true believers in the idea that the purpose of pharmaceutical research is to put patients first and transform the lives of sick people compete in an industry where it was far preferable to develop, say, a marginally better *fifth* statin compound for high cholesterol and market the hell out of it, as Pfizer had done with the bestselling drug of all time, Lipitor? Or bury a $500 million sweetheart reimbursement in the Fiscal Cliff deal, as Amgen did with its army of seventy-seven lobbyists? Or pay a generic company $42 million not to market a cheaper version of your drug, so you can keep selling it at ten times the cost to consumers, as in a recent restraint-of-trade case before the Supreme Court? Could Vertex still be Vertex in our genomic age, when understanding which drugs to prescribe will depend on an ever-deepening biological profiling of individual patients?

What was I seeking? Hope, really. The $325 billion prescription drug business is America's most challenging and one of its most profitable. It's tougher and riskier at nearly every stage than any other business. Yawning biological uncertainties haunt every experiment; the failure rate even after a candidate clears all the myriad hurdles to reach human testing is 30 to 1; the cost of ramping up a successful product typically exceeds $1 billion. Drugmakers operate in the world's most regulated commercial environment, matched only by nuclear power. Small companies face an extra test. Dependent on Wall Street for financing, they must navigate a myopic trading culture that disdains and crowds out long-term thinking and investment. All progress in the pharmaceutical business is backbreaking, freighted with unknowns, takes twice as long as you think it will, and is liable to "blow at any seam," as Tom Wolfe wrote about the endless ineffable peril of staking it all on a lofty high-risk mission.

Mostly I wanted to see what it had taken to prevail against such harrowing obstacles: What had Boger's vision become, and did it represent a true way ahead in our boundlessly promising and still barely comprehended new biological epoch? After he'd resigned from Merck—alone, without first taking anyone with him, and without any assurance that anyone would follow—Boger went home and sketched his goals on a whiteboard: *"Make better drugs, faster. Create the 21st century biopharmaceutical company. Become Merck, only better."* It was almost a haiku. He thought it would take him twenty years and $1 billion. Now, just two years late, at nearly four times the cost, Vertex verged on proving all that he had set out to do.

"One of the most common questions I've gotten lately is, 'Gee, did you ever imagine this would occur?' " he told Vertex's sales troops that spring. It was a few days after Osama bin Laden was killed in Pakistan after a decade-long manhunt, and the company was counting down to launch, primed to go one-on-one against Merck for the richest commercial opportunity in pharma. "My completely unsatisfying answer is, 'Yes, absolutely.' Now, that comes across to some people as fairly arrogant, and to that I say arrogance is only a problem if it doesn't turn out to be true. If it turns out to be true, it's just persistence."

I'd discovered at Vertex that biomedical research emits a high emotional heat. It may be tempting to think that the competitive commitment in other disruptive tech industries is similar but the comparison is slender. In Silicon Valley, you're trying to make a better product, not cure cystic fibrosis or Parkinson's disease. Here the difference between success and failure can be the difference between life and death. Vertex was about to debut not only the first drugs discovered and developed by its own people and commercialized under its own label. It was about to debut itself, an organization of nearly two thousand people sculpted as much by the changing health care economy and the gyrations of its industry over the past two decades as by Boger and the others who joined him. They had had notable early success in the crucible of that new biomedical order—AIDS—but that victory had been pyrrhic because, while it had produced a drug, Vertex hadn't fully emerged along with it. Now the company would correct that disparity.

Boger was right about arrogance. We may not like it in our faces, but it's a problem only when it doesn't turn out to be true.

# PART 1

# Feeding the Beast

# CHAPTER 1

---

## APRIL 28, 1993

Boston's World Trade Center, unlike New York's, was not a beacon of wealth and power but a refitted waterfront mercantile mart. It squatted on a pier above the city's famously ruined harbor, offering a postcard view of downtown. Boger, age forty-two, stepped onto the podium before 250 worried executives at the annual meeting of the Massachusetts Biotechnology Council, a statewide trade group. In four years, his "highly unlikely start-up" had grown to 110 people sprawling among a warren of reconverted lab buildings near MIT—a public company with $50 million in the bank—and its reputation as a leader in structure-based drug design was supported chiefly by a string of impressive publications. The company had competitive research projects in multiple disease areas. Even those who'd thought its long-term goals delusional had to concede its short-term progress.

Across the drug industry, the mood was black. In February, a month after Bill Clinton took office, he announced that his wife, Hillary, would oversee national health care reform, and at an appearance together at a Virginia medical clinic the president denounced "shockingly" high drug prices. He cited figures showing the industry spent more on marketing than on research, demanded that drugmakers change their ways, and suggested that he would propose price controls if they didn't. As one analyst put it, Clinton had decided to "go to war" with the pharmaceutical companies—rich targets in the struggle against spiraling health costs. A few days later a congressional study criticized drugmakers' "excess profits." In the two months since, all drug stocks, big and small, had lost one-third of their value.

"I expect to see frogs raining down from the sky sometime soon," Boger said. He was six foot five, rangy, all angles, with a broad forehead capped by thinning stick-straight brown hair. Tinted aviator glasses did nothing to dim an enveloping gaze and generally ecstatic grin. His cadences echoed a small-town North Carolina boyhood where from a young age he found himself explaining things to his unaccountably obtuse elders. Invited to discuss the industry perspective, Boger (pronounced with a hard *g*) forecast a reckoning. Only the smartest, fleetest, most adaptable companies—companies such as Vertex that were burning tens, hundreds of millions of dollars chasing a broad portfolio of breakthrough drugs for serious diseases—would compete successfully in the new period, he said. Why? "Because we're more motivated, and the fear of death and God is closer to us." Chortling, Boger concluded with a bit of light heresy, a slide of a *New Yorker* cartoon showing an executive telling a lab-coated scientist sketching at a whiteboard, "I like it. Find a disease for it."

Strictly speaking, Vertex wasn't a biotechnology company: it wasn't trying to engineer new products by manipulating genes, so-called biologics. It aspired to make small-molecule drugs—"little pills in bottles with cotton on top," as Boger put it—the time-honored staple of traditional pharmaceutical discovery and development. But with the prospect of launching its own products still many years off, and by no means certain, and with it burning prodigiously through capital, Vertex was a biotech by default. Boger often joked that he didn't know it was a biotech company until Kidder Peabody, the investment bank that took it public, told him it was.

Despite the dire mood, he had earned his cheerful attitude, mostly through backbreaking pursuit, although luck and charisma helped. He had proved he could sell Vertex's story no matter how distressed the market was. Two weeks earlier, Vertex had announced a $20 million deal with Kissei Pharmaceutical Co., the fastest-growing Japanese drug company, to develop AIDS medicines in the Far East. Entrepreneurs forever hustle for capital, but Kissei, an improbable suitor, had initiated the contact months earlier, coming to Boger as if off a cloud. He and Richard Aldrich, Vertex's chief business officer, had been shopping other projects on one of their twice-yearly "death marches" through dozens of executive suites in east Asia in late January. The Japanese had yet even to acknowledge that AIDS existed in their country, so severe was the stigma. Yet Vertex's business liaison urged them to visit Kissei in Matsumoto City, on a misty high-mountain plateau six hours from Tokyo. "Long ride. Switching trains," Aldrich recalls. "We didn't know them. We were resistant. But he said, 'No, they really want to meet you, this could be worthwhile.' I remember we were in the car driving between labs. One of the Kissei guys turned around—Josh and I were in the backseat—and he says to Josh, 'Dr. Boger, at Kissei you are a god.' And, of course, Josh immediately liked Kissei. Within forty-five minutes of arriving there, we went into a side room with the senior guys and started talking money on the deal."

Such good fortune and adulation might have gone to Boger's head if not for the unending pressure, in the face of minimal resources, to get as many projects off the ground and find partners for them as soon as possible—an endeavor that, like drug discovery, resulted in failure more than 90 percent of the time. Right now AIDS and the virus that caused it, HIV, gave the company its best hope for getting a drug to doctors and patients.

Following Boger's speech to the executives, senior scientist Mark Murcko delivered a spirited recap of Vertex's design strategies for a new class of direct-acting antiviral drugs—HIV protease inhibitors—that the company was relying on to catapult it into the ranks of the world's elite drugmakers. Murcko, thirty-three, was the last of a half dozen major Merck defectors to join Vertex and the one Boger had fought hardest to get, platooning the scientists to call him every other night for three months until he agreed to move. More and more, he'd become Boger's technological seer, sidekick, everyman, and commentator: a stocky, precocious, deep-voiced, aside-tossing Sancho Panza. At first Boger refused to go into AIDS, but Murcko and a few others changed his mind.

Murcko peppered his slide presentation of elegant computer-generated three-dimensional molecular models with striking new data and throwaway gibes at traditional discovery methods. He told the audience about a drug hunt of rare effectiveness and speed. Proteases are enzymes that cleave other sections of protein: in the case of HIV, enabling it to multiply. Theoretically, if a small molecule could latch onto the active site, it would block replication. HIV protease offered a promising drug target and Merck and several other companies had compounds in the clinic, although none had yet shown that their molecules were effective in patients. More to the point, the target site was only Vertex's second project; the first had crashed spectacularly when the mechanism for which it was churning out compounds proved to be biologically irrelevant. Merck was spending more on HIV than on any other research effort in its history, throwing hundreds of chemists at the project. Vertex, Murcko said, had just five.

Vertex needed to speed up the process of getting its own compounds to patients, something it had never done before and that would not be possible at this stage without another partner. What the company had was Boger, hubris, and a passionate group of mostly ex-Merck researchers as determined as he was to "do it right"—"Merck's first team making drugs for the hardest diseases with computers," as Aldrich liked to tell potential partners and investors. Structure-based drug design is premised on having a reliable 3-D computer model of the target molecule in the body, so as to be able to visualize the inmost workings of living matter. Vertex had industrialized its processes for isolating and purifying proteins to the point where it could reliably solve complex structures showing how different prospective drug molecules bound to the business end of a target. Murcko's group digitally modeled new atomic interactions and suggested improvements based on predicted changes in activity. This "feedback loop," he told the audience, enabled Vertex's small group of chemists to make much more informed choices than Merck's legions about what molecules to make next.

Not everyone was sold. After lunch, the keynote speaker, Dr. Edward Scolnick, president of Merck laboratories, unfurled himself from a too-small chair on the dais and rose to deliver a talk entitled "Molecular Approaches to Drug Design." Murcko, sitting just to his right, smiled, amused. Scolnick was notably hard charging. He had driven the billion-dollar drug Mevacor, the first of the cholesterol-reducing statins, to market, devoting a significant portion of his toxicology budget to overcoming concerns about statin toxicity. Merck's findings convinced the

US Food and Drug Administration that what seemed to be cancerous changes in animals weren't really cancers and that a resumption of human testing of statins, after a three-year moratorium, was warranted. Now he was driving Merck's AIDS effort. Though the structure of HIV protease had initially been solved at Merck—by yet another of those who would soon grow frustrated and follow Boger to Vertex—Scolnick considered the information it generated just moderately useful for drug discovery.

"I gave you that title," Scolnick said, "but that's not what I'm going to talk about." He spoke instead about the industry's R&D mission and its impact on prices: "what we do compared with what we charge." Merck CEO Dr. Roy Vagelos had frozen some prices as early as 1990, and the company was urging the industry to adopt voluntary controls in an effort to stave off federal regulation under the Clintons. Scolnick touted the industry's case for not controlling high prices: namely, that for fifty years, since the introduction of penicillin and cortisone, both of which Merck played a key role in developing, America's drug companies had produced incomparable value for patients, doctors, society, and shareholders. Their products were thus worth the cost. He finished with a staunch and—given the audience—surprising defense of Proscar, Merck's new drug for shrinking prostates.

With Merck's market capitalization off a staggering $20 billion in the past year due in no small part to Proscar's disappointing launch, Boger thought that Scolnick's road show–style talk indicated how far even the industry's titans had been driven to stoop. Scolnick had enthusiastically backed Boger's efforts at Merck and was put out by his departure, once asking a reporter, "Does Vertex have one single project that wasn't first formulated here at Merck?" even though Merck itself never challenged Boger on intellectual property. Early defectors to Vertex were given up to a month to leave—and perhaps reconsider—but by the time Murcko left, he at first was told to get out in four days.

Boger called Scolnick's talk "bizarre" and then beamed a few days later when he received a handwritten note of congratulations from Scolnick: "Dear Josh, It was good to see you again. Clearly your group is thriving, and the talk by Mark was excellent. I look forward to great things from your group." However much he rebelled against Merck's organization, the *ideal* of Merck still fired Boger's imagination, and he wasn't above gloating when its senior people had to give Vertex its due.

❖

With targets king, the one thing Vertex needed more than anything else were 3-D models of the protein receptors for its medicines. Drugs are molecules. Once they enter the body, they attach at critical junctures in the pathway of a disease, by finding affinities within the folds of the working units in and among cells and binding to them chemically in a way that alters their activity. Since the 1940s, nearly all small-molecule drugs had been discovered serendipitously (brutishly) by screening large libraries of compounds against presumed biological targets and searching for "hits" based on properties such as shape, electrical charge, and either a fondness for, or revulsion toward, water—properties medicinal chemists then tried to tweak through modification. "Monkeys with typewriters," Boger called this approach.

The revolution he intended to lead at Vertex, structure-based design, called for overthrowing this method by vastly increasing the value of precise atom-by-atom information in the hunt for new leads. The most common analogy is a lock and key; know the minute inner contours of a lock and you can design complementary features to trigger it. Proteases, however, are large, heavily folded *active* molecules that work like scissors, with a savage tendency to chew up everything around them. The kinds of small molecules that best inhibit them—gumming up the blades, as it were—mimic short chains of amino acids called peptides. But peptide-like compounds get chewed up by digestive enzymes in the gut, making them unusable in orally taken drugs. A lot of people in the industry thought it was impossible to block all but the scarcest proteases.

From the day Vertex opened its labs, the challenge of producing large enough quantities of ultrapure, active protein to supply the rest of the fast-expanding company fell to a dauntless Australian biophysical chemist named John Thomson. On its ill-fated first project, Thomson had slaved to provide the enzyme that enabled Vertex, from a standing start and many months behind the leaders, to solve its structure in a dead heat with a group at Harvard and ahead of Merck's. He worked hours at a time in a 40-degree cold-room wearing only jeans, running shoes, and a T-shirt. The more difficult the isolation and purification of a target, the longer and harder Thomson went at it. Once, because he'd stuck his neck out, he remained at the bench for eight days straight, isolating protein from calf thymus, on his swollen feet past dawn night after night, washing his own glassware for hours, his hands raw and eyes burning from solvents until he blanked out, sitting upright, Ray-Bans perched on his nose, on a bench in the lunchroom. He went home only to shower.

"One of our chief goals for the next year," Aldrich had said, "is to keep John alive." Proud, hard work—doing whatever it took—was Thomson's key to everything and he exalted the purity of protein science. The wonder of biology is how chains of inert ato.ns receive, via other inert atoms, the operating instructions for how and when to fold precisely into bits of living matter, then interact with other bits to produce life. Every cell is an infinitesimal cosmos, a vast automated forge where proteins are made and interact, performing thousands of precise chemical operations, billions of times per second. Throughout the life sciences demonstrating biological activity is paramount and scientists prize above all else scarce reagents that they can use to mount further experiments. On the day Thomson delivered his first batch of active protein to the biophysicist who would try to solve its structure, Murcko, who shared his extreme commitment to structure-based discovery and his blue-collar work ethic, coined a new measurement: the Thomson Unit, 100 milligrams of ultrapure enzyme.

With the AIDS virus the problem was the production method. If native enzyme can't be isolated from tissue, scientists turn to recombining DNA, putting microbes to work as infinitesimal automated "printers." Genes that carry the instructions to make specific functional molecules are inserted into bacteria and as the microbes reproduce they churn out mature protein, which can be harvested. Not HIV protease. "The more you make the bugs make it," Thomson says, "the more noxious it is to the bug. If you ask a cell to make too many scissors, it punctures itself or makes itself sick." He continues:

So you've got to find a trick. One of the tricks—relatively new at the time—is to get the bugs to make and make and make the stuff with not very high fidelity, so that it doesn't fold properly and falls into the garbage dump of the cell. It's inactive material. You have to then be able to take the inactive material and do something with it to refold it. So you basically brew it up in a harsh solvent—what you call a denaturant—and then you take the denaturant away, so that somehow the thing folds into a nice happy state.

I said, "I don't want to see anyone trying to make this native protein anymore. I want to make it on an industrial scale and rely on ourselves to chemically process it later into an active form." And it worked. It enabled us to jump from making a few milligrams, enough for a handful of crystallization experiments at a time, to batches that took us through a whole program.

A few weeks after the Boston biotech meeting, Boger sat impatiently across from Murcko, Thomson, and several others in a windowless conference room, at an urgent session of the HIV project council. The councils were an innovation of his. He wanted scientists from different disciplines exchanging ideas and sharing what they knew, and he thought the best way to organize them was to give them the power to organize and direct themselves.

The room was scarcely bigger than the narrow conference table, a dozen chairs, and a credenza, above which hung seasonal photos of Mount Fuji, a gift from Kissei. Crystallographer Eunice Kim presented a new structure of a strikingly potent inhibitor wedged deep inside a folded protein. She had determined the architecture, essentially, by taking Thomson's enzyme, coaxing it to crystallize, firing X-ray beams at the crystals, then reverse-engineering a set of atomic blueprints from the resulting patterns when the beams bounced off the lattice and diffracted on a screen. For months, as the chemists made better and better molecules, Kim had been reducing her turnaround times for presenting them with detailed images of how the compounds bound with the active site of the protease. This latest, VX-328, had taken her five days.

"What took you so long?" Boger asked.

"Where's 330?" Vicki Sato joked, as if Kim's accelerating pace might already have caused her to lap herself. Sato was chief of research, accountable for moving compounds from the lab into clinical development. She was referring to a molecule submitted in the last couple of days that was five times less potent than VX-328 but that had looked superior in its ability to survive the gut and remain intact in the blood. It was the ever-lethal difficulty of lasting in the body long enough to get to the virus and deactivate it that doomed nearly all protease blockers, and features to improve VX-330's bioavailability had been deliberately engineered into the molecule based on Kim's structures. The design team *knew* it was sacrificing a little potency in exchange for better properties. The compound differed from an earlier one by adding a single nitrogen atom within a binding pocket measured in billionths of meters.

Six compounds in all were still in contention as the council now set about choosing which one to scale up for human testing—a hugely complex and expensive process requiring quantum jumps of activity at every level of the company. Boger had been pressuring the scientists for months to choose a candidate, but the lead chemist, Roger Tung, resisted him, importuning for more time and data. In June Tung would lead a Vertex contingent to the massive annual AIDS conference in Berlin to disclose the company's work publicly. He wanted to tell a fuller story.

"I'm all for stalking horses," Boger interrupted, "but when the real horse is ready, I don't want to keep him in the paddock because the stalking horse is at the last turn and we want to let him finish so that he doesn't feel bad. I want to go out and shoot him."

To Boger, there was just one issue: getting the FDA to approve the testing of Vertex's compounds in humans before Merck announced its first clinical results. Boger believed Vertex had the better drug: smaller, easier to synthesize, more likely to get into the brain, where the AIDS virus can hide. But if Merck's medicine was highly effective, even in a few patients, all bets were off. The world was desperate for better treatments for AIDS. Other companies—notably Abbott Laboratories, Hoffman–La Roche, Searle/Monsanto, and another structure-based start-up, Agouron Pharmaceuticals—were starting trials with protease blockers. But no one dominated competitive markets better than Merck, and Boger dismissed the others.

"Throw a dart," Sato said.

The council, after arguing for just a few minutes, picked VX-330.

Boger wasted no time. Dispatching most of the scientists with a hail of laughter, he huddled with Sato and a few senior people on the next set of experiments: toxicological studies, animal studies, multidrug studies with other compounds, one-on-one comparison studies, formulation studies to determine how to get the drug into a pill or capsule, blood assays, ultrapure large-scale preparations of the molecule. By Boger's timelines, the molecule had to be available for human experimentation in AIDS patients by the end of the year or during the first quarter of 1994. An immense amount had to be known before then.

Thirteen years into the epidemic, the crisis had only spread and worsened, and hopes for a cure seemed ever further away. At the meeting in Berlin, thousands of grim public health officials, doctors, nurses, academic researchers, patient advocates, and journalists heard a drumbeat of reports about alarming infection rates, especially in Africa. The outlook in vaccine research was dismal. Despite its cautiously rising expectations for protease inhibitors, the drug industry's involvement so far had been suspect, and the most recent studies with experimental drug therapies confirmed the view that the industry was failing to make even incremental progress in fighting the virus.

A new European study, the most conclusive to date, showed that the standard drug AZT, which had raised the first real hope that targeted, effective anti-HIV drugs could be found and developed, did nothing to prolong life for symptomless infected people. They experienced an initial delay before developing the skin cancer and pneumonia that would kill them, but they suffered and died at the same rate as those who didn't receive the drug. HIV is an RNA virus that uses an enzyme called reverse transcriptase to make DNA copies of itself. Because AZT partially blocks the enzyme, it was widely hoped that adding a second reverse transcriptase inhibitor would boost effectiveness. But in the first American study to evaluate the impact of combining AZT with such follow-ons, not one patient showed improvement.

"Only an eternal optimist," the *New York Times* reported in the first sentence of a long article that neglected even to mention protease inhibitors, "would have left the ninth international AIDS conference here last week believing that new drugs will be available anytime soon."

# CHAPTER 2

AUGUST 22, 1993

Debra Peattie, leader of the company's molecular biology group, approached Thomson and Sato about launching a project in hepatitis C, another recent viral contagion where the scale of the medical problem was just becoming clear, and the existing therapies were not good. Unlike AIDS, the epidemic was stealthy, slow moving, and indolent, with symptoms taking decades to show up in most infected patients. The disease was known only since the 1960s, initially as a reaction to blood transfusions, and previously named—"in a less than brilliant foray into nomenclature," its codiscoverer said—non-A, non-B hepatitis. Only now, more than two decades later, was it becoming a recognized public health threat.

Hepatitis C puzzled virologists, baffled doctors, and had no vocal constituency among sufferers, the vast majority of whom didn't know they had it. Three years earlier, scientists at the biotech company Chiron Corporation identified the virus that caused it, making a blood test available, and ever since then doctors were discovering more and more people who were infected but had no symptoms. At the same time, nearly 40 percent of carriers reported that they never used intravenous drugs, never received a transfusion, and had no evident reason for contracting the infection by blood-to-blood contact. The scale of the contagion was unknown and, it seemed, given these mysterious discrepancies, perhaps unknowable.

Sato and Thomson were intrigued. The target was another viral protease. It was a serious unmet medical problem, a wide-open opportunity. Liking the feel of it, they sensed its tractability. But Vertex couldn't initiate a program by itself; there was simply too little published about the virus, and the company, despite its success with HIV, had no virology group of its own. Peattie, scouring the literature, learned that three labs, all roughly equivalent in their abilities, had begun to map the virus, charting the structure and function of its various domains, and were well ahead of the rest of the world. Two of them already had collaborations, with Merck and Hoffman–La Roche. The third was run by an unattached academic, Assistant Professor Charles Rice of Washington University School of Medicine in Saint Louis.

Peattie flew to Saint Louis in late August. She was pregnant with her first child. Having become enthralled at Vertex with the business of research—"the ability to construct negotiations around science"—she recently had been accepted at Harvard Business School, and she anticipated that Rice, like many academics, might be less than enthusiastic about collaborating with a small, unprofitable partner. On the other hand, she knew he had few other options. The hepatitis C virus, HCV, wasn't fashionable like HIV. Universities and federal funding agencies avoided parceling grants to researchers in innovative disease areas.

Most academic researchers fall into their fields of interest, or are attracted to them via some mixture of challenge and circumstance, but Rice was called to hepatitis C. He had been the country's leading expert in yellow fever when an FDA scientist phoned to ask if he could help develop a vaccine against HCV. Because both viruses are so-called flaviviruses—*flavus* means "yellow" in Latin—the caller hoped Rice would find structural similarities. Digging into the biology, Rice's lab determined that HCV started as a larger precursor, a polyprotein, and cleaved itself into at least ten smaller subunits. He and his group identified the protease, found that its optimal activity depended on binding to a small viral protein, and reasoned that since it cleaves at multiple sites, the virus couldn't survive if the protease was blocked. They also predicted other possible drug targets.

Rice was stalled. He couldn't expand his research without funding, and he was finding scant support among the usual sources. "You bootstrap your way along when you want to start something new," he says. "After we'd done quite a bit of work on this on the fly, we applied for a pilot internal grant at Wash U. We got the reviews back, and they said, 'Well, you know this is really important, and you guys have made great progress. But you're really far enough along to write an RO-1 on this [apply for a US National Institutes of Health research grant], so no money for you.' "

Peattie and Rice sketched out a collaboration to determine the structure of the protease, quickly submitted it to their respective licensing teams, and by October the agreement was signed and experiments were being discussed. As Peattie moved to the business side before giving birth to her son, Thomson assumed the lead role on the project. He had recently developed an industry-beating production method for a second protease, ICE—interleukin-1 beta-converting enzyme—a promising target for fighting inflammation. His team's work had enabled Vertex to negotiate a highly favorable licensing deal to develop ICE inhibitors for patients with rheumatoid arthritis and other autoimmune diseases despite having no drug leads of its own. "All we had," Sato recalls, "was another of John's pull-a-rabbit-out-of-a-hat things." To celebrate his and Vertex's achievements, Thomson had just bought himself a terrifyingly sleek naked-frame motorcycle, a Ducati Monster—a "rocket," he called it—and he was ripe to plunge in and go hard against a worthy problem. "All three labs were bogged down, saying, 'We think the protein is like this, but we don't know how to make it or get it active out of the cells,' " he recalls.

"So it's a protein biochemistry problem for a protease specialist hot off HIV and ICE. It was right up my alley. Also we were identifying the attractive opportunity of getting the program kick-started in a collaborative mode with Charlie Rice, and the obvious opportunity for us was to say, 'We can help you move this along and discuss how to get active material.' So I was the sensible person to brainstorm with Charlie on what to do. Charlie wanted to see his fine work stimulate discovery of drugs. He gave us intellectually a kick start into the field. We gave him some money. He gave us some reagents to get started. And that was the start of it."

❖

For every development deal they did, Boger and Aldrich talked with twenty to twenty-five potential partners—"kissing a lot of toads," Sato called it. With HIV, their prospects were limited on the one hand by the small number of companies that knew how to make antiviral drugs and, on the other, by the emerging glut of labs racing to develop pro-tease blockers and numerous other types of treatments. Nestled in the company's sweet spot was Burroughs Wellcome, which sold the AIDS drug AZT.

The British-owned firm had come of age scientifically by creating the first antiviral drugs. Building upon in-house Nobel Prize–winning discoveries about how nucleic acids, the stuff of DNA, are synthesized, it led the way in developing so-called nucleoside analogs: "nucs." These are small molecules that mimic the structure of the building blocks of genes; as a retrovirus tries to make more of itself by assembling bits of genetic material inside a cell, nucs insert themselves, breaking the chain and preventing the virus from replicating. AZT is a nuc and, as is typical of such chain terminators, many of which are used to battle cancers, broadly toxic, especially to blood cells, disrupting the structures that energize them to grow and reproduce. And yet because it was the only drug avail-able, doctors prescribed it at high doses to desperate patients. Though resistance developed quickly, and many patients suffered life-threatening anemia and infections, Burroughs was reaping $500 million a year in sales. Determined to defend its franchise, it had become captivated by its own success, insisting that nucs would remain the backbone of any future treatment for AIDS.

"Somebody [at Burroughs] made the error of saying that proteases weren't very important, so they had been very late to the research party there, where everybody else had jumped on it," Boger recalls. "As op-posed to, say, Abbott or Merck or even Lilly, they were very late realizing that the protease was going to be the mechanism that in HIV would be most effective. They also had some people saying that it was not going to be possible, you can't make drugs against proteases. Of course they were nucleotide guys."

As Vertex cranked up its preclinical experiments in the wake of the

Kissei agreement, it approached Burroughs Wellcome with a novel proposal. A key goal at this stage was determining what a human body would do to a small molecule once an individual swallowed it, a science known as pharmacokinetics, or PK. It didn't matter how potent or specific a laboratory compound was if it wasn't still around in high enough concentrations, hours later, to be taken up by living cells in the body of a sick person. Vertex offered Burroughs, at no cost, 10 grams of a successor molecule to 330—VX-478—so that Burroughs's pharmacologists could feed it to ferrets, dogs, monkeys, and other species and see what happened. Boger recalls: "We said, 'Tell us what you're going to do. Tell us the protocols. If we say okay, you can do whatever you want with it for thirty days and we get the data. And we can use the data if we don't do anything with you.'

"They had better PK resources than we did," he says, "and we wanted to get better information, faster, cheaper, than we could get from the outside. We were pretty sure that this molecule was going to be orally bioavailable but we didn't have any data that it was. So we gave them 478, and they came back and said. 'This is really great. This is better than anything we have.' They had a small protease program, and they said, 'We're shutting down our program if we do a deal with you.' "

Throughout the fall, Vertex positioned itself to partner in HIV while it also sought to raise its visibility. Several other companies had now shown that blocking HIV protease remained the likeliest hope for stopping the virus from spreading. Merck enthusiastically released early clinical data showing that blood tests from four patients taking its inhibitor showed dramatically reduced levels of virus for several months. In the HIV project council, Boger, Sato, Roger Tung, and a few other scientists plotted strategy for the next major medical meeting, in mid-December in Washington, the First National Conference on Human Retroviruses. It was at such meetings that the business of AIDS, as much as the science, was now conducted.

Vertex so far had refused to release the chemical structure of VX-478, raising doubts outside the company both about how effective and unique it was. Unlike the big players, it didn't have its own patent lawyers and was relying on outside counsel, who was proceeding slowly. In Washington, all the other companies would promote their molecules even if they reported no new data. Vertex needed to present something to prove it was in the race. Boger suggested a signature solution. Chemist Dave Deininger synthesized the compound and gave it to Eunice Kim, who quickly solved the costructure. At the conference, after Merck's scientists conceded that they didn't know how their drug worked specifically at the molecular level, Tung concluded his own talk with a slide showing how Merck's drug sat in the active site of the protease, in effect answering the question for them.

Boger and Sato reveled in such in-your-face diversions; they were good for morale, good for the dinosaur-slayer story the company was crafting for itself. But Vertex couldn't maintain its secrecy with Burroughs Wellcome, which, only after seeing how VX-478 bound to the active site of the protease, agreed to collaborate. A few days after the conference, the two companies announced that they would jointly develop HIV protease inhibitors for Europe and North America. The deal, which would eventually bring Vertex $42 million, was in fact worth several times that, as Wellcome agreed to pay the full cost of development— perhaps $200 million—and Vertex won the right to copromote any drug, enabling it to start building a marketing arm. Aldrich negotiated royalties that climbed from the midteens, rare if not unheard-of for a preclinical compound. Vertex's stock rose $2 on the news, to $17.50.

Boger was pleased with the arrangement. Burroughs was the industry leader in HIV. It also had been the target of damaging protests for charging patients $10,000 for AZT, a compound it licensed from the National Institutes of Health, where the molecule was discovered and developed by government scientists. With multidrug therapy sure to be the future—unless one of the protease inhibitors was so superior that it could vanquish HIV by itself—Boger looked forward to bringing his compound quickly to patients in conjunction with the one indispensable agent in the field, AZT; yet skirting that drug's heavy political baggage, which in the current grim environment, and with the threat of price controls looming larger now that the Clintons had uncovered their health care plan, guaranteed controversy and, most likely, a media circus. He could easily imagine enraged AIDS activists resuming round-the-clock pickets and returning to the New York Stock Exchange to pour sheep's blood on investors, chanting "Sell Wellcome!" and "Fuck drug profiteers!"

"I'm glad," Boger said, "our money comes off the top."

❖

Boger and Aldrich agreed that the time to raise money on Wall Street was not when you had to but when you could. One day after the Wellcome deal was announced, Vertex filed a fast-track stock offering that would yield, six weeks later, another $62 million. Though it remained a treacherous period for biotech and drug stocks, the offering built from start to finish, with shares trading at $16 on the day they started their road show in Europe and $18 two weeks later when they ended it in Boston and New York. It would still be years before Vertex might have a drug, but with three deals in 1993, the company was momentarily in the black, declaring a fourth-quarter profit of more than $2 million. It had put $120 million in the bank. Aldrich was promoted to senior vice president, by title and, in fact, the second most valuable person in the company.

Vertex issued hefty stock options to most of the original scientists to keep them from defecting, although with the labs starting to click, and intriguing new targets sprouting across the spectrum of diseases, none of them had any real mind to leave, not soon. Boger deliberately hired people who, like him, craved the chance to compete at the forefront. Yet when the first researchers arrived in Cambridge, they discovered an atmosphere of almost willful anarchy, an antiorganization—"chaos," says Sato, whose job it became to bring order. For two years, the scientists had no offices, lugging their backpacks and briefcases from one communal desk to another to use the phone. There was little rank or hierarchy; decisions were made by the project councils. During its initial public offering, launched during a speculative bubble, the raw, head-snapping speed and intensity of the chase and the relentless pressure to do important science while cutting corners overwhelmed the principles of an idealistic young crystallographer, who smashed a chair in the lunchroom in a fit, screaming, "*You will all be stricken down!*" He left the company soon after to attend medical school.

Boger's "social experiment" was designed to encourage self-selected

leaders who not only would do excellent science but also grasped vis-
cerally that the decline of Big Pharma was due less to cluelessness at
the bench or fecklessness on the executive floors and in the boardroom
than to the immobilizing sludge of middle management in between,
which even at Merck had led to project heads prioritizing how many
compounds a group made over whether or not they did anything useful.
Boger wanted champions, people who would passionately disagree with
each other and with him, who in the end would push groups and projects
ahead because they had smarter ideas, worked harder, generated more
compelling data, and persisted when others would quit.

Mainly through their association with HIV protease, Murcko, Thomson, and Tung emerged from the founding group to become the company's rising stars. Statistically, most pharmaceutical researchers work their
entire careers without helping to produce a drug that makes it all the way
to market. That meant that even if Boger was right and Vertex could improve its odds from 1 in 30 to 1 in 10, most Vertex scientists were subject
to the same sobering reality, a career-long string of failures where you
best found your job satisfaction in something other than success. For
most, it became the daunting challenge itself. VX-478 was not a drug, but
it looked as if it would be, and those who led in bringing it out of the lab
enjoyed a surge in influence, credibility, and prestige.

Each, in his way, spread his wings. A computational chemist and
molecular modeler by training, Murcko stood between the structural
biologists, who churned out torrents of data about the smushy interdig-
itation of atoms, and the chemists, who designed and made inhibitors.
He considered VX-478 a tipping point, absolute confirmation of Vertex's
superior strategy for discovering major drugs. "For me the bottom line
was: five chemists, eighteen months, two hundred six compounds. It all
played out according to script. It was the project from central casting."

Murcko's intellectual curiosity rivaled Boger's and he was often, in
the project councils and at Friday afternoon beer hour, the first and
most effective to challenge him; "Joshua the Indeflectable," he once
mused. Nearly a foot shorter than Boger, he was built like a catcher,
mustachioed. Like Boger, he liked to tweak the mighty, and he thought
as deeply about the iterations of the discovery process and the interactiv-

ity of people and ideas as he did about the forces of atoms. Murcko had
come to biochemistry through high-speed computing. On his first day
at Vertex, he flew to Boston from Philadelphia rather than waste the time
driving and worked that night until three in the morning, hands flying
over his keyboard. In the early days, he often "pinned" the company's su-
percomputer with his experiments; asked how much computing power
he could use to do simulations, he deadpanned, "Infinite."

Awaiting a crystal structure for ICE, Murcko started transforming
his research group into a nascent skunk works, hiring several new people
who knew both biology and computer modeling and encouraging them
to broaden their thinking about what more was possible. He authorized
his scientists to invent new technologies and write their own code if they
couldn't find outside collaborators or buy what they needed. The spark of
innovation is asking questions; Murcko questioned everything.

"The software at the time didn't take into consideration any of the
downstream physical properties of drugs," he recalls. "How soluble is a
compound? If it's not soluble, it can't get into cells. Could you use computers to predict not just the lock and key—how does the drug bind to
the active site—but go a step further, to see how one molecule might
be better than another because it has better physical properties? We ran
computer simulations on hundreds of associated molecules and asked, 'If
I was just going to change one atom, could I increase its solubility?' For
VX-478, one small change predicted to be at least a hundredfold more
soluble turned out to be even better: five hundredfold more soluble."

In January biophysicist Keith Wilson finished the crystal structure of
ICE. It was a big protein for the day, a complicated piece of architecture
with two domains, and it would receive much glowing press when it was
published in the journal *Nature*. Murcko and his modelers went to work.
Sitting at aging Silicon Graphics workstations in a darkened room, they
wore clunky wraparound 3-D glasses. Chugging Diet Coke, unshaven,
they resembled cave-dwelling ancestors of the LeVar Burton character
in *Star Trek: The Next Generation*. On the screen, stick diagrams of hundreds of connected atoms in brilliant reds, purples, and blues rotated
gently, like hair-thin Tinkertoys, in a fathomless black sea. Within hours,
the scientists recognized that although the overall folding was different

from other protein structures, "down deep" ICE resembled a familiar type of protease with a similar cleaving mechanism. Fortified by what they trusted was an original conceptual breakthrough, since no one else had the structure, they started pulling hundreds of scaffolds from chemical catalogs and Vertex's relatively tiny compound library and docking them in the binding pocket. Five weeks later to the day, running round-the-clock simulations, they chose a core for a new class of molecules that, once synthesized, would prove so much more potent than any other inhibitor that it leapt instantly to the lead in the combined drug discovery efforts with Vertex's European partner.

❖

Tung took the AIDS compound VX-478 public. As lead inventor on the patent application, he became the face of what Boger called Vertex's first "real major scientific publicity blitz." Describing himself as "a sort of first-and-a-half generation, mixed background, American-Japanese and -Chinese," Tung was thirty-four, deep voiced and serious. His sobering intensity masked a fiery ambition to aim high, discover drugs, and avoid ever becoming subordinate on a project. In his second year at Vertex, toiling for several months on a grueling synthesis, often until midnight, he started to sprout grey hairs, singly at first and then in clusters. After leaving the lab, he and the other chemists drank most nights until they closed the bars in Central Square—a rite at several local start-ups. Unlike Murcko he wasn't yet entirely sold on the value of structure-based design, though this could also be attributed to a near-universal skepticism of bold claims based on partial data, even when the claims were his. "I live to be proven wrong," he said.

A year earlier in Berlin, Tung was taken to task for not disclosing the molecule's chemical structure, and bravura gimmicks like embarrassing Merck at the Washington retroviral conference, while satisfying, only invited more doubts about what Vertex must be concealing. The issue was its patent position. The delay in filing its US application put the company two weeks behind Searle, which had made a molecule with a similar core that appeared to have, like VX-478, exceptional bioavailability.

More to the point, Searle filed a so-called Markush structure. In the 1920s, chemists, to protect their inventions, sought a way to avoid having to patent individually each member of a class of compounds that would have a similar function. Eugene Markush, a dye manufacturer, filed for and won a patent for a class of compounds with a replacement group of atoms at several positions, meaning it covered potentially thousands of molecules. Based on the alluring but false premise that combinations of different groups of substituents around a common core generate molecules that have the same activity and biological properties, Markush structures amount, legally speaking, to hurling kegs of nails off the back of a moving truck.

Boger delayed revealing VX-478's structure until after the company's European patent application was made public. Merck, meantime, stumbled. At the retroviral conference in December, attendees had crammed into a presentation in the Washington Sheraton Hotel to hear the company report that patients getting the Merck drug had a 42 percent drop in HIV after just two days of treatment, compared with 1 percent for those on AZT. "We were beside ourselves," Scolnick would recall. "We thought we had the cure for AIDS." Then, six weeks later, a molecular biologist examining blood samples from trial participants discovered that the virus had mutated, building resistance to the drug. Another test indicated the virus hadn't developed resistance. Scolnick called Anthony Fauci, the government's top AIDS researcher, to get his read. "You've got resistance," Fauci told him. Scolnick protested, bringing up the conflicting data. "I don't care," Fauci said.

"You've got resistance."

Boger, Sato, and Tung conferred throughout the spring. Formulation and other problems had slowed the work at Wellcome to a crawl. There were so many treatments being tested in humans that it had become difficult to enroll patients for new trials, and it now looked as if Vertex's molecule wouldn't be in the clinic until the end of 1994. Pressured to show their hand, they opted to disclose the chemical structure at the Third International Conference on the Prevention of Infection in Nice, France, in June. The audience seemed skeptical from the outset. Despite hoping for a breakthrough, many were eager to see Vertex come down a notch. The compound was a sulfonamide, one of the class of molecules that were the first antibiotics—the sulfas—and now were also sold as anticonvulsants and diuretics. Novel functional groups extended from a core that shared common features with other protease inhibitors.

Reactions ranged from qualified optimism to mild disappointment. No one felt that Vertex had been blowing smoke, but neither was anyone immediately convinced that VX-478 was all that the company had claimed. There remained major questions: Was it small enough to get inside the brain, something Searle's compound had not been able to do? What of possible allergic reactions? VX-478 had chemical groups similar to those thought to cause some people to react violently to the antibiotic Bactrim. Carl Dieffenbach, the NIH point man for assessing new AIDS drugs, mused to a reporter, "I'm not sure their patent is as secure as they think it is."

Boger weighed these questions only to brush them aside, especially the last. As part of its due diligence, Wellcome had satisfied itself that Vertex's patent claim was clean. Its lawyers had won—and held on to—the rights to AZT through ten years of intense legal strife, providing an intimidating ally. A few months later, during a November conference call with Wall Street analysts, Searle announced it was stopping development of its protease inhibitor. Two clinical trials had showed no indication of antiviral activity, and Searle researchers believed that the compound was soaked up by a blood protein and removed by the liver. Boger was relieved to learn that the problem was unique to Searle's molecule, and he considered himself fortunate to have the threat of a patent war suddenly diminished.

❖

Murcko, as usual, thought well beyond the problem at hand. With the genetic material from Charlie Rice, the main scientific challenge was to discover how the hepatitis C protease worked and how to disable it. But Murcko had a secondary question, one more central to Boger's mission: How do you evaluate new targets without knowing their structures? Could you predict, say, degree of difficulty? When was it wise to invest in new projects and when wasn't it? "Up to now at Vertex, we said we either want to have a crystal structure available or we want to know that we can get there first," Murcko says. "But sometimes maybe you wouldn't be able to get the crystal structure as quickly as you'd like. Maybe if you had a model it could help steer you away from certain projects. Or say this one isn't going to appear so easy."

Murcko did the experiment. Having no new positions but recognizing the uncommon gifts of a recent Harvard postdoc named Paul Caron, he hired Caron as a temp. Caron was advanced in thinking about gene sequencing and physical similarities among proteins, which fold into spirals and loops and cascading sheets according to the ordering of the amino acid residues out of which they're made, following the text encoded in their DNA. In other words, if you had the genetic code for a target you might be able to model its active site *prospectively*, by mapping it against other known proteins. Sitting at a workstation in the modeling room, with a second PC at his side so he could calculate atomic charges and distances, Caron lined up the structures of the few other viral and mammalian proteases that were available and quickly noted that HCV lacked the usual cleft that caused the binding site to be buried in a pocket. "What we saw was virtually a bowling ball," he said. "Very smooth. All the big loops that come around and make the channel weren't there." The target was going to be far more difficult than anything they'd done before.

One of the company's advisors, Harvard structural biologist Steven Harrison, categorically dismissed the model, saying it couldn't be true. Others confessed equal doubts. No one wanted to stop the project, but Murcko and Caron, knowing that the featureless active site presented a steep challenge to design, wanted people to be realistic. "The question at the time, once we had this model, was, 'Do we continue?' " Caron recalls: "It's going to take a relatively large inhibitor to get enough binding energy, and it's not going to be an easy thing. We said, 'This isn't going to be HIV again. This is going to be a long, hard project.' "

Thomson and his protein group also were learning that HCV protease might not be as tractable as the company's earlier targets. "We knew enough about the polyprotein that gets made by the virus that it was not a dead ringer for HIV," he says. "It had some funny new funky ways for creating proteins, including this little cofactor that was essential. It was very conspicuous that there were two adjacent regions that interacted with one another very significantly, and then another third piece that was key to the activity of the protease. And it was a puzzle as to what the architecture was going to look like in the end. More particularly, it gave us this chicken-or-the-egg problem of: we couldn't make the protein, and we didn't have a reliable assay to test whether we had *made* protein. That was really the bind that kept everyone on the planet anchored in the early days."

Throughout 1995, the pressure mounted to solve the structure. Merck, Roche, structure-based rival Agouron, and many others were pouring major resources into the race. Meanwhile, public health officials were reaching the conclusion that many more people were infected with HCV than previously predicted, perhaps twice as many as with HIV. Vertex's attempts at protein production stalled, cycling blindly in a cul-de-sac; the scientists didn't know what material they had made and couldn't test it to determine what to do next. With a reliable assay they could try hundreds of subtle changes in conditions to isolate more protein, but that wasn't an option. "The Dark Ages," Thomson called it.

"Everyone along the way had some different cross to bear," biophysical chemist Ted Fox, who led the lab effort, recalls. "Those project councils were really tough. You got a lot of really energetic people, excited passionate scientists, and you're getting beaten over the head week after week. I remember at one point, we finally had some active enzyme—doesn't look good, not very much; maybe we have an assay—and Josh saying, 'My God. This thing wouldn't survive if it were this inefficient in the real world. Guys, you've got to work harder, or you've got to do more.' "

It was Thomson again who broke the logjam, though others at Vertex were thinking along the same lines. He asked Rice's group to calculate the space between two regions by mapping the interceding genes, then asked the chemists to synthesize chains of amino acids that could be whittled, atom by atom, roughly to that same length. The chemists fashioned tiny artificial spanners, ten or twelve residues across, to form the most intimate molecular "embrace." The exercise worked. "You took your protein from the bacteria that were engineering it, then you mix it with this little synthetic peptide, and bingo, you've got active protease," he says.

As Murcko anticipated, some targets are much harder to isolate and purify than others. Nearly two years after receiving Rice's clones, Fox's group delivered the first Thomson Unit of HCV protease to the crystallographers.

❖

In February 1995, a year later than Boger had predicted to Wall Street, VX-478 was given to eighteen AIDS patients in a Phase I clinical trial, a dosing study of oral availability, pharmacokinetics, and tolerability—not its effect on HIV. Less than three weeks later, US regulators cleared a merger between the Burroughs Wellcome Company and Glaxo, creating the world's largest prescription drugmaker. During the next fourteen months, Aldrich and Boger's brother Ken, the company's lead outside counsel, joined with Glaxo Wellcome's lawyers to try to convince Searle to come to a reasonable solution over its patent claim.

"Searle had this guy, total jerk, totally irrational about this stuff, but by virtue of that, he ended up being quite effective," Aldrich recalls. Early in the negotiations, he and Boger flew to Chicago to meet with a group from Searle. Never one to hide his contempt for the practice among drugmakers of aggressively pursuing patent claims for molecules in areas where they already had shut down their own projects, Boger ridiculed Searle's Markush structure. Searle hadn't made a sulfonamide and had no experimental data on whether it would work. Aldrich told them Vertex wanted to "work something out and get this important drug to patients." The discussions dragged on for months afterward, until Glaxo Wellcome's lawyers also took an unsuccessful stab at it.

In April Boger got a fax from Glaxo Wellcome. "I remember I was on vacation, golfing in Hilton Head, and Josh sent me an email saying we just took a torpedo below the waterline," Aldrich recalls. "Josh is always very positive about things and always pooh-poohs anyone else's science or any problem, but this was clearly a problem that shook him up. Searle was a problem for us and Glaxo."

A delegation from Glaxo Wellcome arrived at Vertex to tell Boger and Aldrich that they were shutting down the VX-478 clinical trials because they couldn't get the patent cleared. The timing—on the cusp of testing whether the drug worked—could scarcely be worse. "We're a public company and HIV is our lead program, so we're looking into the freakin' abyss," Aldrich says. "I said, 'Before we do this, there's one thing we could try.' " Searle's position made no sense: if Glaxo and Vertex shut down the program, Searle got nothing. Aldrich suggested sending a joint letter offering to pay $25 million and a 5 percent royalty in exchange for the exclusive license to all Searle's patent applications in the area of HIV protease inhibition, and warning that if they didn't hear back in a week they would go the *New York Times* with the story that they were shutting down the trial because of an IP dispute with Searle. "A Hail Mary pass," Aldrich called the proposal. Searle took the deal.

"We would have been blown up. We were toast. We wouldn't have been able to raise money. Our stock would have gone from fourteen dollars to two. It was very stressful. It was a very nervous place. People could just see that Josh and I were freaked out. That Friday after we got that yes, I had a few pops."

The dispute was settled none too soon. Two weeks later, at the 1996 international AIDS conference in Vancouver, British Columbia, a cascade of encouraging reports showed that HIV could be suppressed indefinitely by a combination of antiviral drugs built around protease inhibitors. Three years after the despair in Berlin, combination therapy—drug cocktails—transformed AIDS from a death sentence to a manageable disease. The epidemic peaked in the United States with fifty thousand deaths in 1995. With the first of the drugs—Hoffman–La Roche's Invirase, Abbott's Norvir, and Merck's Crixivan—reaching the market during the previous six months, people who had been planning to die sooner than later suddenly and unexpectedly could see futures for themselves. New York, San Francisco, and other urban centers started to fade as "cities of ghosts." Other drugs—better absorbed, with fewer side effects—were still desperately needed.

Vertex, its near-death scare with Searle concluded, remained squarely in the race.

❖

In the labs, the hepatitis C virus continued to pose extreme problems at every stage. A critical challenge in coaxing individual proteins, which are floppy and active, to rigidify and lock themselves into crystals is learning how to subdue them with the right mother liquor. Researchers anthropomorphize enzymes, describing them as "happy" or "perturbed." Organic molecules prefer certain conditions to others—some dissolve better in water, others in fats. The goal is to induce a kind of nirvana: amniotic fluid laced with heroin. And yet the insertion of the synthetic peptide caused the team to have to grind hard for months, substituting endless experimental conditions, amid rumors that another company had solved the structure of the protease and was rushing it into print.

"It's a contradiction," Ted Fox explains. "You want to have your protein in a nice soluble, water-based environment, and then you've got these peptides that are happy only in organic solvents. They don't want to be in water. Different solvents, different buffers. It's a constant game: Can you feed a little bit of it into the protein? Then again, once you've got it inside the protein with the protein wrapped around it, occupying its natural spot, it's fine. A lot of it is fortuitous."

During the summer, the company grew its first diffraction-grade crystals of HCV protease—three years and many millions of dollars since Peattie flew to Saint Louis to meet with Rice. The X-ray structure emerged within weeks, and in a frantic push to get it out the door, Murcko, Thomson, Caron, crystallographer J. L. Kim, and numerous others worked around the clock to submit the paper to the journal *Cell*, which published it in October. To the extreme disappointment of some of them and the disbelief of all, they were not alone. Agouron, which also had a structure of the protease domain in press—though without the cofactor—was claiming a tie, even though its enzyme showed no meaningful activity.

Boger normally avoided public disputes over scientific priority—it was more important to him that information be found and available than who discovered it—but a statement by an Agouron scientist about catching Vertex at the finish line rankled him, and he responded with an incendiary fax. "'First of all,' I said, 'This isn't the finish line. And second of all, your protein is *dead*,'" he recalls. "I was outraged that they were projecting to the gullible scientific press that they had anything. They couldn't use it for anything. It was inactive."

Yet if Vertex once again had come from behind and bolted ahead of the competition, it was not at all clear from the crystal structure itself what it had won. Caron's model proved to be spot-on. The binding pocket of the protease was large, smooth, exposed, and greasy—nothing, apparently, like the cozy nooks of HIV and ICE. As the scientists examined it on their computer screens and started talking among themselves about how hard it would be to design inhibitors to block it, a few preferred metaphors arose: a dinner plate; an aircraft carrier; like trying to land a model airplane on a pizza.

"This protein," Boger observed drily, "was just not well behaved."

# CHAPTER 3

------

APRIL 11, 1997

The hepatitis C virus offered other inviting targets: a motor enzyme called a helicase that can unzip spirals of genetic code, and a polymerase, which spools out new strands of DNA. At Schering-Plough Corporation, an industry leader in structural biology, virologist Ann Kwong had done vital work to characterize the helicase. But Kwong was frustrated, quietly seeking another job. She had come to Schering's New Jersey labs from a postdoc at Memorial Sloan-Kettering Cancer Center, where she'd shown a Thomson-like fortitude working day and night in a windowless cold-room. The company, best known for its blockbuster antihistamine Claritin, sold the first approved drug for hepatitis C, Intron A, a biologic, and it was heavily invested in and publicly committed to using structure-based design against HCV.

Kwong doubted the effort was working and had begun to challenge her bosses in meetings. Now she took the podium to give a plenary lecture on hepatitis C drug discovery at the Tenth International Conference on Antiviral Research in Atlanta. Tung and Thomson, representing Vertex, listened with special interest, then moved swiftly when a couple of members of Kwong's group mentioned afterward that she was considering offers from other companies. Vertex had so far resisted setting up disease groups because it was organized around protein targets, not illnesses. But Sato was impatient with the pace of discovery in HCV, and the company desperately needed better biology, especially in the realm of antivirals. "Roger and J.T. came over to talk to me and said, 'Don't accept. You can't accept. You've got to talk to Vicki,' " Kwong recalls. "I'm like, '*Who* is Vicki?' "

Kwong took the train up to Boston to give a talk at Vertex. The small lecture hall was packed: standing room only. Afterward, Thomson, Murcko, Tung, and several project leaders took her to lunch. She explained why she was leaving her job. "I wanted to do structure-based drug design, work on a team, but the way it was set up was, once the biochemistry group came in, we weren't allowed to talk to each other. I mean, *forget* that," she recalls. "All the guys were just laughing, falling out of their chairs, dying. They were like, 'This is exactly why we left pharma.'

"At Schering, when Vertex published the structure of HCV protease, the head of research went completely bananas—the fact that Vertex could do this, and we had this huge team, and we hadn't done it. At Big Pharma, when you had something like this that was very important, every month or so the VPs would send something down. From the PowerPoints sent up, the PowerPoints would come down, and your direction would be changing, which I think is ridiculous. So I heard Vertex had these project teams, and the people from the different functions ran the projects, and I'm like, 'Yeah, *right*.' But I was really hoping that that was the case. And it was the case."

Kwong was dazzled, but kept her impressions to herself. The problem of designing an HCV protease inhibitor, from a research standpoint, was that there was no clear path. There was nothing to work with, no reagents. No one could grow the virus. There were no cell-based assays to tell you if your compounds were active. That meant, in managerial terms, no clear metrics, no way to measure progress, which in turn meant that the scientists would need to have time and space to figure out novel approaches—time to fail, over and over. And yet Sato didn't seem to want to wait for Rice's group or another academic lab or another company to come up with the tools that Vertex would need to advance drug discovery. She seemed fully committed.

There was no sag in enthusiasm at Vertex about hepatitis C as there was with other less stimulating projects: HCV fired the collective imagination. It was a priority. But at Vertex, speed was the creed, and the project was turning into a costly slog, one that drained money and scientists from other research. With or without middle management, pressure on the team to perform was ratcheting up. Kwong thought piled-up failure was the one true metric of innovation, but coming up empty month after month had strained the organization. Within the terms of Boger's social experiment, that left it to the team leaders to build their strongest case, preferably with data, although now the company's purpose and identity seemed also to be on the line.

"We were living quarter to quarter: 'Should we kill this project? Should we not kill this project?'" Sato recalls. "By now, it was several standard deviations outside of the time that Vertex prided itself on taking to get to a drug candidate. Plus, the clock was ticking on the market. So every quarterly planning meeting it was up for killing, because no one was paying the bill yet. Finally, John just gave one of his inimitable presentations on why we can't give up now: that this was a project that was made for Vertex, and if Vertex can't solve it, nobody can solve it, and we should all just work harder."

It was Sato—her scientific credibility, organizational élan, skill at managing tension, and spirited enjoyment of the whirling loop-the-loop of entrepreneurial science—who took the results of Boger's social experiment and tried to shape them into a successful discovery engine as Vertex added new projects and disciplines. A granddaughter of Japanese immigrants and cerebral only child, Sato grew up in Chicago, came east to study and eventually teach at Harvard, and cut her managerial teeth inheriting a brilliant scientific staff and an empty product pipeline at Biogen. It was perhaps her strongest gift that she could direct Nobel Prize–winning advisors, gunslingers like Thomson and Murcko, and postdocs alike with deftness and poise. It helped, in a world of outsized egos, that she carried herself—back straight, head lifted—like a dancer, and, in fact, found time well into her forties to perform in ballets. HCV, she liked to say, was a "game worth the candle"; the rewards would justify the time, effort and money required.

Sato agreed with Tung and Thomson that Kwong was "really smart" and that Vertex should recruit her to build a virology group. "We didn't have any positions because we never had any positions," Sato says. "So I'm thinking, 'What can I attract her with? What do I have going for me?'

She's going to need a BL-3 level lab to work in. We don't have one of those. I don't know if I can even get a permit to have one of those. Even if I could, it's going to take me eighteen months to build it. I told her we'd build it. I said, 'Yeah don't worry. We want you to come here, we'll do whatever you need.' I think my line was, 'We don't know anything about this, so we need you to come here and help us. We don't have a lot of positions, but we'll figure out a way to help you hire a couple of people.' "

Kwong accepted, prompting colleagues at Schering to doubt her judgment. Had she weighed the downside of leaving one of the world's most profitable drug companies, a commercial and scientific leader in hepatitis C, for a risky future at an untested biotech with a bare-bones program and no clinical apparatus? She told herself, "How can you pass up the opportunity to build your own group?" When she reported her discussion with Sato to one lab mate, she recalls, he lashed back. "He said, 'Well, how many people did she give you?' I said, 'I never asked.' If she's committed to executing this, to me it's stupid to ask her how many people. She's gonna ask me, right? What I wanted to know was how she thought. How does Vertex work? Do they really want to do it or not? You need to be driven by the need, by the question, not by the you-get-2.5-people-because-you're-at-this-level."

❖

Aldrich shopped the project throughout the winter and spring, talking with nearly every major pharmaceutical company. All of them were gearing up for HCV. He squeezed in a tour of Japan—a death march of last resort—where almost all drugmakers had concluded that because blood supplies were now screened, the epidemic would soon go away by itself. Aldrich explained that four million people in the United States and as many as sixty million worldwide were already infected and that if they lived long enough, all of them would develop serious liver disease and would die of hepatitis C. He found no takers. Vertex's ever-escalating terms for doing a deal, as he says, "strained the boundaries of reasonableness. Our stance going into these discussions with Roche, Schering, et cetera, was: 'We've got a program. We've got the best team. We're ahead. We've got the resources and the money to drive the project ahead by ourselves, but if we could find a great partner, we think that we could enhance the value. Also, any deal we do, we keep fifty percent of US rights.' We were asking a lot."

As head of the business side from Vertex's earliest days, and a veteran of other less successful early-stage biotech firms, Aldrich understood acutely its assets and liabilities. He represented it in a rapidly globalizing world that retained a strong belief that leading-edge science would deliver drugs, and he never felt that Boger's exuberance about Vertex's goals was unsupported by the quality of its research. He also knew that if you looked at biotech investments on a rational cash flow basis, they were "ludicrous"—Vertex's less so, but not appreciably. Aldrich's ancestors had arrived in Massachusetts in 1630 and remained in New England, mostly in law and banking, ever since. Behind a casual demeanor, he was a disciplined Yankee, a bachelor who worked murderously, hit the gym, ate at a university club, then allowed himself one chilled vodka and a single cigarette before bed. He wasn't surprised when none of the first-line companies he and his people approached returned any real interest. "What we were asking for was pretty outrageous, since we had nothing, really. We didn't have a clinical candidate, we just had a research program, and they were gonna be footing the bill for everything, yet we were saying, 'Yeah, but we want roughly half the economic value.' "

Wall Street was in the thrall of the Internet boom. It was anticipated that the Next Big Thing in biotech would be genomics, but the patchwork government consortium to assemble and analyze the entire human genetic code was progressing fitfully, and the revolution in breakthrough drugs that it promised remained far beyond the horizon of all but the most patient investors, a vanishing category. Boger intended to be in the vanguard of that revolution but didn't know how, so he joined the board of directors of an early DNA sequencing company, Millennium Pharmaceuticals. Since genes encode proteins, he was eager to figure out ways that the coming flood of genetic information could help Vertex better understand its targets and design drugs.

In March Vertex raised $157 million by selling almost 3.5 million shares of new stock at $45.50 per share. "We caught the price just right," Aldrich recalls. "It spiked up. We were out there. It was an easy offering." The company also began talking more seriously with Eli Lilly and Com-

pany, the Indianapolis drugmaker, about HCV protease. Lilly had nothing in antivirals, but its head of research wanted to get into the area, and
Vertex's protease project seemed like an excellent vehicle. Enthusiasm
built rapidly throughout Lilly's labs.

Aldrich leveraged their excitement by being tough and sticking firm.
R&D still carried a lot of weight at drug companies, and he knew that the
Lilly scientists were pressuring the business side to make a deal. Vertex's
business development person had resigned, so Aldrich directly handled
the "hand-to-hand combat" of getting the agreement worked out. The
negotiations were contentious: "a lot of teleconferences with people
yelling at each other," he recalls. Under the terms of the final deal, Vertex
received $40 million up front; Lilly agreed to pay all development costs,
buy $10 million in Vertex stock, underwrite a hundred-person Vertex
sales force, and pay a steep royalty rate. Any drug would carry both companies' names: Lilly-Vertex.

In other words, on a program that had no drug candidate, and was,
at best, many years and hundreds of millions of dollars from the market,
Lilly would be paying the full cost of development and launch, while
Vertex stood ultimately to get 30 percent to 40 percent of the profits.
An anticipated but still unfortunate collateral price was a pool of ill will.
"They were paying for everything," Aldrich recalls. "Their business guys
were pissed at us. They just felt like they had been beaten up. I never like
to see that in a situation because it's not good for us, it's not good for
anybody to have people on the other side who feel like, 'Yeah, we finally
rolled over and made the deal, but we didn't like it.' "

In early June, as Kwong left for Cambridge, Schering-Plough published the crystal structure of HCV helicase, beating Vertex, Merck,
and the rest of the pack. Ten days later, on the same day the Vertex-Lilly
agreement was announced, Schering signed a deal with another small
biotech company, Corvas International, to codevelop HCV protease
inhibitors.

❖

Boger forever kept one eye on the horizon, his optimism bottomless.
He both dared and enabled others to think beyond their usual ideas and
capabilities. He possessed what collaborators said of Apple Computer

cofounder and recently returned CEO Steve Jobs, a reality distortion field, a rich talent for convincing himself and other people through a mix of bravado, hyperbole, charisma, marketing, and tenacity that almost anything can be done. Boger brought other people around to his way of thinking not through willfulness and manipulation—staring through others coldly if need be—but by an unshakable faith in himself. "I've never made a bad decision," he liked to say. "I've just had bad data." His exuberance was infectious largely because he seemed to embody the idea that the biggest, hardest problems were the most interesting and that an all-out, information-driven mind-set will not only get you the right solution first but also be more fun.

"Josh didn't sit around wringing his hands about the fact that one in three hundred ideas you have in pharma gets to the finish line, and the other two hundred ninety-nine crash and burn," Sato says. "On some level, I think Josh says, 'That *can't* be true. Or if that is true, people don't deserve to breathe.' He has an outrageous sense of possibility coupled with a very deep sense of self-confidence about one's ability to translate that into something that is actually working. And because those were his very strong values, it made it safe for the younger people to think bravely—as long as they delivered. So nothing was too hard for Thomson, as long as you gave Thomson rope, and the reason you gave Thomson rope . . . in the old Peter Drucker sense . . . is that it's okay to be a diva as long as you get up and sing *Tosca* center stage at the Met—to rave reviews. Josh made it okay for people to be a little crazed."

Ann Kwong witnessed Boger's leadership at her first project meetings, where he routinely posed goals that were impossible to reach by traditional thinking and methods. "But that's so freeing," she says. "You know you're not going to take an incremental approach, because you know it's not going to work."

Boger's tolerance for high risk coupled with audacity drove Vertex again and again to exceed expectations, but Vertex could not go it alone— no small company could. Its corporate partners, swept up by dramatic new forces and focusing on profits and market share, meanwhile struggled to reorganize their R&D organizations to be more productive. Ever since Clinton dropped the threat of price controls midway through his first term, the pharmaceuticals sector had strained to hold its place as the most profitable industry in America. It won its long campaign to advertise directly to consumers on TV. It exploited patent extensions and the yawning opportunities beyond the narrow indications for which its drugs were approved: "off-label marketing." It pushed the FDA hard to approve "lifestyle drugs" such as Pfizer's Viagra and Merck's Propecia—a low-dose version of its prostate pill Proscar, now prescribed to prevent hair loss—then pressured insurers to pay for them. The challenge to Big Pharma as it consolidated more and more was to meld cultures while trying to increase profits: not, as at Vertex, to construct a new model from the ambitions of the founders while running perennially, dangerously in the red.

At Glaxo Wellcome, progress on Vertex's AIDS compound VX-478 was overwhelmed by a postmerger hangover that caused Vertex fits. Generally when one company takes over another, the acquirer attempts to absorb or remake the acquired. But Glaxo, facing the loss of 40 percent of its business as its antiulcer blockbuster Zantac came off patent, took the opportunity to remake itself. It hired a human resource consultancy to conduct a global, four-phase, custom-designed leadership and cultural change program to fire up its labs, stoke an unexciting pipeline, and formularize a steady stream of new drugs. The Clinical Process Redesign project stretched on for eighteen months. With lingering internal resistance toward protease inhibition among Wellcome's pro-nuc forces already entrenched, the reorganization only compounded its inertia, costing in all perhaps another year to get VX-478 to market—a year in which Agouron brought a fourth antiviral to patients.

"They were just so unaggressive and internally deferential it drove us crazy," Boger recalls. "I don't know what Burroughs Wellcome would have done, but it wouldn't have been as bad as Glaxo. They were just eternally frustrating. Every other drug in the field was perfect despite the facts. And every little scratch on the paint on our molecule knocked the value down fifty percent. It was this weird opposite of being optimistic and aggressive, such a mismatch to Vertex. We *knew* that we often ignored the scratches on the paint. But they would just stare at them and get depressed about how they couldn't do anything."

As hopes dimmed that VX-478 would launch Vertex into the ranks of profitable drugmakers, the company's aspirations shifted to other areas, most promisingly ICE and its partnership with the French drugmaker Roussel-Uclaf. Here was a wide-open opportunity, an unproven target with a highly motivated partner that shared Vertex's passion and excitement. Anti-inflammatory drugs now more than ever were the industry's grail, as the molecular pathways through which the immune system generates inflammation—and the seemingly endless list of painful, chronic diseases in which inflammation plays a leading part—became better understood. The ICE team quickly leapfrogged over Merck and the other leaders in advancing compounds to the clinic.

In late 1997 Vertex and Roussel picked a clinical candidate, VX-740, derived from the original chemical series designed in five weeks by the modeling team in 1994. Since their original agreement, Roussel had been bought by German pharmaceutical giant Hoechst, which went on to buy Kansas City drugmaker Marion Merrell Dow, morphing into Hoechst Marion Roussel, or HMR, which now gave Vertex $3 million in milestone payments. As VX-740 entered preclinical testing in animals, Vertex pleasantly discovered that the Roussel team was more committed than ever. "Every partnership has its ups and down, but this one had the essentials," Sato says. "It was an area of strategic interest and importance, and it had internal champions who stayed committed through the merger."

For most of the next year, while VX-478 lumbered toward approval and VX-740 sprinted toward testing in patients, Boger kept up the pressure as always to advance more compounds into development. In hepatitis C, progress was tortured and grudging, as ever. One additional intimidating and expensive obstacle for all companies weighing the costs of a full-blown program in the disease was the threat of patent litigation from Chiron, the Northern California company that discovered the virus. Chiron had chosen a novel and risky strategy for identifying the microbe, though it had no drug discovery program itself.

After non-A, non-B hepatitis was identified in the late 1970s, and the search began for the infectious agent, numerous labs had tried—and failed—to find antibodies in the blood or grow the virus. Chiron went after the organism's genetic material instead. Researcher Michael Houghton and his colleagues reasoned that if they could clone one or more genes from the virus, they might be able to make proteins from them, and from the protein, an antibody. The effort took seven years, during which hundreds of millions of bacteria injected with bits of DNA were screened for a putative agent by several different approaches. It was like searching for a needle in a haystack when you didn't know what the needle looked like, only bits of it. New fragments were painstakingly compared with all known human and microbial DNA until a new viral entity was identified. When Houghton's group finally convinced themselves and the scientific world that they had the right collection of virus proteins, Chiron claimed an estate of more than one hundred patents, staking out a vast commercial territory. Any company developing a diagnostic test or a new drug targeting HCV needed to license Chiron's patents, for which it typically charged millions of dollars in licensing fees during R&D alone, and made millions more each year in product royalties.

Vertex and Lilly refused to accept Chiron's terms. "We invited them to sue us," Boger says, explaining how in July Chiron filed suit against Lilly and Vertex, alleging patent infringement.

> We didn't think their patents were valid, so we weren't trespassers on their property—we just didn't think it was their property. I can say that they have acted outrageously in the field. I don't think there's any doubt—and I think you could get a hundred people outside of Vertex to agree—Chiron *retarded* research in the hepatitis C field by their actions. And that's not the purpose of patent law. They went around bullying people, and they scared people away from the field. And people who wouldn't pay their extortion-level demands just got out of the field. I'd have some modest sympathy for them, even if I disagreed with their property rights, if they had a program, but they never did. They basically sat back with their piece of paper and said, "We're not going to lift a finger for patients with hepatitis C, we just want a piece of your efforts." And I just find that an outrageous abuse of the patent system.

Indignation, Boger knew, is a friend, especially to an underdog. Vertex pressed ahead against HCV. Tung and the chemists began de-

veloping new scaffolds and "warheads"—chemical groups that seek and
bind to specific subareas of the active site—against the protease. Kwong
and her colleagues jump-started projects in polymerase and helicase. A
month later, Vertex signed a deal with German pharmaceutical maker
Schering A.G. to develop drugs to help regenerate nerves damaged by
neurological diseases: $28 million up front over five years, plus milestone
payments of $60 million. In October, Vertex and HMR announced that
they had begun signing up patients for the first clinical trial for an orally
administered small-molecule inhibitor of ICE, VX-740. First in humans
with a new pill designed with atomic precision against an untested but
highly promising target, and with a strong global partner and US commercial rights, Boger was right where he wanted to be.

❖

Boger intended Vertex to innovate at every level, in every function. As
the company began to brand itself as the new name in AIDS medicines,
its direct-to-consumer advertising grabbed attention. Almost a year
before it expected to receive approval, it wheat-pasted posters in downtown neighborhoods nationwide alongside pitches for Lilith Fair and
Levi's that screamed "Selling Hope Is Easy in an Epidemic" and "Ambition Will Cure AIDS Before Compassion Does." One ad appropriated
the Silence = Death symbol of ACT UP, the militant patient-advocacy
group that led the protests over AZT. Another featured the NAMES
quilt, the survivor community's tribute to its beloved dead. Bart Henderson, Vertex's chief of marketing, explained that the ads were meant
to get people talking about the need to "push beyond the status quo to
develop better drugs." Merck's ads for Crixivan showed, by comparison,
an African-American man climbing a mountain, reaching the summit,
and gazing at the view below. Its headline: "In the Battle Against HIV,
There's a Change in Outlook."

Boger wasn't just promoting a drug; he was promoting ambition as
a superior virtue. Radical improvement—and the drive and tenacity to
make it happen—was Vertex's paramount value. Boger wanted to infuse
its public image with the idea that a revolution in medicine could be led
by a coalition of advanced drugmakers and patients, together demanding more of the pharmaceutical industry. In using guerrilla marketing to

define Vertex's corporate identity as its first product reached patients, he also was broadcasting his own personal ambition to *make better drugs, faster . . . become Merck, but better.*"

In April 1999, two weeks after the Dow index closed above 10,000 for the first time, Glaxo received FDA approval for VX-478, now called Agenerase (amprenavir). Taken by mouth, an adult dosage was eight 150-milligram soft gel capsules twice a day. Fifth and last to market among the original HIV protease inhibitors, the medicine was well tolerated. With the need for flexible dosing—patients on triple-drug regimens took up to thirty pills a day, at specific times, some with food, some without, often having to wake themselves at night—its long half-life gave Glaxo a commercial advantage, with doctors trying to streamline their patients' regimens in hopes of improving their compliance. The problem was the gel capsules. When Aldrich first saw them tumbling out of a pickle-jar-sized bottle containing a four-week supply, he doubted he could swallow them. Glaxo had been unable to come up with a more efficient way of mixing the active drug with other chemicals into a medicinal product that was stable, so Vertex committed itself to producing a follow-up compound that was smaller and better featured for a drug.

"So we ended up with these horse pills, and we said, 'Okay, we don't have any expertise in formulation but we'll make another molecule that even you can formulate,' " Boger recalls. "They just let themselves get beat in the marketplace even though we had the better compound. They disappointed us. It was more dispiriting than I thought. It wasn't just the money; the fact that the royalties were half of what they should have been to us. But because Agenerase and its follow-up didn't get to be the number one compound, we didn't get the credit for having designed, in the face of the entire industry, the best molecule."

Few at Vertex wallowed in disappointment. After ten years in business, the company was well on track, even if the anticipated shortfall from Agenerase meant it would have to burn more cash and limit the number of new projects and hires. There was too much else to savor. The company threw itself a fitting launch party.

On a warm Saturday in May, nearly three hundred employees and their spouses and dates assembled in the company parking lot and boarded a line of buses that stretched around the block, their windows papered over so the partygoers couldn't tell where they were going. The buses deposited them at the Boston Cyclorama, a hulking brick, steel, and glass rotunda built after the Civil War to house a panorama of the Battle of Gettysburg, where there were potted palm trees and pounding music. The bartenders were drag queens, food didn't arrive until midnight, there were a couple of fistfights involving guests; a short film spoof starring Boger and Sato and directed by the up-and-coming documentarian Nathaniel Kahn, *The Lip Balm Incident*, was shown. Some pot smokers felt encouraged to light up. Many would remember the party as among the best they'd ever attended, though it prompted a small human resources fiasco. The company faced numerous employee complaints: parents of young children unable to phone babysitters to say where they were going because they didn't know; people, mostly from other countries, put off by the vamping queens.

Aldrich felt fortunate to have ushered the company this far, adding more and more revenue streams—real and imputed—to the balance sheet. One welcome consequence of becoming more secure financially while having several programs in development was that the business was in a position to cut its losses sooner than later: it could afford to be truer both to the science and to investors. To Aldrich, this meant not having to prop up less than desirable drug candidates. More and more biotech companies, confronted with onerous burn rates, disappointing clinical results, and slender portfolios were now pursuing questionable therapies longer and longer in the face of ambiguous or even negative clinical data. Such desperation, he thought, inflated expectations ruinously, making the inevitable failures that much harder in the end.

Aldrich's business development strategy centered on building out the company's clinical and commercial arms, a task that became thornier as its original partners were devoured by global players. Renegotiations were expected, and, indeed, made sense for both sides. He was surprised that he still hadn't heard from HMR, Roussel's global parent, about the ICE program, expecting someone there to wake up and realize that Vertex had US rights. He says:

They hadn't connected the dots, and finally they did. I got this nice fax from the senior VP for business development saying that as a multinational company HMR intended to commercialize its products itself. So I drafted a note that basically said, "We're really pleased with the collaboration, and we're happy to go forward. As it stands, there are no problems with it from our perspective." I remember we put it over the fax machine and said, "This is gonna be interesting." Within an hour we got back this stream of consciousness: "We *must* have this, we *must* have that!" Guy totally lost his cool.

We had a lot of leverage and we used it. We actually took them up to the brink of a deal where we would have shared control of North America. They could hardly stomach it but they went along with that for a long time. Then we finally said at the end, "There's an alternative, *but* it's expensive: twenty million dollars immediately, give us all these milestones, pay for our sales force, but you can be in charge of the launch in North America." They jumped at that, since they'd been looking into the maw of shared control, which they hated.

Disputes, lawyers say, settle for the other guy's price. Vertex's business, like its science, revolved around opportunity. With a best-case horizon of three to five years before VX-740 might make it to market, Aldrich settled for immediate cash and ongoing support over any potential windfall if and when the drug would be launched. Such was the innovator's dilemma in biotech: you needed dozens of projects to make it to profitability, but in order to generate more projects you were forced again and again to surrender most of their value before you knew if your drug worked. Without proof of concept, it was anyone's guess what a molecule was worth, so deals were sized to fit the competing needs of the partners. In the end, Aldrich extracted up to $206 million in potential licensing fees and milestone payments—provided HMR commercialized VX-740 for rheumatoid arthritis and two additional indications—but these were so-called BioBucks, contingent upon a staggering array of optimal results, and thus as sketchy and evanescent as they sound.

❖

Vertex had a second approach to rheumatoid arthritis that targeted a different enzyme: p38 mitogen-activated protein (MAP) kinase. Kinases had been discovered in recent years to regulate the molecular traffic inside cells that signals them to function, grow, change, and reproduce. Infinitesimal stoplights, so to speak, they switch on and off billions of times per second, letting some messages through, stopping others. Being so deeply implicated in such a vast expanse of basic human biology, and thus disease areas, made kinases both immensely attractive and daunting as drug targets. As the race to solve the genome accelerated to its conclusion, spurred on by the entry of bio-entrepreneur Craig Venter and his gene sequencing company Celera Genomics, it was now recognized that there were about five hundred kinases, give or take a handful. Labs competed fiercely to determine what they did—which ones made promising targets for which diseases—and how to disrupt them. Their premier screening tool was crude but often effective, a bioengineered mouse with the individual gene deleted so that researchers could observe how not having the gene affected rodents: mouse knockouts.

P38 kinase triggers a molecular cascade that causes acute and chronic inflammation, and inhibiting it had been shown to block disease progression in animal models of both rheumatoid arthritis and stroke. In mid-1999 Vertex began clinical trials of VX-745, the first p38 inhibitor tested in humans. Expectations leapt within the company, not just because the discovery process had been remarkably quick and efficient, but because of Vertex's exceptional commercial position. Its development partner in the Far East was Kissei, its collaborator in HIV. Vertex, advancing for the first time with a potential billion-dollar drug on its hands, owned the rest of the world.

Seeking to leverage what the company had learned from the project, much as it did with HIV and HCV, biologist Michael Su began exploring other protein kinase targets. The traditional pharmaceutical view of kinases held that, like proteases, they were too hard to inhibit without inadvertently hitting closely related members of the same family, causing sufficient unwanted side effects to inevitably—invariably, it seemed—doom them as drugs. But Su argued that p38 showed that Vertex was uniquely situated to strike a blow against that orthodoxy, exploiting kinases as a class by compiling data on all the known subtypes and delineating them on an atom-by-atom basis. Individual kinases now were generating hundreds of research papers each. Publication of the entire genome could only ratchet up the science, competition, knowledge base, and business interest in them all.

"We began to think we could tackle the whole gene family," Murcko recalls. It was a logical next step but also a sharp leap in scale and scope, steep even by company standards. "In typical Vertex fashion, we thought, 'Why mess around with a few targets when you can just do all five hundred?' " They all saw the potential. The real question was how to take what they'd learned from one target and extend that to the next. At the same time, with almost no understanding of the roles particular kinases played in which diseases—much less the cellular havoc that inhibiting them might induce—the idea held substantial downside risk. Not everyone was sold, but the logic for taking a shot was compelling and swiftly gathered momentum. Murcko:

> The other thing we said is, "We don't know which kinases to work on. There aren't that many mouse knockouts in the kinase family. There's not a lot of good cell assays. The biology is very primitive." So you have to be ready when you open your copy of *Nature* or *Science* each week, if there's a paper that says kinase X is implicated in rheumatoid arthritis, that you don't need to start from scratch. You want to say, "Aha, I don't have a crystal structure for kinase X, but I do have a crystal structure for kinase Z, which is very similar to X, so that gives me information right out of the chute. Then I've got compounds that inhibit kinase W, and that's pretty close to kinase X, so I have a good chance that right in my screening deck I've got molecules that will hit kinase X." Instead of wasting a year doing the cloning, doing the expression, the purification, setting up the assays, you could instantly go to town and set up a starting point. You've got a compound—a tool compound—to help you convince yourself that, really, the *Nature* paper is correct, and kinase X really

is implicated in rheumatoid arthritis, without having to go through years of discovery.

Boger was mesmerized by the possibilities. Here was the next rational step beyond structure-based discovery—indeed, a platform for the next stage in the evolution of small-molecule drugs. By positioning Vertex as the company best informed and organized to take the explosion in genetic knowledge and apply it to discovering and understanding new drug targets, Boger staked its future on a new paradigm, wedding genomics to medicinal chemistry. As the dot-com boom on Wall Street climaxed, Celera, Millennium, Human Genome Sciences, and other gene sequencing stocks were swept up in the speculative frenzy. They were the biomedical cousins to e-commerce prototypes Amazon, Priceline.com, eBay, and Webvan, which in July set a new standard for irrational exuberance by raking in $275 million in an initial stock offering with little more than a business plan and some marketing research for a national chain of online grocery stores. Dr. Francis Collins, head of the international genome consortium, issued bold claims about how illnesses would be treated in the postgenomic age: clinicians would subclassify diseases by the genetics of individual patients and adapt therapies to suit them; gene-based designer drugs for diabetes, hypertension, mental illness, and many other conditions would follow suit.

"We saw this bubble come up with the hysteria around the human genome sequencing, and I said, 'We've got to be part of this,' " Boger recalls. "Remember, I'm desperate to get more projects going." Vertex had been considering for years whether there was another way to configure its research, as an alternative to traditional disease-oriented biology, the industry standard. Boger recognized that at some point you have to integrate with disease, but the next step past atom-by-atom design wasn't tackling cancer. Committing the company to a major effort in scaling the success of VX-745 across the kinase spectrum, he decided that an organizing principle that *might* be close enough to the molecular level to be useful would be: "I'm gonna design drugs against certain kinds of targets, and I'm gonna be expert at designing drugs against those kinds of targets. And the targets would be defined not by the disease but by their structural similarity."

❖

On the eve of the new millennium, Vertex recast its core strategy around targeting gene families, what Sato and Murcko, in preparation for a pitch meeting, decided to name *chemogenomics*. Boger imagined now vaulting in leaps—one target family after another—toward his goal of self-sustaining, fully integrated, commercialized drug discovery and development. On Wednesday mornings, he, Aldrich, and Sato huddled over bagels to hash out plans. By industrializing new target identification and screening, chemogenomics would require Vertex to rapidly "blow out" its research efforts in every direction, giving Sato control over what would become an ever-expanding share of the company's resources. Aldrich, meanwhile, started to devise the ways to pay for it, recognizing that the chief risk in scaling up so rapidly was that Vertex would become hostage to a rocketing burn rate, leaving it disproportionately dependent on whatever company financed it—takeover bait, in other words.

Once Agenerase received approval, Boger jumped the list of biotech CEOs who could claim to have brought a drug out of the lab. In pharmaceuticals, past performance can indeed be an indication of future rewards, and Vertex was a far more credible investment opportunity than, say, Millennium, which a year earlier had secured more than $1 billion in potential partnership financing based on an agreement with Bayer to identify, using little more than gene sequencing, 225 targets for new drugs for cardiovascular disease, cancer, osteoporosis, pain, liver fibrosis, blood disorders, and viral infections. Boger thought he would need about that much money to launch chemogenomics and that he ought to have little trouble finding it as long as the delusions and self-deceptions of the genomics boom didn't cause it to implode first.

He began to talk with an attractive suitor. Dr. Daniel Vasella was chairman and CEO of Swiss pharmaceutical giant Novartis, Europe's third-largest drugmaker. Since July, the company had basked in the presumptive success of a new cancer drug that inhibited a kinase and that, based on recently released data from a Phase I clinical trial, promised to transform the lives of patients and, indeed, the treatment of cancer itself.

The compound, which later would be marketed as Gleevec, was given at high doses to fifty-four patients with a deadly blood cancer, chronic myeloid leukemia; fifty-three showed complete response within days of beginning treatment and remained deep in remission. Demonstrating for the first time that highly specific, nontoxic cancer therapy was possible, Novartis's success opened the floodgates of competition, and Vasella was determined to defend—and extend—the company's position. Novartis, based in Basel, was also a market leader in asthma and diabetes.

On a visit to Cambridge, where Novartis was building a new research site in an MIT-sponsored technology park down the street, Vasella sought out Boger. "He called me to talk about . . . *life*," Boger says. "He wanted to buy the company, but he never made an offer, because unlike buying a company with products, he knew that there's no such thing as a hostile takeover of a creative group. That doesn't work. You can't do a hostile takeover of talent. He knew he had to woo us. I didn't see what was in it for us to be taken over. It was sort of uncharacteristic of Dan, because he's a pretty aggressive guy, but he listened to that."

Boger sketched out Vertex's vision of chemogenomics. Vasella quickly saw a potential for Novartis. They agreed to "do something big together," Boger says. "In the background of this, Rich and I were both very aware that any big deal like this ended up being a takeover in the end. They were just gonna be such a big part of what we did. And I think that's what Dan was thinking as well: 'Okay, fine, you don't want to be taken over? I'll just build you a big house, and you move into it.' We realized we had to do this again, pretty soon: find another partnership with another family; otherwise we'd be captive of Novartis."

In February 2000 a senior Novartis delegation arrived at the Fort Washington building, Vertex's hive-like hub, to examine the company and consider what use they might make of its expertise. The all-day meeting started in the East-West conference room, the largest common space in the drafty, block-long, low-slung former trucking depot where labs and business offices jostled for space, sharing floors and hallways. For three hours, Boger, Sato, Aldrich, Murcko, and others gave presentations. "Our job," Murcko says, "was to show them that we had thought it all through deeply." Finally, the president of Novartis research, Dr. Argeris "Jerry"

Karabelas, declared his support for a deal. "Karabelas was behind us," Murcko recalls. "He said, 'This is great. This is what we need.'

"The problem was that also at this meeting was the worldwide head of discovery, who clearly was not happy about this. Essentially, the way we were pitching the deal was: 'You guys, Novartis, you don't have to worry about kinases. We'll take care of it.' Think about it. You're a cancer researcher, you're in Basel, maybe you're humor deprived. And along comes Jerry Karabelas, president of Novartis R&D, and he says, 'I have great news. We're going to do this deal with Vertex, a little tiny company you've probably never heard of; they're gonna do all the kinase discovery. We're gonna shut down all of your kinase programs, and Vertex will do it for you.' And we actually believed that this would work. We were banking—*banking*—on their discovery people partnering with us."

With the deal prewired, preparations raced ahead. Karabelas agreed to pay for 165 Vertex scientists, but the company would have to find them and had no space to put them in. Cranes arrived on the flat parking lot where the scientists played roller hockey. Vertex built in three months, on spec, a four-story, 300,000-square-foot lab building—Fort Washington II— connected like an umbilicus to the main activity by an airwalk. Aldrich, asserting uncommon leverage, directed the negotiations, which at Novartis were handled at the highest level and run out of Vasella's suite in Basel. Under the terms of the deal, Vertex would generate eight drug candidates targeting selected kinases. Novartis agreed to pay $15 million up front and $200 million in committed research funding over six years: cash, no BioBucks. It also pledged up to an additional $200 million in loans to pay for "proof of concept" clinical studies by Vertex, which along with aggregate license fee and milestone payments of $370 million brought the total to about $800 million—precisely the amount that the newest, most authoritative study calculated it took to make one new drug. When the agreement was announced in May 2000, Vertex's share price shot to over $50.

Boger caught the wave on Wall Street just as it turned choppy and then crashed. In April investors got queasy. Amazon, eBay, and Yahoo! lost almost one-third of their value in a month. Dead dot-coms littered cities across America, signaling that the boom of the Clinton years was ending. Yet investors remained intoxicated by siren songs of gene-based miracle drugs, and biotech stocks leapt higher for several more months. After Labor Day, Vertex raised an additional $500 million by selling convertible bonds, hybrid securities that let companies raise cash in return for a guarantee that holders can convert their bonds into stock at a later date for an agreed-upon price. The price was $92.26 per share. Aldrich could see that the current climate in biotech was unsustainable too and that capital would soon disappear. He was happy to complete the Novartis deal and the bond sales, which Vertex would need to fund several upcoming clinical trials, before the gate came down, but, as always, even in flush periods, fretted when the size of the wheel at Vertex got suddenly bigger. "I didn't have any inside knowledge other than my general view that the market was insane," Aldrich says. "I was antsy. NASDAQ was trading at two hundred fifty times earnings. The Internet thing was clearly off the tracks. Genomics was even more egregious in terms of the underlying economics."

The company's stock price spiked throughout the fall, almost doubling by Election Day, when it traded briefly over $100 before closing at $97. More than a few of the scientists and even some of the company's lawyers had become avid day traders, and with the value of their Vertex options soaring, they scrambled to stay ahead of the bubble's gyrations. Boger recognized that he, Aldrich, and Sato needed more executive help and a reporting structure that could translate the Novartis deal into an effective and productive drug hunt. He retained his title as CEO but relinquished the presidency to Sato. The Novartis deal had been all his, an expansion strategy so novel that he considered it "nearly proprietary," and he assumed from here on that Vertex's next collaborations in chemogenomics would copy it and likewise originate from the top. He asked Aldrich to remain chief business officer, reporting to Sato.

The national election hung in the balance, the country consumed by the Florida recount. *Bush v. Gore* groped toward a showdown in the Supreme Court as Washington grew bitter, paralytic. Another long, lightless Boston winter descended. Aldrich, after eleven years of eighty-to hundred-hour weeks, imagined selling most of his founder's stock—worth now about $30 million—purchasing a yacht, hiring a crew, and cruising the Caribbean. Grievously miffed, he doubted that Boger's reorganization could work and considered it wrong, possibly lethal, for Vertex. By the end of the year, he was gone.

His reasons were both personal and practical. He had been with Vertex since the start, working and traveling side by side with Boger. "Josh is the accelerator," he often said. "I'm the brake." He felt that their relationship not only had been highly successful but also would become more and more crucial as Vertex grew. Without financial discipline and restraint, Boger's visions and reality distortions contained real risks— for employees and investors, especially—far beyond just elevating false hopes and a perception of arrogance. Aldrich realized he couldn't tolerate having a limited hand in running the firm in the future—a future that for the first time perhaps since the Searle threat now seemed to him seriously imperiled, not by an aggressive competitor but by a decision to double down on Boger's scientific hubris.

"Josh wanted to reorganize around science and wanted to make Vicki president of the company, and for me, it just seemed a good time to go. I did think it wasn't the best thing for Vertex. I have a lot of respect for Vicki, but that was not an arrangement that there was any chance was going to work for me, frankly. And I don't think that anyone else thought it was going to work, other than Josh."

# CHAPTER 4

## JANUARY 22, 2001

Two days after George W. Bush was inaugurated in a bone-chilling drizzle, the FDA approved Schering-Plough's Pegintron for the treatment of patients with chronic hepatitis C. The drug was the first once-weekly injection of genetically engineered alpha interferon, a biological molecule released to help uninfected host cells resist new infection by a virus. In a study of more than a thousand patients comparing Pegintron to the company's decade-old Intron A, which was shorter acting and taken three times a week, it doubled the cure rate to 24 percent when taken for forty-eight weeks. More than half the patients in the study complained of flu-like reactions: fevers, chills, muscle aches, sweating, exhaustion. A third of them reported being depressed. Roche, too, had filed for approval for a longer-acting interferon, Pegasys, setting up a marketing war as both companies began further studies combining the new medicines with a second broad-spectrum antiviral, ribavirin.

In the collaboration with Lilly, Vertex rediscovered that HCV wouldn't yield to the usual stratagems. The front end of any conventional drug discovery effort consists of a screening assay, wherein a multitude of compounds are tested for biological activity. Most screens are deliberately set up so that about 1 percent of the molecules are hits, but Vertex and Lilly devised one so sensitive that any detectable hit would cause a signal. "Lilly screened their entire sample collection against HCV protease," Boger recalls, "and they had none. Zero verified hits. It was useful to have done that experiment, because even for a company like Lilly, it did sort of cement internally that there was no other way to do this except design."

For the chemists and modelers, the goal wasn't just to invent a molecule that blocked the protease but to fulfill the towering requirements that make all anti-infectives, especially direct-acting antiviral drugs, so hard to develop. The key problem is resistance: the virus, replicating rapidly, evolves variants that allow it to evade a new threat to its survival. Medicinal chemists are skilled at molecular subterfuge, substituting groups of atoms that cause corresponding groups on a target to bind to them instead of their usual partners. But a variety of other properties ultimately rule: the mix of features that make a compound "druggable"— that is, safe, soluble, and stable enough to become an approved medicine. In the end it must be formulated to deliver a pure, precise dose in milligrams to patients while being manufactured—cheaply, reliably, safely, and competitively—in multi-ton lots.

"There are lots of ways to design a potent compound against an active site, and one of them is to put down a lot of grease that touches a lot of grease," Boger says. "You can lob down a lot of stuff that touches a lot of stuff. But the problem is, if you rely on that kind of strategy, you're just asking something like a viral enzyme to mutate and knock your compound out. So we did a lot of dynamics with the enzyme structure to see what kind of flexibility it had. And we did a lot of work trying to ask the question, not what's the best way to make a potent compound, but what's the best way to make a potent compound that the enzyme will have the hardest time kicking out—and, oh, by the way, because of the nature of the active site, that doesn't turn out to be brick dust [about as soluble as sand]. It was a really hard problem. It took a long time."

Vertex and Lilly clashed at every turn. Key to designing a drug is determining where, and in what concentrations, it should collect in the body. The rule of thumb for most drugs—and most drugmakers—is that a molecule should be small enough and soluble enough to circulate in the blood and be excreted in the urine. Large lipid-loving compounds attracted to fats and waxes—grease—are removed from the plasma and gather in the liver, which makes fat and absorbs toxins. No one knew for certain where HCV hid. But a year into the project Kwong had a key insight while sitting in an educational session at a meeting on liver diseases. A transplant surgeon showed a slide comparing RNA levels of the virus in patients just before and after they had their livers replaced. There was a precipitous drop. "*Boom*," she recalls. "Back then it was controversial where the virus replicated. Well, my God, I don't know where else it replicated, but it definitely replicates in the liver. I came back, and I said to the team, 'We need to target the liver.' "

Lilly resisted the idea, passively at first. Its corporate culture was conventional, rigorous, Midwestern, and orthodox, and also, tinged with fresh embarrassment and regret over the dissolution of another recent partnership in antivirals. Lilly had sponsored Agouron's HIV program, then decided it didn't want to develop the molecule and opted out of the collaboration; as a result, Agouron, in partnership with Japan Tobacco, took back worldwide commercial rights to what would become a billion-dollar drug. In discussions with Lilly's scientists, Tung and others from Vertex encountered mounting disagreement over their fundamental goals, even the drug-like profile they thought they had committed to pursuing together. "One of the questions was, is the virus replicating solely in the liver, or are there other reservoirs outside the liver where it's replicating elsewhere and becoming resistant?" Tung recalls. "We had a tremendous disagreement with Lilly over this issue. Their received wisdom is you've got to have significant blood levels of the drug."

Lilly's concept of a druggable molecule was limited by what researchers already knew about all existing drugs and their chemical interactions with the five hundred or so known protein targets in the body—a statistical approach preferred throughout Big Pharma for narrowing leads that's heavily biased toward compounds that are small, soluble, and well behaved. After much intellectual soul-searching, Tung, Thomson, Murcko, and Pravin Chaturvedi, Vertex's head of pharmacokinetics, developed a drug profile that fell outside all its major parameters. "Lilly was arguing for a very mundane, well-known computational algorithm that basically works only within its memory banks," Thomson says. "It can interpolate and piece together successful combinations of the pieces of drugs that we've already tried. But it can't invent new ones. In essence, they were trying to tell us that their new ultraslick horse and buggy was far more sophisticated than our Ferrari because they didn't get what the technology was that made the Ferrari."

Sato, becoming directly involved, insisted that the collaboration focus on molecules that went to the liver. Hepatitis C, after all, was a disease of that organ. More crucially, an amended agreement was needed before Vertex could start to develop relevant animal models to test if its compounds stopped the virus where it needed to be stopped. "Vicki certainly, as she will always do, let them know who's boss, and they responded accordingly," Tung recalls. The Lilly team, already souring on the partnership, continued to make new variants on Vertex's design even as more resistance to the program erupted among the process chemists at Lilly, who were struggling to scale up production of the final class of compounds so they could be tested in animals and humans. Vertex chemist Dave Deininger recalls, "Their preclinical development people said, 'We can't do anything with this. It's got none of the characteristics that say, "Oh yeah, that's gonna be a great drug." It's too big, it's too greasy, it's peptidal. It's a rule breaker of all of the rules.'"

Highly crystalline, the eventual drug candidate—VX-950, made by a Lilly chemist and based on a Vertex design—was "pretty much brick dust," as Boger put it. It was less soluble than marble. Vertex had no formulation group of its own, yet as its faith in Lilly's group faded, Boger could see that HCV might reprise HIV at Vertex: a research triumph but a commercial disappointment, based largely on his partner's difficulty in getting its molecules into pills and down people's throats. It was a problem he recognized but had no way of controlling.

❖

A breakneck expansion strategy, especially in a fast-growing high-tech sector, guarantees a business leader a rare degree of power and influence within the company. As chairman and CEO, Boger had more or less blank-check support from the board of directors, most of whom he had recruited after the original members moved on to newer start-ups, and from its first chairman, former head of the war on cancer and pioneering venture capitalist Benno Schmidt Sr., retired.

His executive team now reported to Sato, who elevated the triumvirate that had made its mark with HIV and ever since then had best anticipated and met the broad challenges Boger laid out—the hits, you might say, in his social experiment. Thomson became vice president of research; Murcko, chief technology officer and chairman of the scientific advisory board; and Tung, vice president of chemistry and head of compound selection. Under the redesign, Tung reported to Thomson, but Murcko, with a relatively small group, retained an independent standing as the company's in-house big thinker, practical visionary, and prodder.

Boger aggressively promoted chemogenomics as the future of the industry, if not a panacea for everything that ailed Big Pharma, something very close to it. With Agenerase on the market, eight drugs in human testing, and hopes of putting five to seven early-stage drug candidates in the clinic during the next twelve months, he told investors that by 2005, Vertex would start submitting two to three drugs a year for FDA approval. The figure, as he noted, was twice that of Novartis and Vertex's newly reconstituted partner in ICE, French drugmaker Aventis, formerly HMR. It was an exorbitant claim, one depending on strong, unambiguous findings from its clinical trials and flawless execution of many functions that the company had yet to incorporate.

Even as the genomics bubble burst amid spectacular flameouts of once-promising single-product biotech companies, and as the slump only deepened in pharmaceutical R&D, Boger's brash evangelism kept the company's share price afloat above $40, twice what it was a year earlier. "My goal is to grow up and overtake Pfizer, not to be Genentech," he told *BusinessWeek*. Pfizer was the world's largest and richest drugmaker, purveyor of Viagra, the fastest-selling pharmaceutical product in history. It recently announced it would spend $5 billion on research in 2001. Former highflier Genentech, the world's prototype biopharmaceutical firm, still had no blockbuster after twenty-five years in business, though recently it had begun selling two breakthrough cancer drugs. Powerful scientifically, lauded for its academic-like corporate culture, a pioneer in off-label sales with its strenuous promotion of recombinant human growth hormone for children whose only disability was that they might grow up to be short, the company had become synonymous with the industry's overly aggressive marketing and oversold hopes. Its stock, which crested at $85 in the bubble, now traded at about $20.

To optimize the research engine it was assembling, Vertex went looking for a company to buy that could quickly do for it what it couldn't do for itself: namely, provide a way to screen large decks of compounds against its fast-growing array of newly discovered targets, then test the molecules in cells to see what effect they had. High-throughput screening was the very opposite of structure-based design: an attempt to focus and accelerate discovery not with precise molecular knowledge but with advanced robotics, miniaturization, and proprietary methods for culling new hits. But in the new era of digital information sharing and genomics—call it *infonomics*—Boger believed the two approaches could complement each other.

One business soon stood out: Aurora Biosciences Corporation of San Diego. Cofounded by future Nobel Prize–winning chemist Roger Tsien, the company led the industry in developing assays, screening, and cell biology. It had drug discovery programs in numerous areas, but its $60 million in revenues came chiefly from providing screening services to more than fifteen major life-sciences companies and research organizations, and its management felt that a merger with Vertex could help it evolve from a screening site into a fully integrated drug discovery company. Aurora's three hundred employees worked in a gleaming white two-story industrial building in a grassy R&D park on a bluff overlooking, to the east, the freeways and metastasizing sprawl north of downtown. A good golfer with a Santa Ana wind at his or her back could stand in the parking lot, facing an uphill lie, and hit the famed Torrey Pines championship course with a strong drive and a long three-wood.

After three months of negotiations and strategizing about how to put the two companies together, Vertex announced on May 1 that it would acquire Aurora for $592 million in stock. "We are going to seize as much of this ground as we can," Boger told the *New York Times*. "This is not something we can come back to in twenty years. The fun will all be over." From Aurora's perspective, the acquisition promised considerable, but not total, independence. After the merger, Aurora would operate as a wholly owned subsidiary of Vertex, but all Aurora partnerships would need the approval of Vertex management. Aurora's chairman, chief executive, and president, Stuart Collinson, would sit on Vertex's board.

❖

A few days after the announcement, Boger and Sato met in Boger's office to review the deal and cut through the remaining issues. Most pressing was Aurora's collaboration with the Cystic Fibrosis Foundation. The foundation was a health care nonprofit started by parents of children with the disease, the most common fatal genetic disorder in the United States. CFF had contracted with Aurora to screen for drug leads against four protein targets that together they had identified as promising.

Cystic fibrosis is a painful and wasting condition, a nightmare for children and for their parents, who watch them suffer more and more, struggling with little hope against coughing and wheezing and respiratory infection and malnourishment, knowing in the end, even with the salvation of a lung transplant, they will almost surely die young. Emerging from deep familial agonies, the foundation drew its identity and character from informed, activist parents faced with an excessively cruel genetic twist; that is, *both* parents must carry the defective gene. They don't have the disease themselves, but in coming together, they give it to their kids.

Founded in 1955, when children born with CF seldom survived into their teens, CFF grew into a radical, determined, hard-charging, and highly successful pioneer in what's called venture philanthropy. It raised and invested nearly $600 million to advance life-sustaining therapies, using the proceeds from one to leverage the discovery of others. These symptomatic treatments had nearly tripled life expectancy, and patients now lived on average until their midthirties. But there still were no medicines for attacking the underlying genetic defect that causes the cells that line the body's cavities and surfaces to become congested with thick mucus secretions, blocking up the pancreas, the gastrointestinal tract, and, most lethally, the lungs and airways.

CFF President and CEO Robert Beall had approached Tsien, among others, about searching for new avenues, to get at the underlying cause of the mucus buildup. Tsien was the only one interested. In May 2000, after enlisting William Gates Sr., head of the Bill & Melinda Gates Foundation, to donate $20 million toward the idea, Beall and CFF pledged Aurora $47 million over five years to screen for novel therapeutic agents, the largest contract ever awarded a for-profit business by a nonprofit health organization. Sato was scheduled to meet with Beall the next day to discuss the future of the program. A biochemist, he was a former division chief in endocrinology and manager of most of the CF work at NIH, a heavily networked activist-scientist with a strong medical staff, excellent contacts in high places, and ferociously committed donors. Bald except for a ring of white hair, a visceral, bespectacled bullet of a man, Beall shared the urgency of patients and families and was indeflectably determined to press his collaborators to feel the same way.

Boger would present his recommendation the following week to the board. Unlike the partnerships with, say, Aventis in rheumatoid arthritis and Lilly in hepatitis C, this was a nonobvious alliance and an unusual business model. The disease, the foundation, and Beall alike all raised numerous issues complicating his and Sato's thinking. Though they were enthusiastic about the opportunities and drawn by the challenge of chasing a genetic disease with a small molecule—disabling a normally functioning protein is one thing; fixing a broken one quite another— both of them first had to weigh the economics, which at a glance looked prohibitive.

They knew they could very reasonably shelve the project. CF affected only thirty thousand people in the United States and about seventy thousand worldwide. Could Vertex justify to investors passing up potentially far more lucrative opportunities if it meant having to fund clinical trials for so limited a disease? (With so few subjects and a small cartel of specialists controlling access to them, the cost of CF trials would reach $100,000 per patient.) Would the CFF, with its own interests and its emotional urgency about bringing new treatments to patients, meddle in Vertex's research? Did Beall, with no experience in industry, even know what a real drug candidate looked like? Sato recalled their line of reasoning in a Harvard Business School case study:

> This deal with the CFF made great sense for Aurora as a screening company, but can we craft a deal that makes sense for us as a drug company? We already have several compounds in Phase II with greater market potential, so will continuing a partnership with the CFF just be a distraction? It's the opportunity cost that's key here:

If we commit to CF, are we limiting our upside on drugs with bigger market potential? And obviously, the terms of the Aurora partnership don't begin to address the investment we will need to make to launch a real drug discovery effort.

As they went back and forth, Boger's phone rang. It was Aldrich, calling to congratulate them on the merger. Boger and he had remained friendly, continuing to share season tickets to the Red Sox and, on occasion, swap business advice. Aldrich said he favored extending the agreement with the foundation, provided the funding approximated what Vertex might get from a corporate partner, and that the CFF partly fund some early-stage development as well. Though the market for CF drugs was comparatively modest, he urged Boger to keep the program going. "Working with pharma," Aldrich says, "we were always caught in this tug-of-war over who gets the economic value of the drug. Seventy-five percent of the negotiations center on that. When working with charities you just don't have to engage in that tug-of-war. It's true, Wall Street gets more excited about deals with big corporate partners, but in the end, this is a less expensive form of financing."

It helped that a small group at Aurora was passionate about the work, which outside the company was largely dismissed as both a long shot medically and, given the relatively small upstream rewards, just not worth the investment—a black hole, scientifically and financially; a hobbyhorse for Beall. Academic research in the field had bogged down; Vertex would have to do almost everything itself. The area was saturated with disappointment. For a decade, hopes had soared that a normal version of the CF gene could soon be delivered to affected cells in the lungs, pancreas, and other organs of dying children, and that those cells would begin to churn out healthy replacements for the defective protein at the root of the disease. But all such "gene therapy" experiments had been discontinued in 2000, a casualty of oversold hopes and deaths among the first test subjects.

Boger and Sato weighed their decision as dispassionately as they could. Rare hereditary diseases, particularly those caused by lone genetic mutations, represented just the sort of scientific challenge that Vertex and Aurora coveted, far more tractable and open to a highly focused research effort than, say, most cancers, where a dozen things might be happening to make cells go wrong and metastasize. The rewards for groundbreaking science in one such disorder might open up related possibilities in other areas. "We're always a bit of a sucker for exciting science with a mission," Boger told the HBS authors. More to the point, the bottom line was, insistently, the bottom line. "Simply and clearly put, without the CFF funding, Vertex would not be in CF."

No company would be. Not even the richest could afford it.

❖

Many people from around the world would remember the moderate and uneventful summer of 2001 as the fluttering end of a daydream. America stood at a time and place in history where its top news stories for two months running were the mounting suspicions surrounding a California congressman about the disappearance of a young female aide and a spate of shark attacks up and down the East Coast. After the tech-led bust on Wall Street and the disputed election, the national mood was—*how to say it?*—hungover, postcoital, dispirited.

In August, before heading for his West Texas ranch, George Bush announced a new federal policy governing research that used human embryonic stem cells. It was Bush's first prime-time address to the nation from the Oval Office. He decided that federal money could be spent on such experiments, provided the cells were already derived from an embryo that was created for reproductive purposes and was no longer needed, and that the donor had given informed consent and held no financial stake in the research. Balancing the resistance of antiabortion forces against urgent calls from those for whom such innovation was a last hope, Bush crafted a careful compromise that opened the door, however slightly, to a new and controversial field of biomedical discovery. His aides hailed the decision as Solomonic, a preview of a humble, temperate White House. Half the country still regarded his presidency as illegitimate.

Dr. John Alam, age thirty-nine, assumed a pivotal role in recasting Vertex from a lab-driven boutique—"a creative group," as Boger called it—to a prescription drugmaker. As senior vice president of drug evaluation and approval, Alam had overall responsibility for preclinical development, clinical development, and regulatory affairs: that is, the full gamut, from figuring out what toxicological studies to do and determining what doses to prescribe for what indications to having a product approved for sale by the FDA and other countries' regulators. Within the free-floating group of executives swirling around Boger and Sato, none had more corporate obligations, or access, or sway, than Alam.

He was striking, angular, intense, sharp-featured; slim and dark-skinned with straight black hair and rimless glasses. Thomson had a lot more people reporting to him, Murcko had more strategic influence, but Alam was a company officer, attending board meetings and traveling widely with Boger and Sato to reassure investors that Vertex's clinical sights were set as high, and in as competent hands, as the research labs. "It was always then Josh and Vicki and some number of senior people," Alam would recall. "But if there was a two and a half, I was perceived to be the half person."

Alam focused throughout the summer on the anti-inflammatory compound VX-745, the company's p38 kinase inhibitor. At his previous company, Biogen, he'd been the sole physician associated with the approval of a biologic molecule, Avonex, another anti-inflammatory now on the market for rheumatoid arthritis. With such drugs, once you got clearance to begin human testing, you determined the pharmacologic effects empirically, based on drug levels. The varied effects of a biologic within the cosmos of the human body—good and bad, wanted and unwanted—are revealed over time by experimenting on sick people. Small molecules are different. You need to be able to predict what they'll do in humans before you administer them. "Small-molecule drug development is more rational and also just fraught with risk," Alam says. "It's toxicology, toxicology, toxicology."

Vertex had known since 1999, when the company announced it was starting human trials with VX-745, that the compound passed the blood-brain barrier, a semipermeable cellular screen that allows very few substances to cross into brain tissue. In itself, this was no cause for alarm. Short-term studies at ten times normal therapeutic concentrations in rats, dogs, and other species showed no adverse effects. A twelve-week

Phase II study in patients with rheumatoid arthritis launched in January demonstrated that the drug knocked down inflammation, was well tolerated, and had no serious adverse side effects on the central nervous system. As Vertex awaited longer-term toxicological results from an outside vender, and Alam prepared to press the FDA for permission to expand the trials, Boger strained to contain his fervor. "Everyone could taste it," he says. "It was the first kinase drug, and it was outside the Novartis deal. We were getting Novartis to pay for all of our kinase research, yet our lead drug was outside the agreement—it was, like, how can it be better than this? We got early clinical results from arthritis that said it was working. And we were just getting ready to configure a Phase III trial—an approval trial. The drug would have been on the market in 2006, and it would have been our drug, 100 percent, worldwide."

In August Alam got the complete results of a longer-duration study in dogs. The data, unexpected because the initial studies had been clean, showed irreversible organ damage. Not only did the animals show symptoms of neurotoxicity but histologic changes as well; there was some degeneration in the nerve fibers of the spinal cord. "I said, 'Wait a minute, this isn't a go decision,' " he recalls.

Boger was stunned by Alam's report. He expected—wanted—the tox study to show something. Dogs have "leaky brains." Their blood-brain barrier is more porous than other animals', including humans'. Though most neurotoxicities found solely in dog brains prove to be harmless in people, they at least give a clue what to look for. And if there's any possible risk to patients, you'd rather know sooner than later, even if your doubts are based on a famously poor predictor. "Drugs kill dogs," Boger often said, "and dogs kill drugs." On September 2 Boger, Sato, and Alam went through all the data and agreed that they couldn't start the drug in the next cohort: they would submit all the data to the FDA and tell the agency they were stopping the study. Consulting their calendars, they set a timetable. They would take the decision to the board and then announce it publicly on September 12.

Boger recalls: "It's something in people that's hard to determine—a bad problem in the brain. You don't know whether it has cognitive readouts because the dogs ain't telling. All you know is that their brains look bad when you cut them open—kind of a hard assay to do in the clinic. It's the only case that I can recall knowing about directly in which there was no way forward. And so we had to stop this thing. And that was not pleasant."

❖

If 9/11 "changed everything"—as voices from all over began predicting even as the buildings still smoldered—Boger didn't notice. Anticipating market chaos, panic selling, and a disastrous loss of value in the wake of the attacks, the New York Stock Exchange and NASDAQ remained closed until September 17, the longest shutdown since 1933. When trading resumed, the market declined the first day by 7.1 percent, the biggest one-day drop in history. Vertex shares slid into the upper twenties but held firm. Boger thought the toxicity issue was "extremely bad luck," but the company was moving ahead with second-generation compounds that didn't cross the blood-brain barrier. He remained upbeat and confident as Vertex readied a press release announcing the VX-745 reversal.

"Internally, I don't think that anybody thought that it signaled anything wrong," he says. "I think it did signal to us that we had to take our own word seriously. If we thought the industry odds were one in three hundred, and we could do one in thirty, then we'd better be prepared for one in thirty. It wasn't going to be one in two. It just hammered home that design wasn't going to be the magic key that unlocked every problem, that there still was going to be the unpredictability of whole animal biology."

After a year of interviewing candidates to fill the holes in Vertex's executive ranks, Sato hired as chief financial officer an ambitious and handsome, young British expatriate, Ian Smith, a partner with accounting giant Ernst and Young, who had vetted the Aurora acquisition. Smith would try to do for the company what Aldrich had done: finance Boger's ever-widening ambitions until Vertex made money.

As general counsel, Sato and Boger decided, despite raising some eyebrows, to bring in Boger's older brother Ken, a senior partner in the national law firm of Kirkpatrick and Lockhart. Ken Boger, calm and genial in negotiations, towered like his brother but was broader and more rumpled; he was a Vietnam veteran with an MBA from the University of

Chicago and a law degree from Boston College. He had an accountant's eye for irregularities, a boxer's sense for opposing weaknesses, a minister's calculating gaze, and a deep appreciation for human foibles. Ken had been legal strategist, enforcer, and rabbi on all Vertex's development deals since 1989. He and Smith were scheduled to start in late September. When Sato called to tell him about the next day's press release concerning VX-745, she half-joked, "I assume you still want to show up tomorrow."

Vertex's most senior corporate lawyer had resigned, apparently to clear the way for Ken's arrival. "The department was very thin at that point," he recalls. "So the decision was made by Josh and Vicki to do something that I probably wouldn't have done, which was to let the chief patent counsel be the guy to sign off on press releases." As the highest-ranking lawyer on staff, Andrew Marks had primary responsibility for intellectual property issues, but he was also the person employees consulted regarding whether or not they could sell shares under Securities and Exchange Commission (SEC) rules. On September 20 he wrote to Boger in an email, "I guess that I am troubled about any employee-trading prior to that release because it is likely to have an effect on the stock (looks like I can't sell any shares) and, depending on the degree of that effect, could create the perception of insider trading."

The next day Marks received a draft of the press release. The timing would be consequential, as on that day he also sold 20,900 shares of Vertex at an average price of $22.81, netting $476,765. "The Thursday before I started was when Andrew saw the press release on the 745 program termination and was asked to review it," Ken recalls. "Well, Andrew was a compulsive day trader. A lot of people were back then. And he had a big margin account with a brokerage, and he had Vertex stock in his margin account. All of his dot-com shares were falling. So he gets a call from his broker, saying, 'We're gonna have to liquidate some of your shares to meet the margin call.' And he told them to liquidate the Vertex shares. This is when he knew about the release."

Vertex's stock price dropped 20 percent on Monday, sliding to $17.74 and tripping simultaneous alarms. The SEC, which flags all stock transactions by employees when shares fluctuate wildly, brought charges against Marks for insider trading. After pleading guilty, he was sentenced to a year and a day in jail. Bells also went off in law offices in San Diego, Detroit, San Francisco, Boston, and New York, where plaintiffs' lawyers, having discovered a lucrative new opportunity in securities litigation, joined together to bring suit against Vertex in federal court for defrauding investors. The suit claimed that the defendants—Boger, Sato, Alam, Murcko, and Marks—concealed what they knew about VX-745's neurotoxicity long before the announcement; and that by making "false and misleading" statements, they'd been able to artificially inflate the price of shares. "Delaying making this news public," the plaintiffs alleged, "not only permitted Vertex to receive milestone payments on VX-745 from Kissei but, as importantly, allowed it to continue to sell the myth of its enhanced drug delivery process and sign collaborative deals with other drug companies and, finally, acquire Aurora."

Conflating Marks's desperate trades with two years of optimistic public comments by Boger and Sato and a talk by Murcko at the American Chemical Society about predictive modeling, the case tapped into a new sense of frustration among some longtime Vertex investors: a feeling that they'd been misled, that structure-based design was supposed to anticipate these kinds of risks and fix them, that Boger had sold them a chimera. Murcko recalls in an email: "The argument was basically, 'Look, you Vertex guys are such geniuses, you know how to predict *everything*, you give lectures at international scientific meetings on the subject, you're widely published and cited authorities, yet are saying you *had no idea* that VX-745 was toxic. Is *that* what you're trying to make us believe?' "

Within a month after Vertex announced it was terminating its trials in rheumatoid arthritis and a blood disease called myelodysplastic syndrome, the stock regained almost all it had lost in late September. The lawsuit would proceed anyway. Lilly, meanwhile, accepted VX-950 for development against hepatitis C. Though the announcement would be delayed a few months, Boger was already looking ahead, considering what moves would be necessary as he, Sato, and Alam reshuffled the development portfolio in a down capital market. On the day he and Murcko learned they were being sued, they were standing in the open space where Vertex held its beer hour, looking out over the grassy spot across the street where George Washington once ordered his troops to build a battery against the British. Now it was a small, neglected park. In the middle was a towering flagpole, and one of the cement-filled cast-iron cannons had in its trajectory the Prudential Center—the Pru—which, like other skyscrapers, now looked like an eerily easy target.

"I guess we've made the big time," Murcko joked.

"You only get sued," Boger nodded, "if you have something worth going after."

❖

For CEOs of small, unprofitable biopharmaceutical companies, the annual J. P. Morgan H&Q Healthcare Conference at the Westin St. Francis Hotel in San Francisco during the first week in January is, as Boger once described another investor cavalcade, "a meat market." Indeed, it is *the* meat market, several days of ongoing presentations and meetings and dinners and mingling where five thousand biotech executives and financiers gather to compete, compare, schmooze, drink, and hook up. A CEO with a hot story or stock can't walk down the mobbed corridors without dozens of people pressing in and proffering their multilanguage business cards.

It's rare that the announcement of a clinical candidate would merit much attention here. But in the six years since protease inhibitors revolutionized AIDS treatment, expectations that they would soon do the same for hepatitis C had risen—and sunk, thwarted by the grinding pace of HCV research and the scant breakthroughs reported by the companies themselves. Excitement at H&Q about the field had faded, resembling if not equaling the moroseness that poured out of the 1993 Berlin AIDS conference, where the promise of new cures had all but gone out.

And so it was perhaps no surprise that the *Times* devoted a twenty-inch story to Boger's routine announcement during his presentation that Vertex and Lilly had chosen to develop the HCV protease inhibitor VX-950, and that Vertex was receiving a $5 million milestone payment for the achievement. Boger was careful to note that the first patients wouldn't be tested until at least 2003 and that further animal tests were needed before clinical trials could begin. VX-950 was nowhere near being a proven hope against disease, much less a compound that could be put in a pill and swallowed. Yet amid the general pessimism, and joined with the fact that no other company had yet disclosed a clinical candidate, this was something to crow about: a lead public position against a major epidemic for which the existing treatments were grueling and mostly ineffective and where Big Pharma's pipelines appeared empty. Boger regaled his audience with a new metaphor profiling the scientific challenges and risks posed by the virus.

"Instead of stuffing a bomb in a cave, which is what the HIV protease inhibitor does, it's like climbing a sheer rock face," he said. "In my nearly twenty-five years in the industry, this is the most difficult drug design problem that I've ever encountered."

With two landmark successes in designing protease inhibitors—three, if you counted the more easily formulated follow-up to Agenerase that would soon get FDA approval, Glaxo's Lexiva; possibly even four, if you also included pralnacasan, the new preproduct name for VX-740, the ICE inhibitor that Aventis was testing in almost three hundred patients in Europe to see if it relieved their crippling arthritis—Boger could foresee Vertex's next expansion. It got deep in discussions with Glaxo about the more than five hundred human proteases. Glaxo might have been an unlikely partner given the ongoing dismay over its corporate diffidence and failure to dominate the AIDS market despite having several strong compounds, but its R&D and business people had bought into the idea. Again Vertex would be getting paid off the top, and Boger professed no worries as long as the checks cleared.

In Wall Street's eyes, Vertex was one of the few early-stage biotechs to survive the bubble unscathed. It had enough programs in the clinic and cash on its balance sheet to weather sudden downdrafts in its business. Vertex shares climbed back nearly to $30, but once again in April lost a fifth of their value when Aventis announced disappointing results from the pralnacasan study. This time no one sued. The anti-inflammatory failed to ease symptoms significantly. But true to the enthusiasm that had driven the collaboration from the start, both companies said they would continue working on the medicine, since as many as 40 percent of patients in certain groups, including those who took the highest doses, showed some improvement. Although hardly transformational, the molecule was active. Pralnacasan might not stop the body's joints from breaking down—might not be what was coming to be called almost reverently in the industry an "oral Enbrel," referring to the breakthrough biologic that was winning the market in rheumatoid arthritis and was well on track to get labeling approval for several other illnesses—but pralnacasan warranted further interest. What else it might do could be discovered only by performing larger, longer, costlier, more varied trials.

❖

Thomson managed Vertex's drug discovery machinery in Cambridge even as he built it; promoting from within where possible, shifting around responsibilities, "picturing problems in a more structured manner and making that a template for solutions," he recalls. As he tried to arrange the pieces to give fabric to an organization that was moving on two tracks—Novartis, and everything else—he also continued to drive the Lilly collaboration, endeavoring to identify and convince key managers in Indianapolis that VX-950 was a developable compound. "We were in their faces, and that made them nervous," he says. Thomson stopped smoking at work, striving to be less wild man and rogue, more exemplary.

Sato assigned Murcko overall responsibility for bringing Vertex and Aurora together and assigned Tung to teach Aurora to hunt for drugs. Each welcomed the assignment. Murcko wanted to ensure that Vertex kept expanding its possibilities and didn't bog down in one approach to drug discovery, even one as powerful as design. Tung remained skeptical of Vertex's boldest claims for design and was eager to test his own theories outside the mother ship of the Cambridge labs.

Tung and Murcko alternated weeks in San Diego while Aurora's senior management team, absorbing the impact of the merger, reshuffled. Landing on top after the shake-up was biologist Paul Negulescu, formerly head of research, who now ran the site for Vertex. Negulescu was even-tempered, deft, strong willed, enthusiastic, and patient, a true believer in Aurora's high-throughput screening who had joined the company at age twenty-seven fresh from a postdoc in biophysics and immunology. One of Aurora's first five employees, he'd been hired as director of biology, "which meant director of myself," he recalls. Now he resolved to "stay through it, make the acquisition a success" by helping make the site fully capable: bringing in PK, pharmacology, and medicinal chemistry to help improve and advance the hits emerging from Aurora's "cell-based stuff."

Negulescu reported to Murcko. "We had a lot of projects, most of them very, very early, and Mark guided us through the culling process," he recalls. "The one thing I remember about all of those interactions, whatever the purpose of the visit or the topic, there was always the sense that we would make something good happen. I think that's part of Mark's genius. He's a glass-is-half-full person, and you shouldn't underestimate how important that was. The support, and confidence he gave us just by being here and saying 'I believe in you guys' was probably as significant as anything else."

Cystic fibrosis was an ion-channel project, one of several types of screens Aurora had developed. After the gene for CF was identified in 1989 by a group co-led by Francis Collins, speculation rose about what it did in the body. Researchers discovered that it encoded for a protein that they chose to name—still with no insight about its function—cystic fibrosis transmembrane conductance regulator (CFTR): literally, the pore-forming protein on certain cell surfaces that lets through charged atoms and molecules and that we know about only because it is broken in people with cystic fibrosis. Soon after, they determined that CFTR channeled chloride ions and water, explaining, ex post facto, how for centuries folk healers diagnosed the disease by licking babies to see if they tasted salty.

Knowing what CFTR did made it only more daunting as a target. Hedging its strategy, Aurora proposed in its initial agreement with the foundation to pursue alternative pathways as well. But Beall and his medical staff pressed the company hard to focus on the disease at its heart. "I credit Bob with that," Negulescu says. "He was going around looking for companies not just to work on CF but on CFTR—*for* CF. He was frustrated by the fact that we knew that CFTR was the defective protein, and we knew that it was a channel, and we knew what was wrong with it, but there was nobody looking at how to fix it. They were looking at alternative channels that one could try to activate to control the symp-

toms, say, with inflammatory responses. Or how does one interrupt the
bacterial environment in the CF lung? Everything but the real issue."

Passion is an underrated virtue in drug research. Big Pharma R&D
leaders, forever imposing processes and metrics, believe that strategy
ultimately prevails; pick an approach, stick to it, be meticulous, you'll get
there. But Beall's deep conviction that only a drug that hit CFTR could
transform the lives of people with cystic fibrosis had inspired a growing
faith at Aurora, if not yet at Vertex.

Negulescu flew to Cambridge to lay out the goals of the CF program
before the scientific advisory board. Murcko, anticipating resistance to
Aurora's approach, cautioned him not to expect a warm reception. What
Aurora would do, Negulescu explained, was find compounds to correct
each of the two main types of defects. In some cases, there's sufficient
CFTR at the cell surface, but the channels don't remain open long
enough to let the ions and water pass through—so-called gating mutations. In many more patients, the problem is that not enough protein
gets to the cell surface, because the protein is improperly folded. When
Negulescu said that Aurora was screening for molecules to activate the
protein regions responsible for both problems, an advisor muttered,
"Fantasy science." Boger and Sato, sitting in back, said little.

In April microbiologist Eric Olson took over the CF project. He was
unlike Kwong and others who, inspired by the company, chose to join
Vertex because that's where they thought they could do their best work:
in Olson's case, Vertex and CF seemed to come together to select him.
A lean, blond, soft-spoken Minnesotan, Olson had worked in antibiotics for sixteen years at Upjohn and Warner-Lambert, recently acquired
by Pfizer. He'd become interested in CF through a colleague whose
daughter had the disease, and they'd collaborated against pseudomonas,
the most common bacteria to attack the CF lung. Living in Ann Arbor,
looking a year earlier for a job, he'd interviewed at Vertex for a position
as program director, spending an hour with Sato and coming away
impressed. Meanwhile, he also interviewed with Negulescu at Aurora,
before the merger. He got an offer from Lilly, but heard that Lilly might
be soon getting out of anti-infectives.

"I almost just about didn't join Aurora because even though they could set up screens, there was no way they could ever make a drug," Olson recalls. "But I got wind when I was out there that they were in deep discussions to solve that problem. They said, 'Don't worry about that.' I did worry about it, but I really trusted Paul. A couple of weeks later, I was in the airport, going to Iowa, when the thing crosses the news: 'Aurora bought by Vertex.' I was so glad, since I'd already gone to Vertex. They each had a piece that I felt could make a big difference."

Olson's group had developed two related approaches to measure CFTR activity. Both employed high-throughput "patch clamp" assays in which swatches of cell membrane are isolated using a micropipette tip, and then hooked up to microelectrodes to gauge the opening and closing rates of individual channels. Screening for molecules to activate CFTR with gating mutations—so-called potentiators—technicians looked for changes in the electrical flow over a two-minute period. For those that might fix the folding problem and get the protein to the surface—correctors—they ran the same experiment but overnight, giving time to allow more protein to activate and get to the membrane. After sifting through tens of thousands of compounds, in June Aurora recorded its first validated hit for a potentiator. The molecule wasn't potent or specific; much less was it known to be effective, safe, or easy to formulate and manufacture. But it showed that a small molecule could enhance the functioning of CFTR.

Olson, Negulescu, and the CF team regarded the hit as an important moment, but they neither expected, nor found, real corresponding enthusiasm in Cambridge. Having a hit meant the true start of a project, and Tung and others did what needed doing to move the molecule along. But as seasoned drug hunters they remained skeptical. "They came out and set up all the stuff that had to be set up, but it wasn't clear what the feeling in Cambridge was," Olson recalls. "My sense was, 'Look, the foundation's covering most of the cost; you guys want to keep doing it, go ahead.' In the meantime, they were trying to advance all these clinical programs anyway; they weren't ready to start a bunch of new projects.

"Let's put it this way," he says. "Nobody stopped it."

❖

The first annual Liver Meeting® at the John B. Hynes Veterans Memorial Convention Center auditorium in the Back Bay during the first week of November was an attempt by both its sponsors and the city to rename, rebrand, and trademark the half-century-old yearly get-together of the American Association for the Study of Liver Diseases (AASLD), where until the last few years the reigning topic was cirrhosis. They hoped to make the meeting—and the expensively refurbished three-tier conference hall—an important venue in the business of biomedicine. It wasn't Vertex-Lilly's VX-950, but another company's molecule, that stole the show.

In four papers describing the discovery, safety, and early antiviral activity of a small-molecule inhibitor of HCV protease, Boehringer Ingelheim's BILN-2061 proved beyond a doubt what Vertex, Schering, Roche, Merck, and many other companies had claimed on faith for a decade: that by blocking the protease with a selective inhibitor and making it orally available, you could dramatically reduce the amount of virus in patients. In a group comparison with a placebo, an inactive pill, all ten patients taking the drug demonstrated after two days more than a hundredfold decline of HCV RNA in their blood. When they stopped taking the medicine, the numbers returned.

Whatever the sting of losing an apparent lead in the public race for a cure, Sato, Kwong, and the HCV team were buoyed and influenced far more decisively by Boehringer's affirmation of the target. They now knew that the first company to bring approved protease inhibitors to the millions of infected and sick people with hepatitis C would usher in a new age of treatment, just as with AIDS. Kwong's group, still struggling to bioengineer an animal model to resemble the human liver, doubled down. Many, including Kwong herself, also drove "midnight projects": independent side experiments encouraged and supported by Vertex.

The rumblings Olson heard about Lilly turned out to be true: after several fitful years of on-again, off-again commitments, and the debacle of losing Agouron's AIDS drug, the company was pulling out of anti-infectives. Vertex's main champion in Indianapolis was transferred. Lilly began requesting amendments to its licensing agreement that would put more of the onus—and costs—on Vertex for clinical proof of concept. From Aldrich's initial hard bargaining to the widespread belief inside Lilly that VX-950 wasn't druggable, there was little but bad feeling about the program. As Lilly went through its portfolio in the weeks after the Liver Meeting, VX-950 was discontinued.

With protease inhibition validated by Boehringer's results, Boger might have felt more liberated and relieved to have the molecule back had he, Sato, Alam, and the scientists been better prepared to meet the costs and risks of going it alone. Pharmacology studies and patient trials were expensive enough; VX-950 cost $2.5 million per kilo to make. The compound was extremely tough to work with. Vertex knew it could be improved, though not, specifically, when, how, or how much. The road ahead would be long, dark. "I don't think we were being overly optimistic in terms of extrapolating how far we could improve the situation both in the formulation and synthesis," Thomson says. "But they said, 'No, given where we are at the moment, we want a calculation that says this has profitable margins, without extrapolating any further progress.' The program didn't explode, but we had a stalemate on whether and how to proceed. So the lawyers stepped in to find a solution."

Investors and analysts, examining what appeared to be Lilly's rejection of the molecule, found little upside in their restructured settlement. In exchange for worldwide rights to compounds identified during the collaboration, Vertex granted Lilly a small royalty on future sales, a gesture acknowledging that a Lilly chemist had made the molecule that now was a Vertex asset and would-be product.

"One of the things that Lilly said in their parting shots is: 'This is not a negative to you, this is not a negative to the program.' They told us all the things that we could repeat back to our investors," Boger says. "But they did make it clear that they did think that 950 wasn't the drug. It was on its way to a drug. It was a good molecule to go in and test a hypothesis. But it would never be a drug because we would never be able to formulate it, and we'd never be able to manufacture it."

# CHAPTER 5

JANUARY 6, 2003

The idiom of the biomedical business shields the user from the grittier realities of the trade. A disease, for instance, is not an illness but a "market opportunity" or else simply an "opportunity." So is what used to be called a condition, something lamentable but not a sickness: say, erectile dysfunction or wrinkles or baldness. The choicest word in the lexicon is *value*: the full measure of a product's usefulness to patients and society, although *value* also codes for profits, including obscene ones. Speaking with investors you might tell them, "In the coming year, we will add strong value to our business by enlarging our opportunity in erectile dysfunction" so that you can avoid having to say, "Next year we expect to make billions flogging boner pills."

The subject on everyone's mind at this year's Morgan health care conference in San Francisco was the frenzy in mergers and acquisitions. The M&A cycle in biotech was cresting as companies stepped up their efforts to enhance and expand their pipelines and commercial capabilities and to build up their patent estates. Millennium, hopscotching from gene sequencing to pharmaceuticals, had bought four other companies in the past five years. Boger addressed the trend in his presentation. "As we completed our first decade," he said, "we found ourselves needing to scale the Vertex discovery engine across multiple gene families. The merger with Aurora Biosciences gave us product buildout capability, cash generation, and additional technology capability. While the deal resulted in an eighteen percent dilution for Vertex, we will recoup by moving the price-earnings [ratio] when the company becomes profitable."

When and if Vertex eventually would earn more than it spent remained anybody's guess. But that only encouraged someone like Boger to make it sound as if the march to profitability was advancing on schedule, firmly under control. "In the year ahead, we are looking forward to continued progress with our late-stage pipeline, including the launch of 908," he continued. Glaxo had filed for FDA approval on the reformulated version of Agenerase, and Boger remained dimly hopeful that the molecule formerly known as VX-908—delivered in two small pills twice a day—would stimulate Glaxo to move more aggressively in AIDS, where numerous players were jumping in and experimenting with different combination therapies. "We also expect our partner Aventis to begin a Phase II-b rheumatoid arthritis study of our oral ICE inhibitor pralnacasan."

He went on: "We have set aggressive goals for product development in 2003, which we believe will position us to succeed with our long-term corporate and commercial strategies. Specifically, in the coming year, we will commit to two drug candidates to move forward on the path for approval, launch, and commercialization by Vertex. At the same time, we will focus on maintaining a high level of momentum and innovation in our drug discovery organization to generate a continued flow of novel drug candidates in our pipeline. In addition, we will continue to maintain a strong financial profile as we pursue our goals."

Boger didn't discuss what was pressing most on *his* mind: how to build off the progress in chemogenomics; more specifically, the end-stage negotiations for a second gene-family collaboration in proteases with Glaxo. As with Novartis, Vertex had to show that it was ready to hire another couple of hundred scientists and put them immediately to work. It had leased a new six-floor, 300,000-square-foot facility a mile away in Kendall Square, epicenter of what was now called, a decade after Boger's talk to the state trade group, the Cambridge "biocluster." Vertex's cash burn would soon grow notably larger. Part of Boger's talent was making investors feel that the hugely speculative, tortuous, open-checkbook process of bringing drugs to patients could be brought under safe, orderly, and predictable restraint.

CFO Ian Smith was by character and training resistant to Boger's reality distortion field. Someone needed to be "the brake," as Aldrich had put it, and Smith, thirty-six, managed the company's business as he had audited others' as a partner with Ernst and Young, keenly attending to the underlying hydraulics. He was slim hipped, athletic, over six feet. His flashing smile, clipped accent, frequent out-of-season tan, and toned, dark-haired good looks resembled those of the actor Hugh Jackman. Smith had come to Vertex via an accelerated and incongruous route. As a teenager in England, he grew up above the Queen Victoria Pub in a gritty part of Manchester, preferring soccer and cricket to school. A quick grasp of shapes, numbers, and patterns propelled him to a business degree at a polytechnic. Fearing a return to the rough life he'd escaped, Smith became a chartered accountant. By the time he was thirty, he was living in Boston and advising top executives at Reebok and Staples on their expansion plans. E&Y made him a partner soon after.

The high-throughput, gene-family approach was proving to work better on paper than in practice, but Vertex was committed to it, and it was Smith's job to raise and manage the capital that would enable the company to broaden it. He had no background in science, but, having served as outside manager on Millennium's four acquisitions and been skeptical regarding the Aurora buyout, he doubted the platform's sustainability as a business model. Like Aldrich, he recognized that the reliance on Big Pharma to fund the biotech industry was coming to an end, and the gene-family model was more of a hypothesis than a proven method for finding drugs. Boger's enormous claims for it and his thoughtful arrogance, so winning with investors and analysts, challenged Smith to try to rein him in while also teaching himself to show Wall Street he was Boger's match for projecting confidence in Vertex's vision. "It was beautiful working with Josh," he recalls. "You ask Josh what keeps him awake at night, and he goes, 'We're not going to achieve everything I think we should achieve.' I'd look at him and say, 'Really?'—because I took care of the contingency planning."

In early spring, Boger's expansion strategy struck a shoal. Construction on the building in Kendall Square—a $45 million architectural challenge to its neighbor Genzyme Corporation, with terra-cotta and linear panel-glass accents, a six-story skylighted atrium, and glass-enclosed elevator cabs—was completed. Ken and his lawyers put the final touches on the multivolume protease deal with Glaxo, bigger than the Novartis agreement—even as the fashion for high-tech, target-based solutions as the answer to plummeting productivity was fading fast across the pharmaceutical industry.

"The agreement had been negotiated down to every last detail," Boger remembers, "but the CEO had not been involved. The deal got up to his desk, and he just pushed it to the side and said, 'I don't think we'll sign this right now.' That was not a good day. It was crushing. The building was completed and empty and we were paying rent on it, and because of accounting rules we were having to write it off as a loss and take these horrendous losses on our P&L."

Smith confronted Boger and Sato. The balance sheet was badly out of whack. The convertible debt secured during the last days of the biotech bubble still loomed. Vertex, like the rest of the sector, had lost four-fifths of its value but was expanding as if the correction had somehow not applied to the company. Something had to give. All three of them understood implicitly that of all the optimistic promises Boger had made at the Morgan conference, the most urgent was to begin to bring its own drugs forward, as that was the only route toward profitability. Yet with the failure of the p38 kinase inhibitor VX-745 and the uncertainty over pralnacasan, that horizon, too, seemed further away than ever. Smith likened the company's situation to a "perfect storm" and told them that the only way to keep afloat was to restructure the balance sheet: hunker down, focus inward, ride it out.

"We had no productivity out of research; molecules were not coming out," Smith recalls. "We were struggling with the Novartis collaboration. We were losing way too much money. We didn't have a strong cash position, and we had no visibility of being cash-flow positive. We had drugs failing. Yet we had a debt repayment sitting on the balance sheet

of about three hundred million dollars. The business was basically struggling on all three fronts: research, development, and finance. So we laid
a small portion of our research efforts off. We consolidated down."

❖

Vertex was a fifteen-year-old company with more than 850 employees at four sites, having added labs in Iowa and Oxford, England. Its
research budget for the fiscal year was about $200 million—less than
one-twentieth of Pfizer's. The company anticipated an annual loss of
$140 million to $150 million. By Smith's math, the business needed to
cut its workforce by 15 percent, or about 110 people. These layoffs would
come chiefly in research, where a full-time employee, or FTE, cost on
average $375,000 a year in salary, benefits, lab equipment, and supplies.
No one in senior management had laid off anybody before.

Thomson bore the brunt. From the beginning, he'd warned of the
risks of chemogenomics, arguing that without an equally committed
partner, a diffuse, structure-and-target-based approach to finding and
treating the cause of a disease was less likely, in the end, to be as productive as the therapeutic-area paradigm favored across the industry.
Companies like Merck, once they specialized, say, in heart disease or
anti-infectives, assembled research campuses focusing just on the biology of the illness. Vertex was generating vast databases of information
about every known kinase but few drug leads because its own biology
was weak. What's more, it was getting little help, unsurprisingly, from
the people at Novartis, who resented having their insights outsourced.
As Novartis R&D head Karabelas quipped, "Data, data everywhere, and
not a drug, I think."

Thomson and his organization had put in place the industrialized
platform that Vertex had promised, but the effort had stretched him,
the scientists, and the company's research in uncomfortable and untenable ways. Senior biophysicist Jon Moore remembers the tempo:

> We hired so many people so fast because we needed bodies, basi-
> cally, to do the work. Every single day, chemists would come in, and
> by lunchtime, you'd have a yes or no. If somebody came in and gave
> a poor seminar or didn't seem like they had it to fit in, I don't even

know if they got lunch. We were looking at so many targets at the same time. We were trying to do hits-to-leads on them. We were making proteins, assaying them, screening them, doing chemistry on them. We were as organized as we could be given the scope of what we were trying to do. I felt we were doing lots of things but not particularly well. Things like structural biology and biochemistry and enzymology can be made into high-throughput processes, but especially in the structural world, you're only gonna get the low-hanging fruit. The easy things will fall, but it's always inversely related. The hard things, the interesting targets, are always going to be more difficult. That's just how it worked.

When Sato told Thomson he would have to scale back research by 20 percent, he approached the problem with his usual rigor, intensity, candor, and doleful appreciation of the unique constraints he faced. "Being head of research for the Cambridge site is not like being at a site where the big cheeses don't hang out," he says. "You've got to be able to bite your lip. You've got to be able to show decisiveness—take a free rein. But sometimes the reins are pulled tight." Thomson and his managers searched for an algorithm to decide how to refit the labs and select the scientists Vertex could best afford to lose, without disabling projects. They arrived at a matrix by first considering the effect on the future of the company. "There was a need to rebalance the workforce," he recalls. "Asymmetric release of people is often a vital part of it, so that rules out that it's a blunt culling of the weakest across the board. Our decisions had to be legally defensible, ethically defensible—also economically and scientifically. Then there were elements of longevity. Emotional? No, I would argue ethical."

A few married couples worked in the labs. Thomson, Moore, and others felt strongly that Vertex had a vital obligation to their families not to let both individuals go. In April—midway between the "Shock and Awe" and "Mission Accomplished" phases of the Iraq invasion—Sato called a Saturday meeting of all Thomson's direct reports. Thomson laid out his criteria. Everyone was emotional—including Ann Kwong, who Sato now promised wouldn't lose any of her own people—igniting a loud free-for-all as lab chiefs and project heads fought to protect their own scientists. By Monday, when those who were being laid off were summoned to the East-West conference room, Thomson had lost control of the process, though he still held himself to account.

"These people don't know why they were being brought together— and then smack in the audience is this couple, sitting next to each other looking terrified, and I'm up there to say, 'You're all being let go,' " he recalls. "I had to take it on the chin for the company."

As Vertex restructured its research division, Thomson, Moore, and those others who were there from the beginning "grew up fast," Moore says. Thomson learned that Vertex, which prided itself on its humanity, under certain pressure could be as capricious and unscientific as any other company. Vertex thrived on its own exceptionalism, but he found nothing distinctively heartening about the way people who had given themselves to the company—"bled purple," as others said of him—were being treated. Nor did he always bite his lip. "I was a boat rocker by that stage," he says. "Boat rockers in the early days were what we wanted. But around this time, reminding people of a need for a conscience was not appreciated."

❖

Boger gave himself and Sato until the first week in November to decide which two programs Vertex would take forward on its own. That the company needed to advance two projects had become an article of faith and was a blunt admission both of the risk that neither one was likely to get approved and of the fact that they had no presumptive favorite. Vertex couldn't afford three projects, yet choosing only one would be putting all its eggs in one preposterously risky basket. No other decision in the company's history was likely to prove as fateful.

As usual, they set in motion a process to maximize their inputs. The central pillar of Boger's social experiment was the decision-making process itself, which called for as many voices and as much data as possible, but in the end was the opposite of consensus; that is, Boger, reserving the final decision, would leave things indeterminate until all the information was in and he could make the most informed choice, even if it totally reversed his earlier statements and positions. It was this fluidity that had allowed him, after writing into the company's charter that Vertex wouldn't work on HIV, to reverse himself when Murcko, Tung, and their colleagues convinced him that they had a conceptual advantage. "Success in drug development is usually tied to two or three people who are passionate about their opinion beyond explanation," he noted at the time. "Vertex has a long history of ignoring my opinions."

In May, just after the layoffs, Vertex senior management met off-site to discuss overall strategy. Joining their ranks was the company's new chief scientific officer, Peter Mueller, who formerly ran R&D in North and South America and Japan for Boehringer Ingelheim, the German firm developing the HCV protease inhibitor that Vertex hoped to catch—and beat—in human trials.

Son of a Bavarian banker, Mueller, forty-eight, was a polymath, having been sent at age ten to study at a Benedictine monastery outside Munich, an intellectual "boot camp" where four out of five students failed to make it through. After receiving his doctorate from Albert Einstein University in Ulm, he joined its faculty as professor of theoretical organic chemistry and then migrated to pharmaceuticals via astrophysics. Mueller was relentlessly inquisitive, equaling Murcko in his hunger to innovate. He pushed people hard—himself harder. More to the point, he had vital experience in getting drugs across the finish line. Regularly dressed all in black, his curly blond hair greying but not thinning—with a jolly disposition in the lunchroom, an impatient severity in meetings, and always a rambling Bavarian syntax—Mueller was what Vertex most needed. He was a closer.

"Peter had more personal drug discovery and development experience than the rest of us put together," according to Sato. "I hadn't developed small-molecule drugs: HIV was the first one I was sitting at the table for. Josh didn't have any drug development experience at Merck. John Alam was the closest because he had taken Avonex all the way, but it was a biologic. Peter had been in the business. He knew big drugs, he knew good base hits, he knew things that looked promising but ultimately were gonna die. So his experience and judgment at the table were very important."

Two months later, senior management met again, this time with program executives and departmental chiefs, to review the company's portfolio and try to prioritize their opportunities. Their challenge, as described in a Harvard Business School case study, "was to compare drug candidates at different stages of development, with different technical properties and different potential therapeutic applications." In other words, Boger and Sato initiated a scientific process in which fifty or so managers from across Vertex would develop criteria to determine which of the company's molecules to gamble their future on. As part of their analysis, participants used "real option valuation" (ROV), a modeling tool that stacks up the costs and risks of a drug's clinical evaluation against its estimated commercial value.

They focused on four types of risks: *target, mechanism, molecule,* and *market.* Boger believed the goal was to diversify the kinds of risks Vertex would encounter. If you had a choice, you didn't want to place all your bets, say, on anti-inflammatories; or on unproven targets like p38 MAP kinase and ICE; or on molecules like VX-745 that might be neurotoxic, or, like VX-950, went straight to the liver. You wouldn't choose two close follow-on compounds whose performance would be wide open to competitive drubbing, as Agenerase had been. You wanted to vary the type and number of land mines, anticipated and unanticipated, you would surely confront during a period of progressive investment stretching up to a decade into the future. "Companies tend to have biases in how they evaluate risks and which risks they are comfortable with," he told the HBS researchers.

"Some companies systematically underestimate target risk, some underestimate molecule risk, and some underestimate market risk. And the interesting thing is that when you're inside the company, you're probably not even aware what your biases are. So, to protect ourselves from these hidden biases, we deliberately want to make sure that we're taking different kinds of risks in our portfolio. By balancing our risks, we can avoid being blindsided ten years later."

Four molecules emerged as front-runners. Two held reasonable promise as "oral Enbrels": VX-765, a second ICE inhibitor chemically distinct from pralnacasan, a so-called fast follower into the huge rheumatoid arthritis and osteoarthritis sweepstakes; and VX-702, a second-generation p38 inhibitor that didn't cross the blood-brain barrier, now being tested in a Phase II-a pilot study against another inflammatory response afflicting almost two million people a year in the United States alone, acute coronary syndrome. A third compound, VX-148, was one of a number of inhibitors Vertex had designed against the enzyme called inosine 5'-monophosphate dehydrogenase (IMPDH). It was nearing the end of a midstage study in patients with moderate to severe psoriasis, a scabrous and painful skin disease. Blocking a validated immune system target, IMPDH inhibitors were also being tested in the treatment of multiple sclerosis and cancer, and Vertex's first IMPDH inhibitor, called merimepodib, was in Phase II trials for treating hepatitis C.

The fourth portfolio candidate was VX-950, which because it cost so much more to make and was last in development, for the least certain of markets, lagged severely by every measure. Mueller's first reaction to the molecule was negative, and he was initially skeptical that it could become a drug. "We kept going to these planning meetings where people would do the ROV, and hep C routinely came out at the bottom," Sato says. "It was four standard deviations away from any program that was ever going to deliver any value to Vertex. Josh and I would go, 'Wrong answer.' The analytics kept saying that hep C was a disaster. There were a couple of meetings where poor Steve Lyons, who was program executive, I'm sure felt like Saint Sebastian on a bad day."

Sato and Boger refused to give up on the molecule. So did Mueller, who without their support couldn't kill the project even if he wished to. Mueller's departure from Boehringer—and his familiarity with its protease inhibitor—restricted him from asserting himself either way at this point. The process lurched ahead throughout the summer and fall, even as he cautioned against putting too much stock in the company's assumptions. "ROV models are more valid for late-stage development compounds, when you have a pretty good feeling of the potential market ahead in one or two years," he advised. "Everything else is pure specu-lation. I'm not aware of any prediction for early-stage compounds even close to market outcomes."

The deadline for a decision—the third-quarter earnings call with Wall Street analysts—loomed. Boger began to exhibit his own core bias, a blend of romance and risk-reduction. As Thomson put it: "HCV is a profoundly important medical area where Vertex can make a difference. We have a locked-down target with a low biological risk." What Boger had learned about drug development he had learned at Merck, which is that if a drug has the right concept, and there are no physical reasons why it can't be scaled up—and, above all, there are no toxicology problems—all the rest is capital, execution, and competitive commitment. Sato admired how Ed Scolnick, hell-bent on getting to market at the same time as the other HIV inhibitors, persuaded chairman Raymond Gilmartin to build a $150 million manufacturing plant a year before Merck could expect FDA approval. Vertex would have to commit to as much effort or more *before* it ever made a profit—yet another strike against VX-950.

As Boger and Sato emailed back and forth on managing the portfolio, hepatitis C kept bobbing to the top of the discussion. "Nobody else was winning," Sato recalls. "I would have felt differently if somebody had come up with a kick-ass molecule, but nobody had."

❖

Kwong and her group weren't let go by the layoffs. Her midnight project for years had been to develop an HCV protease animal model, and she had hired an Indian virologist named Raj Kalkeri to work with her on it. They were trying to engineer a mouse that would produce active HCV protease specifically in the liver, so that Vertex could tell if its—and its competitors'—molecules were reaching and blocking the enzyme where it counted. They had stepped up the pace and were closing in on a solution before the cutbacks. Yet with human testing of merimepodib and Boehringer's protease inhibitor overshadowing VX-950, management showed little enthusiasm. Kalkeri was laid off. "People were saying, 'What's the point?' " Kwong recalls.

Kwong and Kalkeri had come up with a novel strategy and, having nearly proved the concept, weren't prepared to drop it. First they had fused the genes for HCV protease with those of another enzyme that they reasoned would be cleaved, broken down, and released into the bloodstream once the protease was activated. Then they inserted the genes for this fusion protein into a type of virus that, as Kwong knew from earlier gene therapy studies, all went to the liver when injected into the tail veins of mice. Eventually they began to produce animals whose livers churned out both HCV protease and a toxic enzyme that could be measured in the blood as a marker of protease activity. Under the microscope, the mouse livers looked like a ravaged human liver from someone with late-stage hepatitis C.

Neither of them felt they could stop now. "Raj actually begged to come back, on his severance, to work on the proof of concept of this model," Kwong recalls. "There are two things that are astonishing about this. Number one, that he did this. He came in day and night. We had to kill ourselves that summer. This is not a two-hour-a-day thing. This was really exhausting. Second thing is, Vertex let him do that. What company would allow somebody who was laid off back in the lab to work day and night? Only one that knows that, whether you're inside the company or outside, you're consumed by the project and won't be stopped."

Kalkeri finished the work in September, and Vertex began feeding mice VX-950 and Boehringer Ingelheim's protease inhibitor at a range of doses. He had made a version of the mouse in which the toxic enzyme was expressed only if the HCV protease was active, not if it was blocked. Facing the portfolio decision, Boger was enthralled by the data. "What we could do was administer compounds to the mouse and actually see the liver not get damaged, because we were blocking the production of this toxic enzyme," he says. "We could do whole-animal PK and actually get livers to look better. We're getting molecules down the mouse's gullet, into the body, through all the ways in which molecules can get trapped, and the molecule had to get to the liver and shut down HCV protease in the mouse liver, and if it did that, it saved the liver."

Impelled by slides that showed that VX-950 could not only stop the virus but also visibly repair sick livers in infected animals, Boger put to rest whatever remaining doubts he may have had from the ROVs. VX-950 *worked*. The rest was straightforward, or so it could be made to seem. "I went and talked to Peter; I showed him the data," Kwong recalls. "He was blown away. He said, 'We've got to do this and this and this and this.' I said, 'Peter, the guy who did this was laid off last May, and he

worked all summer with me to get this data, but he's laid off." Peter went
straight from his office to HR and got him rehired."

❖

Millions of hearts across New England sank in unison at sixteen minutes
after midnight on Friday morning, October 17, as the Red Sox, in the
eleventh inning, lost the seventh and deciding game of the American
League Championship Series, 6–5. The loss kept them from advancing
to the World Series, which the team hadn't won in eighty-five years.
Boger was among the bereaved. After the game, Boston Manager Grady
Little slipped immediately into lore, joining the team's other scorned
ghosts. Little had left on the mound his tired and battered pitching ace,
Pedro Martinez, who, five outs away from victory, gave up three runs in
the eighth, launching the game into extra innings "There is no reason to
blame Grady," Martinez said in a postgame interview. "Grady doesn't
play the game. If you want to point the finger, point the finger at me."

Several hours later, before the stock markets opened for trading, a
new Wall Street biotech analyst, Geoffrey Porges, initiated coverage
on Vertex by issuing his first research report. Analysts handicap stocks,
which makes them especially vital to "story" companies, with no sales
and earnings. Boger was vocal about his frustrations with their trade.
The previous year New York State Attorney General Eliot Spitzer, celebrated by *Time* as American capitalism's "top cop," persuaded the SEC
to impose rules requiring that Wall Street research departments be kept
at arm's length from investment banking operations. Accordingly, most
analysts now were paid by their firm's trading desks, meaning that (a)
they earned less than before, and (b) their fundamental interests were
tied not to how well a company performed but to how much volume it
generated for their employers. In Boger's mind, the first change meant on
average they were less talented; the second, that their self-interests were
tied to promoting volatility.

Porges, forty-three, had joined Sanford C. Bernstein & Co., a premier
research firm, the previous year. He was a lanky, probing Australian with
a medical degree from the University of Sydney, three years of postgrad-
uate training in internal medicine and pediatrics, and an MBA from Harvard Business School. A former vice president of worldwide marketing

in Merck's vaccine division who went on to operate a technology and investment firm in London, Porges knew the industry from all sides and had well-established contacts with—and spoke the language of—every subgroup, from doctors and nurses to sales reps to hedge fund managers, on four continents. He talked with them all, frequently. He titled his analysis "Vertex Pharmaceuticals: Still Floundering."

Porges described Vertex as having "a relatively broad early stage product pipeline and excellent long term prospects but no immediate risk of profitability." He didn't think Lexiva, the successor to Agenerase, would do well in the marketplace, noted the company's "relatively unattractive near-term pipeline, which does not meaningfully contribute to the company's financial position," and criticized its "lack of clarity and definition" about its long-term product portfolio. "We rate the stock underperform with a target price of $10.60," he advised Bernstein's clients. The stock price had closed the day before at just above $13.

Striding the halls that morning among the labs in Fort Washington II, Boger brimmed as usual about Vertex's plans. Saying he was excited about the possibilities, he told the Harvard Business School group researching its portfolio process, "The portfolio is playing out exactly as we had hoped. We've got a stream of revenues from our partnered projects that will help fund our development costs. There are multiple paths for us to become profitable. We're in a position to choose."

Financial value and commercial potential were major criteria in choosing a molecule for clinical development, but these were the first compounds Vertex hoped to sell in North America under its own name, so issues of personal and corporate identity also surfaced. "I started out to make an important drug company, not just one that makes me financially comfortable," Thomson said. Boger aspired not only to be successful but also to change the standard by which success in pharmaceuticals was measured, both within the industry and in society. He liked to invoke George Merck, who in the 1940s and 1950s transformed his family's fine chemical company into a drug industry paragon by emphasizing science and virtue over commerce. In August 1952 Merck was pictured on the cover of *Time*. "We try to remember that medicine is for the patient," he told the magazine. "We try to never forget that medicine is for the people. It is not for the profits. The profits follow, and if we have remembered that, they will never fail to appear. The better we have remembered that, the larger they have been."

Pfizer, by comparison, had always displayed an opposite bias. John McKeen, its president during the same era as George Merck, told *Forbes*, "So far as is humanly possible, we aim to get a profit out of everything we do." As Boger and Sato discussed the portfolio decision further with the executive team, it became progressively clearer that arguments favoring a quicker turnaround time to market or a larger disease population were no match for the Merckian logic that said that the way to be most successful in making drugs was to transform the lives of people with consuming medical needs. As Sato told the HBS researchers, "When choosing between going after the fifth beta-blocker or the first or second something else, what do you *really* want to do? Are you going to be more excited about making a drug for X or Y; X and Y being otherwise equal? Having medical impact is important when picking a candidate. 'Dollars in' is a legitimate proxy for medical need as well as an independent marker in its own right. However, you need to be careful when assessing medical need using commercial success. Did Lipitor, the fifth statin to reach the market, address significant medical need? Maybe, but maybe not as much as its sales suggest."

The FDA announced approval the following Tuesday for Lexiva—VX-175. For Smith, the notion that Vertex would be able to fund the development and commercialization of its own drugs through royalties, milestones, and research collaborations, as it had done up to now, had passed. Porges was right. The only way to survive was to own everything: research, development, commercial. Costs rose exponentially the closer you got to product approval. Vertex's crop of late-stage candidates, fifteen years in, was anemic—"relatively unattractive." Floundering was an alien concept to Boger, but when biotech investors looked five years into the future and couldn't see a path to sustained profits, they blamed management. As long as the balance sheet remained weak, Smith was in no position to go to Wall Street to finance the company's buildout. He had nothing to sell. He hoped Glaxo would at least do better in Europe than it had done with Agenerase.

The day before Halloween—five days before the earnings call—Vertex received discouraging news about pralnacasan, the anti-inflammatory ICE inhibitor. Late-stage animal studies revealed potential liver toxicities; its partner Aventis was suspending human testing until the issue could be sorted out. Not an outright discontinuation, the decision might as well have been final, since Vertex was relying on pralnacasan royalties to fund the development of its portfolio. Halloween was a Friday. Going into the weekend, the stock price nose-dived from $17 to $7, losing 60 percent of its value. That night, senior management huddled in a conference room long after almost everyone else had left.

They had the weekend to develop a strategy. Smith had scheduled a debt refinancing directly after the earnings call, hoping to benefit from the pipeline announcement, but now that was out of the question. On top of everything else, they had to rethink their entire budget.

❖

As Boger had promised investors in January, Vertex planned to announce two candidates for late-stage development. It would hedge its risks by pursuing parallel projects. Throughout the weekend, senior management combed through the most recent clinical results from all the trials. The picture among the four favorites was clouded both by the pralnacasan data, which several members feared had dimmed the outlook for Vertex's fast follow-on, and by disappointing news from a recently completed trial of its psoriasis drug, which was closest to market and the compound that analysts were focusing on to put the company in the black.

Not all the news was bad. Another IMPDH inhibitor, merimepodib, was showing unexpectedly strong results against hepatitis C. A dark horse in the portfolio race up to now, the compound didn't attack the virus directly; instead, it boosted the effectiveness of the current standard of care. Six-month results from a midstage study showed promise for hard-to-treat patients who'd failed a previous course of treatment and thus had no other options. Strong safety data for the acute coronary syndrome drug and substantial gains in formulating VX-950 rounded out the latest updates.

"A key point in the weekend came when Josh reframed the decision," John Alam recalled. "We had been focusing on which two to take forward. Josh stopped, and said, 'Why are we stuck on choosing two? Why don't we focus our resources on one and keep the others going?' It was then just a matter of setting the priorities."

There were many reasons, of course, not to focus on just one, risk chief among them. How could any company—or any investor—rationalize a decision to bet everything on a 30-to-1 shot? Boger crafted a strategy that he believed would enable Vertex to press ahead to market relatively soon—say, in 2007—while sustaining its most promising efforts, all the while diversifying most of the risk. The greatest risks for VX-950 were *molecule* and *market*. It remained uncertain if Vertex could formulate and manufacture the drug, and how many doctors and patients would rush out to use it; some analysts considered the disease a niche opportunity. Merimepodib, on the other hand, worked through a weak pathway—it had considerable *mechanism* risk—but Vertex already had identified a commercially scalable process for drug synthesis and could begin its first comparison study within a year. By combining two products into a single vision, Boger committed Vertex to a franchise and a new corporate direction: oral therapy for the treatment of hepatitis C.

"Very quickly," he said, "we came to understand that merimepodib and VX-950 could give us a strategy. We could go into the market with the existing treatment. We could then follow with VX-950. And then we might have the possibility of using them in combination. No one was paying attention to merimepodib before because they didn't see it as a combination treatment with VX-950."

Wall Street analysts, expecting to hear about Vertex's strategy for rescuing pralnacasan and developing its psoriasis candidate, were dumbfounded by his message during the conference call: the company would stop trials on the latter, continue development of four projects, and make merimepodib its top-priority program. "It was just awful," Chief Commercial Officer Dr. Anthony Coles would recall. "There was a long silence on the other end of the telephone. It was as if a relative had died."

Boger attempted to explain the company's reasoning, but the analysts wanted to hear upbeat forecasts for the coming quarters, not speculation about the possible synergies of two molecules they'd scarcely heard of—

much less one rejected by a partner. They, too, use ROV and other models to try to value companies, and it was simply impossible for them to do anything more than guess what Vertex was worth. Understandably, they were displeased, and it showed in their reports. "We have a very different model than most biotech companies," Alam told the HBS interviewer. "Most emerging biotech companies focus on one project. They bet the company on that one project. Analysts like that model because they can value a one-product company much more easily. That's not the strategy for Vertex. We think there is value in having multiple options in our portfolio. We can prioritize among the projects based on the best information we have available. But we don't get credit for having a portfolio."

As the annual Liver Meeting concluded at the Hynes Convention Center, Vertex began to reposition itself as a fully committed player in infectious liver diseases, no longer one that would just report research breakthroughs but one that aimed to become a dominant company: a rival to Schering, Roche, and Merck. Kwong couldn't make the case that Vertex had a better compound than Boehringer Ingelheim, still the leader among those chasing protease inhibitors, but she presented compelling preclinical data, including experiments suggesting that VX-950 could knock down replication ten-thousand-fold in forty-eight hours and could clear the virus in less than two weeks. Ken Boger settled the company's five-year-old patent troubles with Chiron, which dropped its lawsuit and granted Vertex (for an undisclosed sum) a nonexclusive license to develop VX-950. The company announced that it would be able to start testing the drug in people in early 2004.

In December Vertex held its annual investor day in New York. The occasion gave Boger, Sato, Alam, Smith, and others from the company their last forum before next year's Morgan conference to make their case to Wall Street. Of the biotech analysts who attended, only Bernstein's Porges reacted positively. The next morning in his note to investors, he wrote that although Vertex was among the worst-performing stocks in the sector in the two months since he'd started coverage, and was terminating three of its most advanced pipeline programs, "much of the portfolio and financial uncertainty we identified has now been resolved, and, we believe, in aggregate, that the company will benefit from more positive news flow going forward, as well as a more aggressive and realistic commercial strategy."

Porges noted that the launch of Lexiva was ahead of expectations, already having surpassed three other drugs and Agenerase itself in prescriptions, and that it would be likely to generate significant income to pay for clinical trials. Smith, he wrote, had signaled clearly that the company was committed to changing strategy and lowering its burn rate while it actively sought more partners to out-license programs and bring in more cash. More crucially, Porges identified hepatitis C as a sizable commercial opportunity, suggesting that if Vertex won approval for merimepodib in 2007, the market potential for the compound could rise to $1 billion before newer agents arrived. Porges kept his $10.60 target price but, as the stock was trading 20 percent lower than that, upgraded his recommendation to "market perform." He now thought the stock would do as well as the Standard & Poor's 500 and other leading indicators.

In Porges's estimation, Vertex had done much of what it needed to do to regain its footing. But his endorsement also contained a disclaimer, saying, in effect, caveat emptor—buyer beware—as well as a blunt challenge to Boger, Sato, and the executive team. He wrote:

> The risks to our thesis about Vertex are scientific, financial and operational. Scientifically any setback or delay on merimepodib would be disastrous for the company, given the singular focus they have now given the product. Financially, failure to achieve a meaningful reduction in expenses in 2004 and beyond would be a disappointment to investors and could compromise the company's ability to secure additional financing. Operationally should the company fail to advance one or more meaningful product outlicensing/collaboration in the first half of 2004, the outlook would again turn negative. Each of these initiatives, and their associated risks, is essential to the financial survival of this company and deserves laser-like focus and complete commitment from management.

# PART 2

# Game Worth
# the Candle

# CHAPTER 6

FEBRUARY 14, 2004

VX-950 had been revived—barely. It remained the darkest horse in the portfolio race: the molecule that Lilly had rejected, eclipsed by Boehringer's compound, a distant follow-on to merimepodib in Vertex's strategic arsenal. HVC research in general had crept steadily forward, a study in frustration. More than ten years after Deb Peattie flew to Saint Louis to enlist Charles Rice to help solve the structure of the protease, Rice, now a world leader in infectious liver disease at Rockefeller University, gave a grim accounting to virologists at a meeting in San Francisco. No one had yet been able to grow the virus, he said. As a work-around, scientists had devised a so-called replicon system in which pieces of the genome were used to produce particles of HCV, which served as a model for testing drug candidates. Boehringer's pilot candidate, BILN-2061, had toxicology problems, and the company was going back to the drawing board. As ever with HCV, breakthroughs were halting and elusive.

The larger problem was to understand—and then show—how a drug or combination of drugs could achieve sustained viral response (SVR): in other words, a cure. The key with killing off viruses is not how to exterminate them (you can use bleach) but how much of a substance to dose people with that will be both effective and safe. To rate the effectiveness of a molecule, researchers traditionally test how much of the substance it takes to knock down a targeted biological activity by half (IC50). Since trillions of HCV particles are made daily in infected people, numerical drops are measured in "logs"—each log being a power of 10. A 2-log drop is a 100-fold decrease; a 4-log drop, a 10,000-fold decline.

Inside Vertex, there were well-founded qualms about VX-950's potency, and Kwong made it her mission to dispel them by developing a better test of effectiveness. "The reason that we were in trouble is that BILN-2061 was 350-fold more potent than we were in a certain assay," she recalls. That assay compared effectiveness at a 2-log drop. But when Kwong ran a multi-log drop essay comparing 2061 side by side with VX-950, "we weren't in spitting distance of each other," she recalls.

Kwong and Vertex now realized that the longer you incubate with the drug, and the more drug you put in, the more effectively you could clear the virus. While the hepatitis C team pressed ahead with animal studies to establish an optimal range of dose regimens for human testing, Ian Smith and Ken Boger launched timely financial and legal maneuvers aimed at reducing the company's exposure while squeezing more value from its assets. Smith went back to some of Vertex's significant bondholders and proposed a private restructuring of their notes, offering, essentially, a 6-to-1 split. "Typically, if you're a bondholder, you hold the company hostage," he says. "You don't go, 'Yeah, I'll take a risk on your equity.' They say, 'Give me the damn cash. I'll take sixty cents on the dollar instead, as opposed to taking your stock, which might be worthless.' "

Since the crash of pralnacasan, Vertex had seen perhaps its most precious asset among investors—the benefit of the doubt—dry up. "We were being treated like everyone else," Boger recalls. Of sudden and equal concern was brewing internal doubt. John Alam, in particular, was facing the realization that Vertex's target-based discovery engine might be steaming down the wrong track: that outside of anti-infectives, a few cancers, and some rare genetic diseases, the idea that you could inhibit a single target with a small molecule to stop a disease was "a false belief." Still, Smith found that there remained sufficient hope in Vertex's business. He was able to round up enough investors willing to convert their shares, removing $320 million in debt from the balance sheet.

Ken pressed to retrofit the terms of the Novartis partnership by arranging for an earlier, more rapid stage transfer of Vertex's drug candidates to Novartis for clinical development. So far Novartis hadn't accepted any of Vertex's molecules, claiming that Vertex hadn't shown sufficient clinical relevance. Ken, after persuading Sato and Boger that the present arrangement was unsustainable, convinced Novartis to come to terms, splitting the research efforts to give Novartis first-option rights on Vertex compounds, while allowing it to push ahead in kinases on its own. He recalls:

> I proposed a way to redo the deal to knock out this proof-of-concept stuff, and that said to Novartis, "You have to pick the compound when we propose it for development, and if you don't pick the compound that we propose, then we not only get the compound, we get the whole target." I was pretty sure that I could get Novartis to go along with it, because I knew they had a problem that was ongoing.
>
> They were violating the agreement, big-time, and I had made a decision from a business development and legal standpoint not to call them on this. I was in a "Don't ask, don't tell" mode. I knew they were breaching the agreement by doing work on the side with kinases that were supposed to be exclusive to us, and the people who were doing the work were really influential inside Novartis. They were the people who were supposed to be killed off by the deal with us, but they didn't get killed off, they grew. I figured, let's just let this sit out there and use this someday.

Novartis agreed to the amended deal in early February. Vertex now was free to shop those kinase projects that Novartis had rejected, and Boger approached an old friend at Merck who'd just been put in charge of a new cancer effort, proposing a collaboration in Aurora kinases, a group that Vertex had first shown to be a potentially important class of targets. The company had a potent inhibitor that profoundly reduced tumor growth in cancer models. At the same time, Sato, Kwong, and Tony Coles's business development team took Kwong's data on the road, meeting with dozens of potential partners in HCV. As Porges had warned, the company had until June to line up another collaboration, or the view on Wall Street, still scraping near its nadir, would again turn negative.

❖

Boger had told the Harvard Business School team, "A lot of biotechs are founded on the German academic model, with a couple of principal investigators and their closest troops. As the firm expands, later employees are considered less important. In contrast, we believe that the last person in the door is just as important as the first. We consciously reject the German model for the Silicon Valley model." But what if the last person in the door is an actual German academic? And what if he's handed the keys? The gradual immersion of chief scientist Peter Mueller into Vertex's social experiment was bound to roil Boger's hypothesis.

At an off-site, all-day Saturday meeting—on Valentine's Day— Mueller proposed a new strategic vision for the company, attempting to galvanize Boger's and Sato's fervor for hepatitis C into action. Sizing up Vertex's situation from the perspective of getting a drug out the door and becoming a sustainable pharmaceutical company, he argued that it first had to free itself from its dependence on Big Pharma. "With those types of arrangements, there are some obligations linked to it, and you have not the freedom to operate that you would otherwise have," he recalls. "You're more or less a slave of those partners. They provide development support, but at the end of the day it doesn't drive a small company to the next stage of evolution.

"The other thing was we had in the pipeline molecules that I would say were so-so. We had to maintain those activities, but to move forward, you have to demonstrate that you can do something on your own. Given the small amount of dollars that we had, those activities were very, very limited."

The issue was how to advance the company. The group was restive and large—maybe fifty in all—more or less the same body that had conducted the portfolio review. The turbulence, shocks, reversals, and doubts of the previous nine months lingered, and feelings were raw, conflicted. More than a few regarded Boger as not quite the same Pied Piper he had been, a casualty of his own inflated expectations. Merimepodib was advancing but was "crippled," Alam says—the best of an uninspired lot.

Mueller made his case. "I strongly came forward with the idea that we have to develop the anti-infective drug VX-950, because it was probably the highest chance of success from a *clinical* development point of view," he says.

There was a huge debate. Can we afford that? It's very difficult. It's long trials because standard of care is forty-eight weeks. Cost of goods of the molecule were outrageous, when you have to produce a couple of kilograms, and a kilogram costs $2.5 million, and it's cumbersome to make. People get nervous because it takes away money and capacity from other projects.

It also meant that we had to make a commitment to build a development organization. That was actually the more significant piece, because then you have to go and do financing. We didn't have the development capabilities that you need to make a drug all the way through the end. You had to basically build this more or less from scratch, with all the functions that are needed underneath. There is chemical development, analytical chemistry, formulation development, a quality environment. We also had to think about a manufacturing buildout, because we couldn't afford to put a plant in the ground for a couple of hundred million bucks. The clinical end was the same thing. You need a set-up in clinical that has all the components that you need. We're going to need a strong regulatory environment, strong clinical development environment, strong clinical operations that can handle bigger and complex trials across the globe.

Boger agreed with Mueller that VX-950 was Vertex's best hope for a medicine that transformed the lives of patients. This was a decision he had positioned himself to face ever since he'd committed Vertex to getting as many projects as possible out of the lab and into patients. He was not without other options, which complicated his choices. He told the group that Vertex would mount a full-scale development effort with VX-950, effectively freezing some IMPDH projects that were further advanced. "That was a tough call," Boger says:

It wasn't a company killer. It wasn't we have to keep this going because this is all we have. But that made it even tougher. This was a
program that was being run at the thirty-five-million-a-year level for
which suddenly the whole bill was ours—unexpectedly. It doesn't
matter if Lilly says publicly that they've canceled all their infectious
disease programs. Everybody thinks that they wouldn't cancel
anything they thought was valuable. So you have to convince Wall
Street, convince the board, convince everybody, that it didn't matter what Lilly said, this was worth finding thirty-five million a year
to keep going, knowing that it was going to go to a hundred million
real soon. And why *this*? If we had that kind of financial resources,
why didn't we put it behind something that wasn't tainted that way.
The typical biotech thing to do was to say, "Well, we're putting this
program on hold until we find another partner"—which would
have happened. Instead, we just went full speed ahead.

Mueller prides himself on his Bavarian understatement. He knew
better than anyone else, including Boger, the full implications ahead for
the organization, and he admired the courage it took to take on the risks.
Boger often told people that he wanted Vertex to be the most feared
pharmaceutical company and the most fun pharmaceutical company:
feared because of its fearlessness, fun because it was more exciting to
dare, excel, and win against all odds than simply outperform your rivals.
Sato thought the combination derived from the fear of falling short of
one's sacred ambitions, of not being good enough, and the deep personal pleasure and raw thrill one takes in an almost psychedelic level of
hard work and engagement in several major challenges all at once—at
life, exultant, breathing hard but evenly, on the cutting edge. Even at the
bottom of Vertex's fortunes on Wall Street, she recalls, "The Kool-Aid
was strong."

"So we made this commitment that we could solve all those problems
and do it," Mueller says. "That was actually the fundamental moment
when the decision was made: 'Okay, we do this drug, and along the line
we build an organization to make it all happen.' This is one of the boldest
moments that I have seen from a corporate decision-making point of

view. I must say I give kudos to Josh and his guts. He stood up and said, 'Yeah, we do this.' "

❖

As the second quarter advanced, Vertex management regrouped while making the sort of partnership deals that the company, now more than ever, needed to do to survive. Although it had more than $460 million in cash, Smith expected a net loss for the year of $140 million to $150 million, meaning its balance would soon drop near $300 million. Despite better-than-expected royalties from Lexiva, at its current burn rate the company had two years or less before it ran out of funds. Losses as far as the eye could see, the founding promise of all biotechs, had started to add up to real money, ratcheting up pressure on the executive team.

In May the company announced that the Cystic Fibrosis Foundation would pay $21 million in direct research funding for the next two years. Bob Beall, pleased that Vertex had proved it could find small-molecule compounds that corrected the defective ion transport in the lab, was eager to subsidize its late-stage drug discovery effort. As Aldrich had predicted, Vertex's full commercial rights in CF—a so-called orphan category since it has so few patients, and which affords, among other inducements, accelerated time frames to approval and superpremium pricing—looked more valuable now that development deals with big companies had lost much of their allure for both sides.

Sato scrambled hard to find a limited partner to subsidize the work in HCV, one that would deliver near-term revenue and supply badly needed development support while the company geared up VX-950—and itself—to conduct large-scale human trials. She was under the gun to bring in something fast: the first human study was scheduled to start in June. With Boger uncharacteristically staying out of the discussions, she approved a $33 million collaboration with the Japanese drugmaker Mitsubishi for development and commercialization rights to the drug in the Far East. It wouldn't cover the bills but it was something. Vertex retained exclusive rights to the rest of the world.

In mid-June, the day the agreement was signed, VX-950 was given to the first of thirty-five healthy volunteers in Europe to test its safety, tolerability, and pharmacokinetics in an early-stage dosing trial. That morning in Cambridge, chemical engineer Trish Hurter, forty, joined Vertex, taking over its formulation group. Hurter, compact and hugely energetic, a spirited horsewoman, was from South Africa, having gotten her PhD from MIT before migrating into and out of the paper industry, having two children, and managing, most recently, formulation development at Merck. At Vertex, where most discussion galloped along like the dialogue in *The Social Network*, Hurter was instantly in a speed class by herself: "Walk fast, talk fast, ride fast, drive fast, type fast," she explains. She adds, in another context: "Shit happens, move on."

What the start of human testing signified is that VX-950 could be coaxed, despite being less soluble than marble, into a state where it could be taken orally, get through the digestive track, and be absorbed, three to four hours later, through the lining of the small intestine and into the bloodstream. What the experimental subject got was not a pill but a cloudy suspension made in a beaker from ingredients stable for only a day or so. Concentrations detected later in the blood were a measure of how much reached the plasma—the exposure rate.

VX-950 "loves being crystalline and loves hanging on to itself," Hurter explains. After two years and many millions of dollars, Vertex chemists had developed a process to keep it from locking into a lattice by melting it at high temperature, then infiltrating it with polymers as it flash-cooled. The polymers, like infinitesimal chains of pop-beads, mingled with the drug, diverting and delaying the atoms from reconnecting long enough for the substance to be absorbed. The process plainly worked, but yields were scant and unreliable and results erratic, swinging from batch to batch.

"They had done this tox study in Arkansas in July, and the exposures they got from dogs were really bad; way worse than they expected," Hurter recalls. "Then, a little later, in one of the multiple-dose panels in humans in Germany, they got almost no exposure, even though they'd been getting good exposure before. They'd thought they had stability for twenty-four hours, because that was the data that people had generated in February in Boston. There had been a heat wave in Europe, so the general hypothesis was that the drug was crystallizing. We didn't know what was going on, basically."

Hurter dug into the chemical process, an amorphous system like glass in which the drug substance is kept in a jumbled-up state that prevents it from finding itself and crystallizing. She hadn't worked with an amorphous system before. All Vertex could spare for her experiments was two grams of the drug. Her group was supposed to have thirteen people, but the previous director had left during the winter, taking a couple of chemists with him, and within two weeks after her arrival, it was down to five. Unwilling and unable to wait until she could conduct a more thorough search, she hired two temps.

"The chaos was intense," she recalls, "but fun. Nine-fifty is definitely way up there in terms of challenging molecules. Merck didn't have anything like that in their repertoire at the time. It's insoluble in everything; it's not even soluble in solvents. All the normal things you do just aren't enough for it. It's big, so it's hard to get through the GI membrane, but to get it through the membrane, you need high concentrations; and yet it's hard to get it soluble—you've got to work on both aspects. It's also very difficult to make and very expensive to make. They had twenty-two steps with really low yields. All kinds of hazardous process steps. A real bear."

On a site visit to Aurora, she found herself in a car from the airport sitting next to physical chemist Pat Connelly. One of Vertex's early scientists, Connelly had come to the company straight from a Yale postdoc, and within five years he had left to start his own structure-based discovery company specializing in antibacterials. After selling that business and starting another, he'd returned to Vertex earlier in the year. Thomson and Murcko had asked him to help develop its CMC (chemistry, manufacturing, and controls) environment. He and Hurter discussed the problems with VX-950 for five hours during the flight back from San Diego.

Connelly appreciated her urgency, especially regarding the minimal human exposure during the Phase Ia trial in Belgium. It wasn't hard to imagine the sheer unacceptability of Vertex heading into its first trials with VX-950 in infected patients, scheduled to start on November 1, with a drug that couldn't be guaranteed to show up in the blood. Connelly had spent much of his early career investigating the effects of heat on the bonding of proteins and small molecules, using a supersensitive calorimeter that can measure temperature differences of a millionth of a degree. The machine was still in the lab.

"I said, 'Listen, we don't have a good way to detect how this amorphous suspension is going crystalline, but it's a phase change; there should be heat involved. Let's put it in this calorimeter and see what happens,' " Connelly recalls. "So that's what we did. We knew it would crystallize; the name of the game was how long would it take. Sure enough, we saw a little blip. It took four and a half hours to complete."

Investigating further, Hurter learned that the Belgian clinicians had waited almost the full twenty-four hours before mixing the suspension, meaning that the VX-950 had crystallized long before the patients drank it. Reassured that the compound was stable long enough to make the transit reliably from mouth to small intestine to bloodstream, but was highly susceptible to heat, they designed a temporary solution: VX-950 needed to be kept refrigerated, and then mixed with water and shaken, not stirred, in an air-conditioned room right before it was administered. "The James Bond protocol," it was called. By early fall, Hurter's group had a formulation sufficient for the Phase Ib trial: a few dozen patients, fourteen days. But getting from there to making a pill that could sit on a pharmacy shelf anywhere in the world until some far-off expiration date remained another order of challenge entirely. Adding two more chemists in October, she bolted ahead.

❖

For a year, prompted by Smith's and Ken's dire warnings, Boger had strengthened Vertex's business. The company added two more knowledgeable, experienced directors to its board: a former vice chairman and president at Pfizer and a former US assistant secretary for health, whose combined expertise, contacts, and profiles might help steer the company through the regulatory process. The week after announcing the Mitsubishi deal, it signed a global collaboration with Merck in cancer research, delivering $34 million in up-front development support for its lead kinase inhibitor, $350 million in BioBucks, instant vindication of the Novartis restructuring, a sense internally that the company was emerging from its doldrums, and deep personal benediction and satisfaction for Boger.

Merck remained the industry gold standard, however tarnished by the recent reversals to pharma's image. With drug prices skyrocketing, the sector had fallen further, faster in public esteem in recent years than any other industry in history, according to polls showing that Americans now viewed drugmakers as on a level with tobacco companies. Former Merck chairman Roy Vagelos, noting the "exorbitant" costs of new medicines and "galloping" annual hikes of old ones, predicted a reckoning, again raising the specter of government price controls, as he had done during Clinton's first term a decade earlier.

With pharma spinning, Boger could see more clearly the opportunities for changing not only the way drugs are discovered but also how they are developed and sold. The awareness startled him, coming as a kind of secondary revelation, since when he started out, it hadn't crossed his mind that Merck and its peers could topple so rapidly.

Ken recommended that Boger meet with a former client of his (Ken's), an independent Boston management consultant named Bink Garrison. Most pharmaceutical CEOs confronted with questions about the direction of their business in a changing market rely on McKinsey & Company, the consulting giant, which had helped build most of the top ten prescription brands and supported nearly all the largest mergers within the industry. Garrison, a former advertising executive, counseled McKinsey when it wanted to develop *its* business, and he worked primarily with Fortune 500 companies. After graduating from Princeton University in the 1970s and a stint as a nuclear weapons officer on a navy submarine, he'd started his career as a copywriter. Tall, lean, soft spoken, he favored bow ties and other WASP accessories, though his everyday speech combined what he calls "the adman's bag of tricks" with an incongruous, hipsterish patter. Garrison dressed like George Will but sounded like Don Draper channeling Allen Ginsberg.

"I went over there and met with Josh and Vicki," Garrison recalls. "They kept talking about this communication problem they had with the board. I said, 'Well, what do you mean?' They said, 'Well, they don't really get what we're doing.' I said, 'So, okay, what *are* you doing?' Twenty minutes later, I had a little bit of an idea, and I said, 'I have good news— and good news for me. The first good news is that you do not have a communications problem. You have a strategy problem. You can't express what you're doing. And the second good thing, for me, is I do strategy.'

"I said, 'Who are your folks?' My proposition was, 'Let's put them in a room together for a day, day and a half, and find out not only what you think but what they think. And what you'll create is alignment between you and your team, and you'll have something to tell the board.' "

Boger leapt at Garrison's offer. He believed that what Vertex needed wasn't management consulting but a homegrown process for navigating the challenges of becoming a world leader while drawing on and renewing its culture—to discover what sort of company it truly could become not by reviewing others' best practices but through rigorous self-examination. "I was explicitly trying to systemize, institutionalize something that was already there but that was somehow going to get lost unless it had a more explicit verbalization and home," he says. "When I met Bink, I said, 'This is the guy. I need *this* person because he understands how to drive that in a nonhokey way, how critically it's lacking in large companies, and how you can't retrofit it easily.' "

Garrison met one by one with all the members of the executive team. His framework derived from Jim Collins's breakout bestseller *Built to Last: Successful Habits of Visionary Companies,* the classic investigation into the norms and practices of innovative companies that management gurus and CEOs globally regard as a data-proven tool of what defines corporate culture. It was based on yin and yang: complementary opposites. "On the one side of the yin is core ideology—which is core values, what do we really believe in? it's not negotiable—and on the other side is core purpose: Why are we on the planet?" Garrison explains. "In a company that's been around for anywhere longer than twenty-four months, that's not something you create. It's an archaeology project, not a creative project. You discover it. The other part of it is an envisioned future—a ten- to thirty-year ridiculous goal. Collins calls it a BHAG: Big Hairy Audacious Goal. What would it be like, the top of the mountain? Describe it."

On a day in July, Garrison met off-site with senior management and a handful of others, including Murcko and Thomson. His goals were to generate a clear, concise, easily understood description of what Vertex is:

a Vertex "vision, mission, and differentiation" statement, an acceptance that corporate positioning is based on reality and substance and not fluff, and a "perceptible rise in energy based on newfound clarity." To get them talking, he had them use analogies: If Vertex were an animal what would it be? Or what brand of car would it be? A Ferrari, a Lexus, a Prius? "The whole purpose was to till up the garden," he says, "to get some action."

Boger had told Garrison after their first meeting, "I'm gonna hire you away." Garrison, whose greatest gift may be that top CEOs appreciate him almost as a personal trainer, said he wasn't looking for a job. Boger replied, "I'm gonna make you an offer you can't refuse. I'm gonna cut your salary by ninety percent." Throughout the fall, as Garrison began to probe deeper into the company, Boger showed that he was serious. He made the Vertex Vision Process—a grassroots, companywide exercise to find out what people really believed about who they were and what they hoped to achieve—his top priority.

"This is something I was desperate for," he says. "We ran it, as usual, as a complete experiment. The intellectual content is, first of all, values are immutable. They're not something that you can decide to change. If your values are that you're evil, you can't decide to be good. Once your value is being evil, that's your value, and it doesn't matter how many consultants you bring in, you can't change it. When is that set in stone? Probably much earlier than most people think. I would say in most social and organizational structures it's set within six months. So we actually went on an exploration. We asked across the entire Vertex world, independently, what were the values. It came down to basically, on our best days, how do we act?"

❖

When Sato first explained the concept of fear and fun to the researchers at Aurora, many recoiled. It troubled them that fear was an articulated component of Vertex's culture. "It's not fear of being chained to the galley oars," she explained. "It's about fear of not rising to the challenge of what the world needs. It's the fear of being mediocre." In that sense, there was no distance between Cambridge and San Diego; aspirations were equally high at both sites. Tung had moved from Massachusetts and now co-led with biologist Paul Negulescu a Left Coast version of Vertex, one where

the ambition and drive were the same, but half the scientists also partic-
ipated on weekends in CF walks, and more than a few took a couple of
hours at dusk to surf at Torrey Pines State Natural Reserve before grabbing some fish tacos at a stand and returning to the labs. Project leader
Eric Olson, a clear-thinking and adept strategist, uprooted his family to
Cambridge, where he could represent Aurora and drive ahead the CF
clinical program. At his send-off in the cafeteria, his colleagues gave him
an enlarged photo of the panoramic view from the boardroom, a snow
shovel, and other necessities for winter in Massachusetts.

With the added funding from the CF Foundation, Negulescu, Olson,
and the project group pushed more deeply into the problems of trying
to correct broken CFTR in the epithelial cells of patients. These are the
cells that line cavities in the body and also cover flat surfaces. The new
team leader was Peter Grootenhuis, a Dutch medicinal chemist and deft
scientific manager, a recent veteran of a rapid M&A mash-up at another
firm that left him working at the same site but for four different companies in four years. Grootenhuis, a part-time virtual professor at Dutch
universities, guided the incorporation of Aurora's cell biology into the
overarching methodology of making drug compounds, which is improvement through modification. He inherited a familiar standoff. The
chemists needed better assays to tell if their molecules were working;
the biologists were impatient with the chemists for not delivering more
potent molecules. Though the group had several hits from three different
screens, progress had stalled in developing leads. Without better assays,
there was no way forward.

"I had to first make sure that we work well together as a team," Grootenhuis says. "The thing about CF is that there was no animal model, so
all we had were assays that we ran in different cell lines. We expressed
human CFTR in mutants, but there was no way of knowing how reliable
or predictable of a read that was. The view in biology was that if we can
base our assay on human bronchial epithelial cells from CF patients, that
was about the closest we were going to get."

As leader of CF biology, Fred Van Goor set about conceiving of a
model system, one that would convince them that they were on the
right track. A deliberative thinker with a laid-back quality belied, at age

thirty-six, by a mane of gray hair that seemed to balloon when he was deepest in thought, Van Goor had done his PhD thesis on ion channels and then spent another five years at NIH studying endocrinology. He recognized that everyone with defective CFTR had the same array of symptoms, but CF was not a singular disease. It resulted from a myriad of different mutations causing two main types of defects: either not enough properly folded protein gets to the cell surface, or there's sufficient protein at the membrane but with channels that don't remain open long enough to allow salt and water to pass through. In the lungs, the trachea (windpipe), and the sinuses, this retarded flux allows mucus to build up and cake, providing a fertile field for bacteria and gunking up the cilia, the fine hairlike projections that sweep away detritus and help keep the airways clear.

Van Goor's group had run the screens that yielded molecules that increased both the amount and functioning of CFTR, but not within human cells, much less cells that might provide a clue as to whether they would work against disease.

Sabine Hadida, who headed the CF medicinal chemistry group, had joined Vertex from the same orphaned San Diego outfit as Grootenhuis, though they hadn't worked together. As a postdoc at the University of Pittsburgh, Hadida had invented a widely adopted synthesis, and she had an ability, like Tung, to go into areas where the chemistry was thought to be well understood and discover new angles of approach. As her group tried everything it knew to improve on its hits—and as she sensed that the biologists, unaccustomed to working with chemists, weren't helping—her dismay at not having a more representative assay flashed over. "Number one, we had to convince Fred that we knew what we were doing and that this assay was not really helping us," she recalls.

"Number two, we had to convince the whole biology group that those hits were not really going anywhere. We were able to eliminate the activity, but we were not able to improve it. Everything we would do would keep it the same or worse."

Van Goor reasoned that human bronchial epithelial (HBE) cells from patients with CF could provide a solution. A few academic labs had cell lines, and with the help of the foundation, he recruited collaborators to help test several groups of compounds. While they were looking both for corrector molecules to fix misfolded CFTR and for so-called potentiators to open the channels longer, the team found itself reckoning with a startling new reality: that far more CF patients had folding mutations and thus would be helped only by correctors, which were much harder to find and develop. "The rate of finding a corrector is 0.002 percent," Van Goor says. "Then you have to make it a drug."

Here was the emerging dilemma of twenty-first-century drug discovery and development, of medicine itself. Among CF patients, the most common defect is the deletion of three bits of genetic code in the F508 region of the gene for CFTR—delta-F508. Half of all patients have two copies of the delta-F508 mutation and up to 40 percent have one, meaning that the great majority of patients—those who by and large, because they have almost no functioning protein, suffer the severest symptoms—would need a corrector for any hope of relief. Meanwhile, only about 4 percent of patients—around three thousand worldwide—were known to have the G551D mutation, a gating defect that had yielded Vertex's most promising hit for developing a potentiator. How did you decide what to pursue, the greater need/opportunity or the path likelier to succeed? For sick patients and their families, personalized medicine couldn't be more personal.

"In CF you're talking about eighteen hundred different mutations," Van Goor says. "That's like eighteen hundred different diseases. It's a real research, then development, then regulatory, then marketing challenge. Everybody says they know all about it, but how are we going to develop medicines for that? Because that's the future. The future is to know somebody's genotype—modifier genes or mutations that cause disease—and then make drugs that are tailored to those. How do you work in that new world? How do you evaluate it? It doesn't cost any less to develop a drug for five people than it does for a hundred thousand: same preclinical studies, clinical studies, it's all the same. What we're trying to do is use biology to provide a reasonable rationale for how different classes of mutations are distinguished from each other, so it's not eighteen hundred, it's three."

In August Hadida's group made a potentiator that Van Goor personally took up to the lab of a collaborator at Stanford, who had developed a line of delta-F508 HBE cells. The compound was ten times more potent than the starting point. "We had not seen that in all these years," Hadida recalls. The goal and philosophy of everyone on the project remained to go hardest for a corrector, but the gain in activity was a milestone, affirming their approach. It convinced them to press harder on both fronts. "We said, yes, now it's real," Grootenhuis says. "The problem with that molecule is that it was a brick. It was completely insoluble. It had bad pharmacokinetics. A bunch of things were wrong, but we loved the activity."

❖

On November 1 Vertex reported at the Liver Meeting (AASLD) that VX-950 was well tolerated and had favorable PK properties in healthy volunteers, according to Phase Ia clinical results; Alam announced that the company would begin testing the drug in infected patients within a few weeks. That morning's *Wall Street Journal* featured a front-page investigation into Merck's four-and-a-half-year rearguard action to conceal what it knew about the safety of its blockbuster painkiller Vioxx, which it recently had pulled from the market after studies tied it to heart attack and stroke risk. "By 2000," the paper reported, "one email suggests Merck recognized that Vioxx didn't merely lack the protective features of old painkillers but that something about the drug itself was linked to an increased heart risk."

> On March 9, 2000, the company's powerful research chief Edward Scolnick e-mailed colleagues that the cardiovascular events "are clearly there" and called it a "shame." He compared Vioxx to other drugs and wrote "there is always a hazard." But the company's public statements after Dr. Scolnick's e-mail continued to reject the link between Vioxx and increased intrinsic risk.
>
> As academic researchers increasingly raised questions about Vioxx's heart safety, the company struck back hard. It even sued one Spanish pharmacologist, trying unsuccessfully to force a correction of an article he wrote. In another case, it warned that a Stanford University researcher would "flame out" unless he stopped giving

"anti-Merck" lectures, according to a letter of complaint written to Merck by a Stanford professor. A company training document listed potential tough questions about Vioxx and said in capital letters "DODGE!"

Reputation, vital in any business, counts even more so in medicine, where it correlates to a rare degree with trust, market muscle, and influence. In Boger's analysis, Vioxx was a case study in how to destroy an organization's image. First comes a problematic product, then executive hubris, then foot-dragging, and then finally the thuggish self-deception that you can beat the rap with cover-up and intimidation. One decade you're on top; the next you're acting like a tobacco giant. Boger was more saddened than surprised by Merck's spate of reversals: its new-drug pipeline was nearly bare, and within two weeks, three hundred personal injury lawyers would gather in a ballroom at the Ritz-Carlton Huntington Hotel and Spa in Pasadena, California, for what the *Times* called "a combination strategy session and pep rally on Vioxx claims" estimated at up to $10 billion. As always, he believed the only way to avert such a downfall—beyond having a robust portfolio and high wattage across the functions—was through organization and culture.

He was trying to build an organization that was more capable of carrying out midstage and late-stage development and the transition to commercial operations—without killing off the spirit in research. He had accepted early on at Vertex that the people who seem most essential during one phase of a company's evolution are not always the same ones to propel you on to the next. He and Sato had structured the company in such a way that program executives reported up the chain to him, while everyone else, including the heads of functions, reported to her. The system was stretched now and would surely be unsustainable once Vertex had several molecules in late-stage trials.

After bringing in Garrison to probe what made the company tick, he launched several more preparatory moves, starting at the top. A few directors were alarmed that Vertex had done no succession planning. "They kept asking, 'What if you get hit by a bus? This organization is too big; we have too many bets on the table,'" Boger recalls. As he stepped up his efforts to explain to them where he was trying to take the company, they were distrustful, skittish. "I was worried one more jolt and they might panic." Boger saw no need to groom a successor, but to placate them, he brought on a new non–executive director who he thought shared his vision and could show them the kind of person they should consider if and when he could no longer run the company.

Matthew Emmens, fifty-three, was chairman and CEO of Shire PLC, a midsized English firm rapidly building on a diversified roster of specialty drugs to become a successful, research-driven multinational. In the mid-1970s, after graduating from Fairleigh Dickinson University with a business degree, Emmens started in sales at Merck when it had great products to sell and a reputation for integrity, and its six hundred reps included only one woman. Learning science and medicine from the doctors who were his customers, he absorbed their concerns and affinity for R&D, and after rising through the ranks to lead a new joint venture to promote a heartburn drug called omeprazole (Prilosec), one of the biggest-selling drug franchises of all time, he ran Merck KGaA, known as "German Merck," and established EMD Pharmaceuticals, its US prescription drug business. Medium built, fit, white haired, circumspect, a mechanic since high school who flew his own turboprop, Emmens had a passion for taking things apart, fixing what's broken, and reassembling them for optimal performance. He knew how to set goals and build a team.

"I introduced Matt as our get-hit-by-a-bus guy," Boger recalls. "I wanted a smart commercial guy on the board, but even more so, I told them, 'This is the guy who can be the temporary CEO while you take a year finding a new one.' He was my archetype. He was a guy I knew and trusted."

Boger also started to identify experienced, high-level veterans of Big Pharma to run Vertex's expansion and restructuring. After searching for more than a year for someone else to manage the program executives and move projects to a commercial footing, he hired Dr. Victor Hartmann, a former head of business development at Novartis, giving Hartmann the title of executive vice president for strategic and corporate development and the clear impression that he expected him, as Boger put it, to "shock the system" and "get development to deliver." Boger recalls, "When Victor came in, he could deliver the message, 'I don't care how you work it out internally in your development function, but here's what you need to deliver. Here's what a clinical package needs to look like. Here's what the documentation needs to look like. Here's what the backup needs to look like.' But he wasn't in charge of development. In classic Vertex fashion, we were trying to get people to change their behavior voluntarily."

A week after AASLD, Vertex announced that it was beginning a Phase Ib study with VX-950 in patients infected with hepatitis C. The study would evaluate three different doses of the drug over fourteen days of treatment. Results were anticipated in the first half of 2005. Meanwhile, Hurter's formulation group got another 20 grams of the molecule—"floor sweepings," she says—and began experimenting with different polymers and conditions to try to improve stability. Garrison traveled to Oxford and San Diego to run focus groups with employees, to get them talking about what drove them ahead.

At a meeting with the board, Boger put up an enlarged photo of a bus. Borrowing from Ken Kesey and the Merry Pranksters, he invited them one by one to post a picture of themselves on the bus. For Boger's generation, "You're either on the bus or off the bus" was a handy metaphor for commitment, for being all in or else left behind. The directors all put up their pictures. For Boger, the best way to avoid being hit by a bus was to be driving it, fast and supremely sure of his sense of direction, like legendary Magic Trip driver Neal Cassady, who also inspired the Dean Moriarty figure in Jack Kerouac's On the Road. Boger neglected to make much effort to ensure that the board was happy with him. Nor did he pay much attention to managing them.

"Completely not," he says, "but it's a two-way street. They hadn't added much value in the last few years. So why should I manage them well? And what am I getting out of this? In a list of problems of trying to build a pharmaceutical company, managing the board because you serve at their pleasure can't be high on your list because it doesn't solve any real problem. It can create a problem if it gets out of control, but it doesn't solve a problem. You toss the lions some meat so they don't cross over the fence, but it doesn't build anything."

❖

Sabine Hadida's chemistry group produced another tenfold leap in activity by improving the PK, formulability, and synthesis properties in a potentiator, VX-770. A senior researcher named Tim Neuberger, frustrated by the difficulties their collaborators were having trying to isolate and grow HBEs, pressed Van Goor to bring the model in-house. Over the years, everyone on the team had gotten to know patients, many of whom had died or were dying, and the team felt a degree of urgency that seemed to go beyond the normal bonds between researchers and the people they hope to cure.

Their efforts had proven the concept that a small molecule that targeted CFTR could reverse the damage from mutation. Van Goor had made comparison videos, taken through a microscope, of live cells from the lungs of CF patients. In untreated cells, the cilia rotated aimlessly; slow and uncoordinated, they failed to move mucus off the cell membrane. The second video showed cells treated with a corrector. The cilia, synchronized, beat faster, like wheat raked by a breeze, showing how in the body they potentially could sweep bacteria and mucus from the airways. The images were powerful, but without its own cell lines, Vertex was stalled. With lung transplants having become a last resort, Beall brokered an agreement in December for Van Goor and Neuberger to receive a lung from a CF patient with the delta-F508 mutation.

The lung arrived at midnight from the East Coast. Neuberger drove in an hour later to begin the work of harvesting cells. He had never done a dissection before. Having read a couple of papers describing extraction methods, he assembled forceps, scalpels, tweezers, scissors, and buckets of liquid in a biosafety cabinet. Taking the package from its cooler, he teased apart the wrapping to discover an organ both healthier and smaller than he'd anticipated. "There was just this tiny little lung in there," he recalls. He could think only of his own children. A thirteen-year-old boy had died of an anaphylactic reaction to a new antibiotic. Lungs from very sick patients are scarred and saturated with mucus and pus as dense and black as tar balls. This lung was pink and smooth.

Neuberger examined the lung with an eye toward cutting away tissue he was certain wasn't part of the branching system of airways. Grabbing the bronchial tube with a forceps, he sliced down deeper and deeper to isolate the airways as much as possible. He carefully cut away the stiff cartilage tubes that support the upper bronchia and the softer rings that surround them as they branch into the lobes. Then he snipped the airway tissue into smaller pieces, rinsed them out with enzymes to get rid of the gooey mucus, performed a series of washes, and then treated them with a bacterial enzyme called pronase. Active cells like HBEs are held together and supported by an extracellular matrix, which also prevents bacteria from spreading through the body. Pronase evolved in bacteria to chew through that barrier.

After thirty-six hours, Neuberger pulled out the pieces of tissue and put them into petri dishes with a medium fortified with vitamins, minerals, and amino acids. After slicing open the airways, he scraped and filleted sheets of cells, collected the sheets under a microscope with a pipette, transferred them to another vial, and spun them down. He added another enzyme that clips intercellular adhesion molecules, and the sheets fell apart into smaller groups of differentiated HBEs. Under a microscope, he could see the cilia still beating. He transferred the smaller groups to flasks, where, he hoped, he could coax them to reproduce, grow, mature, and form cilia of their own. The first set of experiments would take six weeks. Though the process would work the first time, it would fail the second and third times, and Neuberger would work to find optimal conditions for growing HBEs for the next three and a half years.

Mueller visited Aurora in January to review the site's progress. Negulescu and Tung, adjusting to the recent lurching in Cambridge, were determined to advance a compound into clinical development, and Tung had led the chemistry effort against sodium channels. They had made promising molecules against chronic pain caused by injuries to the nervous system. Negulescu recalls:

In those days, the program that we always expected to bring forward first was our pain program. We thought this was the one that was gonna deliver for the site. So all our agendas, for all our meetings, always had sodium channels first, then a couple of other things we were doing, and then CF, trailing—so that if the meeting went long, it got cut, or pushed to the very end.

The meeting with Peter was a classic case of that. We had an all-day meeting. Around four o'clock, he started to get antsy; he had to get a car to the airport. And so Fred gets up and starts talking about the CF program. Peter's mumbling. He's starting to put his papers away. Fred's looking at me: *"What do I do? What do I do? I can't go through my whole forty-minute presentation."* So Fred showed his movie, and out of the corner of his eye, Peter saw it. He started watching it. He stopped fiddling with his briefcase. And he sat down. And he said, "This is freaking amazing." It convinced him, like it convinced many people, that this was a real effect.

For Sato, the progress in CF affirmed that the Aurora buyout was working much as she and Boger had hoped. It also signaled a threshold for her and for Vertex's research organization. Its AIDS medicines were discovered before she arrived, but HCV and CF were diseases she had championed. Now Mueller and Hartmann would build the company as it moved into late-stage development with compounds brought out of research under her lead. As Vertex took shape as a commercial organization, she could only be drawn further from science if she stayed. Sato retired, returning to Harvard to teach. What she'd found in fourteen years of leading Vertex in an I formation with Boger—besides the necessity of fear, fun, and passion—was that the perseverance required to make a drug was being strangled by the controlling view on Wall Street that the only measure of a company's success is the extent to which it enriches investors. "The mantra of shareholder value is putting at risk the kinds of qualities that competitive innovation really requires," she says. "There was a time when people were looking at Vertex and saying, 'Another example of two billion dollars in shareholder equity flushed down the toilet.' Innovation has lots of parents. It requires not just capital. It requires certain personal attributes. It requires patience."

In late February 2005, a couple of weeks after Vertex announced Sato was leaving, a federal district court threw out what the company's press release called the "purported class action lawsuit" filed against her,

Boger, and others soon after 9/11, the crash of the p38 kinase inhibitor, and lawyer Andrew Marks's ill-timed margin call.

❖

Even before the Valentine's Day meeting a year earlier, clinical virologist Dr. Robert Kauffman had become VX-950's most ardent champion; also, elliptically, its most reserved. Kauffman, fifty-seven, was dark browed and meticulous, an inward-seeming physician who throughout his career had alternated between studying viruses in the lab and studying them in patients, at Harvard's teaching hospitals and with other companies. He'd known since second grade that he wanted to be a research scientist, then discovered in his late thirties that his unique talent wasn't in the lab but in clinical experimentation. At Syntex Laboratories, he had driven the development and approval of an immunosuppressant, CellCept, a major organ transplant drug. With a low-key, just-the-facts affect and a serious, inquisitive, and respectful manner, he brought a seasoned maturity and bedside equanimity into the company's deliberations: ballast to Alam's cerebral self-assurance.

After joining Vertex in 1998, Kauffman had fleetingly believed that merimepodib would enable him to retire in less than a decade, but as he'd grown more and more aware of its shortcomings, he grew more forceful in pushing VX-950 ahead. "To give an immunosuppressive drug to someone with a viral infection is a little bit of a tricky thing to do because there's a fine line between making the infection worse and curing it," he recalls. "You had to thread the needle, and it's pretty hard to do that without getting into trouble. The company was under a lot of pressure to bring a drug into Phase III. I basically said, 'It's fine to develop merimepodib, but you can't do that and not develop VX-950.' Nobody would understand why you failed to develop a direct-acting antiviral and yet you were developing this mechanism that a lot of people didn't believe in. It just wouldn't make sense as a company."

Since the start of the Phase Ib trial—the first study in a small number of patients diagnosed with hepatitis C, and designed to confirm the hypothesis that Vertex's protease inhibitor could lead to improved clinical outcomes—Alam had largely stayed on the road promoting Vertex's combination approach to controlling HVC. After Eli Lilly dropped out, it had become much harder to sell the idea to investors. Boehringer was steaming ahead with its compound and Alam worried that VX-950 might not be competitive—another too-weak molecule like pralnacasan. Perhaps more than anyone else at Vertex but Boger, he badly needed a success.

"We spent the six months from an investor relations standpoint between November and May having meeting after meeting where everyone said, 'There's no way you're gonna match BILN-2061 because it's a hundred times more potent. 950's a dog,'" he recalls. "So we'd have to go through the whole rationale why 950, dose for dose, was, in fact, going to work. There was a whole bunch of people we just had no impact on. Internally, the team set up a baby hurdle—a 2-log drop at two weeks. That was the hurdle for going forward because no one felt that we could come close to BILN-2061, even internally. People were worried. The team didn't want to put it out that we would stop the program if it wasn't as good as 2061. They didn't trust management to make a wiser decision, so they set low expectations."

Alam's father, a retired FDA pharmacologist, was gravely ill, and most weekends Alam flew to Washington to join his mother at the hospital. His father's medical history was of more than incidental or merely personal concern. Now seventy-two, he'd been diagnosed with non-A, non-B hepatitis in the 1960s and then was confirmed to have hepatitis C in the 1990s. He also had type 2 diabetes and atherosclerosis, leading to a three-vessel coronary artery bypass surgery in 2002. Typically, when he was in his late sixties, his liver disease had worsened. He'd considered being treated with interferon and ribavirin, but, according to Alam, his doctors told him, "First of all, you don't need to be treated, and second, you're not gonna tolerate twelve months of therapy anyway, because you have cardiovascular disease." Now he was back in the hospital for a "redo" and an aortic valve replacement, and his liver was failing.

Clinical trials aren't run by drug companies themselves but by independent investigators. They're conducted in stages. Before a drug is studied in large groups to see if it is effective (Phase II), then compared with commonly used treatments (Phase III), you must prove that the idea behind it works and that what you're seeing isn't an artifact. Patients are

assigned randomly to different groups and investigators are "blinded"—
kept from learning any information that might bias results. The VX-950
trial was again being run in Belgium, following the common practice,
especially among small companies, of testing products overseas before
approaching the FDA.

Kauffman faced a worsening dilemma. He was sensitive to the fact
that Alam couldn't have an inkling of how patients were doing on the
drug, since it would put him in an untenable position of having either
to lie to investors or breach securities laws. Kauffman also knew that to
move ahead with the project, Alam, his boss, needed to authorize major
new expenditures as the company accelerated into the next phase.

"We didn't know who was assigned which treatment, but it was aw-
fully obvious after a very short time who was *getting* which treatment,"
Kauffman says. "We would look at each other and say, 'Oh my God.'
There was nothing we could say. It got to the point where we could see
viral resistance. Ian, of all people, was interested in these findings. It was
very difficult at times to maintain a completely neutral appearance. But
we did, very successfully. Nobody knew anything."

At Vertex, as at other companies, trial findings are first revealed internally at confidential disclosure meetings. Legal, ethical, and scientific
safeguards are crucial. Ken ran the meetings, deciding who, for corporate
reasons, should attend, and managing the struggle over how to inter-
pret new data. On a day in April, he convened a group including Boger,
Hartmann, Smith, Mueller, Coles, Alam, Kauffman, and others around a
double-wide table in the Frankfurt conference room in Fort Washington
to review the Phase Ib two-week dosing data with VX-950 in patients
infected with HCV.

The factual turning point was one slide. At the end of two weeks,
patients receiving 750 milligrams every eight hours had a median drop
in HCV RNA of more than 4 log—a 25,000-fold reduction in viral levels. The compound drove the virus down hard and fast, and then kept it
down; the viral load in some patients was below the detectable limits of
the most sensitive assays. BILN-2061 had been given for only forty-eight
hours, after which viral levels rose, but this was the first demonstration
anywhere of, as Kauffman says, "a really potent antiviral agent that could

make viral loads just go down, amazingly and dramatically." Hartmann, the most experienced drugmaker in the room, said the last time he'd seen a result as profound was with Novartis's revolutionary cancer treatment Gleevec.

Boger could see the way ahead, and it was both daunting and massively expensive. Most often clinical results have to be subjected to statistical analysis to determine incremental value, but the effect of data like this was deep and visceral. He instantly reordered the company's priorities around the slide's swooping silhouette, which was dubbed, for its steep dive and hockey-stick leadout, the "Vertex swoosh."

"Up until we saw that, that wasn't the number one program," Boger recalls. "We were still trying to keep pralnacasan alive. It was our other HCV program which was going to lead. As soon we saw the swoosh, it took over, since then you knew you were on the fast track. You knew you had all that work to do, and you knew you had the molecule to do it, so it changed things a lot. And it became substantially the solution to Ian's problem."

Smith relished the swoosh as salvation, nothing less than the key to the company's future, which it well might be, as well as for the magnitude of the story he could now tell to Wall Street. He announced within days a fast-track stock offering that would yield, less than a month later, $175 million. As Smith, Alam, and others went on a hastily assembled road show, confidently touting VX-950 as the molecule that would transform the treatment of hepatitis C, several analysts raised their ratings to "outperform." Porges was the most bullish, despite the dilution of stock and management's boosterism, advising in a research note: "We . . . believe the stock is likely to continue to strengthen as investors recognize the size of the opportunity and the significance of the Phase 1b data recently released." Boehringer finally halted development of its protease inhibitor due to cardiac toxicity, leaving the biggest outstanding question, Porges warned, a possible challenge from Schering-Plough.

Alam reeled from the reversals. In late May he flew to Chicago for the Digestive Disease Week conference, where Dr. Henk Reesink, author of the European study, presented his findings. "We blew it out of the water," Alam recalls. He and Kauffman discussed what the data revealed about viral dynamics, the complex interplay between virus and host, and they began to model treatment regimens to improve SVR rates in patients. Several weeks later, Alam joined Smith on the road show. He called his mother from an airport and learned that the low blood flow during his father's heart surgery had caused his liver to fail. "I finished the road show, flew back to Washington, and buried my father," he says.

❖

Hurter's group reworked the VX-950 formulation over and over with different polymers and conditions. Anticipating the start of new and bigger trials in the months ahead, she drove her fast-growing team to emulate her own breakneck pace. By May, they were able to make 60 kilograms for advanced animal toxicology studies; with a better powder and a better suspension vehicle, the drug substance remained stable for two days. They turned immediately to figuring out how to make the new material into a tablet by combining it with other inert compounds that acted as stable carriers for the active ingredient. By July, they had a pill formulation for the Phase II clinical trial scheduled to begin in October.

"After we figured out what the amorphous formulation looked like, we got some stuff made," she says. "It took us less than a month. We made the stuff in England, where they have a place where you can manufacture and dose at the same facility. You can literally make the tablet upstairs and take it downstairs and give it to patients. We were waiting on the data, and Josh was being all gung-ho. I was getting worried because, 'Okay, I think it's going to be good, but it's not a slam dunk.' I was saying something to Josh at the beer hour, like, 'I hope you don't think this is totally bomb-proof. And I'm glad you have faith.' And he looked me in the eyes and said, 'Trish, I have the *appropriate* amount of faith in you.' "

Boger at last had a molecule powerful enough to foil the "constant headwinds of doubt" that had buffeted the company for the past four years. His ebullience soared. After Sato left, he took back the position of president, and with the organization maturing around him, he plunged further into making Vertex the model twenty-first-century life sciences company he'd dreamed of creating ever since he decided to leave Merck. Soon enough, that would mean engineering and running a competitive commercial organization built for expanding success across the globe.

But first Boger felt he needed to ensure that Vertex was prepared to become such a company by cementing its culture, building its reputation with doctors and regulators, and raising its—and his—visibility in the industry and internationally.

Garrison's work captivated him as it progressed from fifteen focus groups involving about a third of the company's employees across all the sites, to a smaller team that would distill all the discussion and testimony and draft the core purpose and values. Figuring out a distinctive shorthand for the company's unshakable mission was not hard, relatively speaking, and the team soon arrived at the prosaic "Innovate to redefine health and transform lives with new medicines." It sounded exactly like the statements of hundreds of other companies, but with three disruptive verbs—*innovate, redefine, transform*—they thought it made the point. Developing the values was harder. After asking them to find out what inspired them, Garrison now pushed them to focus on a few taut phrases: "It's a giant regression, basically. A lot of companies do values that are aspirational: we wish we were more collaborative, so we're gonna make collaboration a value. *Unnh. Error. No good.* If you do values that aren't true, the employees say bullshit. The other thing we said is, 'Here's how many you get. Three.' You go to companies all the time, and they have nine, ten values. They might as well have none. When there's three, there's nowhere to hide."

In July Hurter's team was deep in the final challenge of a commercial formulation. "We were really in the thick of things," she recalls. "I was going completely nuts. We were called to some all-employee meeting at Le Méridien hotel. I remember I was walking down there, *speeding* down Sidney Street. I'm ranting and raving, 'I don't have time for this!' I get down there, and there's all this loud music playing, purple balloons everywhere, and it's very festive."

The packed scene was the introduction ceremony for Vertex's values, a mini-launch that within weeks would be reprised at the two other major sites. Determined to build on Garrison's work, Boger devoted major resources to the effort, putting key people in charge of Vision into Practice (VIP) teams and initiating a three-phase implementation plan. The company adapted the familiar branding elements of clinical trials, where phases are named in series with slogans. Boger now explained Phase I—AWARE—six months of internal messaging that would include the inevitable banners, desk tchotchkes, and coffee mugs but also off-site brainstorming sessions to ensure that everyone "knows it and gets it," Garrison recalls. Boger put up a slide of the three values, arbitrarily ordered.

### Fearless pursuit of excellence

### Innovation is our lifeblood

### "We" wins

Here was an expression of corporate excellence that thrilled Boger. Not only were the values the actual values that had driven him and the other founders out of Merck and into the marketplace, and had coursed through the organization for sixteen years, but they were also cleverly turned to differentiate themselves from the commonplace themes of excellence, innovation, and teamwork that most companies adopt. " 'We' wins," for instance, despite its slight grammatical awkwardness, is more than the sum of cooperative endeavor. It claims that at Vertex if you have a job that you know you can complete on your own with 100 percent certainty, you'd rather enlist others even if your chances of success diminish slightly, since the benefits of interdependence always outweigh the risks. Fearless pursuit of excellence, Boger says, was "a demand for excellence always in the face of something that had to be scary. It almost had to be scary before it would warrant our attention.

"Innovation is our lifeblood," he says. "What we found was, it wasn't a choice. We didn't choose to be innovative. What came up from the Vertex conversation was that to not be innovative would be to have the blood sucked out of you. It was a physical need. It is what keeps us feeling alive. And so, it isn't that we're innovative because we care about others. We're innovative because we care about ourselves. We would die if we couldn't be innovative. We can't stand it. It makes us ill if we don't innovate."

Hurter was far from alone in becoming an instant evangelist. "I'm gullible," she says. "I soaked it all up." Her chief job now involved hiring people for entirely new functions during a growth spurt, and she found it invaluable to have a succinct statement of what sort of place candidates would find if they joined. "We can say to people, 'This is what we're about. If it sounds cool to you, you should be here. If this sounds dopey to you, you should not be here,' " she says. "I also tell them it's a very chaotic atmosphere. Ultimately, we still change our minds a lot, and yank people around, and change priorities, and change our minds to try to do something better—and, oh yeah, it's due last week. And if you find that frustrating, you really shouldn't work here."

❖

In December Vertex announced it was initiating a program of midstage clinical trials with VX-950. The first twenty-eight-day study would test the safety, tolerability, and pharmacodynamics of the compound when combined with pegylated interferon and ribavirin, the existing standard of care. A few days later, the FDA granted the molecule Fast Track designation—expedited review, since HCV infection was a serious, life-threatening disease with an unmet medical need for shorter, more effective, safer regimens. Garrison, after struggling for a final few weeks against Boger's reality distortion field, joined Vertex to see the vision process through. Boger convinced him that only by coming inside the organization would people take him seriously enough, and trust him enough, to help them do what needed doing, which was to weave the values and vision into the fabric of a breakneck expansion as Vertex "scaled."

Garrison was first given the title of senior vice president for organizational development—E-level standing. But as people started meeting with him, it was clear to everyone that building a chain of command wasn't his field. After a few weeks, he realized it would be useful to have a more descriptive title. He went to Boger, who told him to come up with one. A friend made a suggestion: senior vice president and catalyst. Boger, his social experiment now morphing into intensive culture building, told Garrison he couldn't imagine a more "pregnant" description.

# CHAPTER 7

## JANUARY 9, 2006

Boger began his talk on VX-950 at the Morgan conference by show-ing a picture of the Apple iPod. "Every so often," he said, "there's a game-changing product—one that transforms a product category, one that transforms a company, and one that transforms an industry." He went on to compare Vertex to Thomas Edison's Menlo Park invention factory. It was impossible to be too bald here, and Boger deployed a prized talent for predicting a future that at once seemed over the top, yet sensible and within reach. And yet as he now stepped onto a broader stage, he carried the extra weight of sixteen years of making grand claims that hadn't panned out, a reputation for overselling, and losses approach-ing $1 billion. An investor leaned over to a reporter and whispered, "He certainly doesn't lack for confidence."

He'd reached the point where, from now on, the company's—and his own—evolution could be classified pre- and post-swoosh, and he foresaw seismic changes at every level as Vertex ramped up to deliver the drug. "It's a tremendous responsibility to live up to a molecule like VX-950," Boger told the audience, adding, "This astounding data can actually be a bit terrifying." As if to hammer home the message, Ver-tex simultaneously issued a press release announcing interim findings from another small Belgian study done by Henk Reesink. In a blinded, placebo-controlled trial of just twenty patients testing the combined effects of VX-950 and pegylated interferon, half of those taking both drugs had no detectable virus after two weeks. They had a median 5.5-log drop—a 300,000-fold reduction—in viral RNA.

The speed of the viral response was breathtaking. But as with HIV, the central issue was resistance; there would be variants that VX-950 wouldn't hit, and another drug, or drugs, would be needed to achieve a cure. Alam—driven in equal parts by the data and a grim certainty that a shorter-duration treatment against HCV could have saved his father— believed beyond a doubt that the interim findings made a convincing case for shortening the course of treatment from a year to as little as three months. That all the patients in Reesink's study had the most resistant, hardest-to-treat genotype of the virus—genotype 1—ratcheted up company, and investor, interest in launching such trials as soon as the FDA would approve.

Publicizing the results from a tiny European trial to coincide with a key investor event helped Boger grab added attention on Wall Street, where the share price had nearly tripled since May, and where even small studies in carefully selected patients can spark what Smith calls "the magic" of soaring expectations among investors around a promising stock. It also was guaranteed to irk the FDA, which, more than any other federal watchdog (with the possible exception of the Nuclear Regulatory Commission), defines and controls the process by which a product gets to market. Emerging out of the Progressive Era mission to "civilize capitalism," the agency defends very seriously both its responsibility to protect the public and its prerogatives, which include deciding how much, how long, and under what circumstances an unproven treatment can be tested on experimental subjects. "Because we sent kind of the wrong message," Alam recalls, "when we got on the phone with the FDA examiner, he just screamed at us, 'You guys believe in your drug too much. You're irresponsible.' From that day on, we basically severed any kind of relationship."

If Boger, Alam, and Vertex had a fatal blind spot, this was it: a faith in themselves and in data so blazing that it eclipses other sensibilities. Getting on the wrong side of regulators just as Vertex was at last moving up to become a true drug company so deeply troubled some board members that they began to discuss among themselves the need to rein in Boger. But he and Alam saw no other course. "In every division of the FDA, they're completely rabbinical; it's all about precedent," Boger recalls. "It has nothing to do with any information that you bring them. So we—Vertex, who they don't know from Adam—come with a drug and we say, 'Look, we know there's an existing cure for this disease. We want to add our drug to the existing drugs. And we think we can show you—the way this is going to be used is—we're going to shorten the treatment of those drugs that are on the market'—in the FDA's language 'change the label' of those drugs. And ours is a new chemical entity that goes by a new mechanism.

"They go, 'No no, no. That's not the way you do it. You're going to dose your drug for twelve months with interferon-ribavirin and see what improvement you get. And then we'll talk about shortening treatment.' And we say, 'There's no reason to dose our drug for twelve months. Its effect will be one hundred percent done in some number of weeks'— we thought eight weeks—'at which point we need these other drugs to sort of sweep up.' And they were just completely against that."

In the realm of clinical trials, a lucrative hybrid industry—part academic, part commercial—had lately started to ply the gulf between drugmakers and regulators. Companies now outsourced the management of large worldwide clinical trials to outside investigators and contract research organizations (CROs), committing significant portions of their R&D budgets to specialty services that line up doctors and patients, publish studies in medical journals, and add a patina of independence and prestige. For its Phase II trials to show how VX-950 could be an effective drug, Vertex had hired as its principal investigator Dr. John McHutchison, internationally recognized for coordinating major clinical trials in liver and gastrointestinal diseases and associate director of the Duke Clinical Research Institute in Durham, North Carolina.

A droll, wiry Australian, McHutchison had moved to California in the late 1980s to study liver disease, began working in HCV, and never returned Down Under. With a large operation and a panoramic view of the treatment landscape, he consulted with dozens of companies focusing on numerous different approaches—anything with a biological rationale. But Alam had also hired McHutchison to help educate investors and analysts about hepatitis C, and since 2003 he'd assisted Vertex by giving an overview of the disease and the market opportunity at major investor events. Hepatitis C was still poorly appreciated even by doctors, most of whom seldom saw a case. When they did see any, they tended to refer their patients to specialists before they developed cirrhosis or liver cancer or went into full-blown failure requiring a transplant, the hallmarks of the epidemic. In most cases hep C presented as an invisible, symptomless, slow-motion buildup until it flashed over into a racing downward spiral in late middle age. The disease confounded investors, who could only guess at the size of the problem. McHutchison told them that in the United States alone, the costs would be equivalent to those of asthma. Around the world, where up to two hundred million were infected, the ramifications were geometric.

From McHutchison's perspective, hepatitis C drug development was a "graveyard." But the swoosh sped up the life cycle for testing new agents in patients. In December he and Alam quickly designed and pulled together a twenty-eight-day trial of VX-950 combined with peg-riba. "We did that four-week trial because we didn't have the twelve weeks chronic tox data to dose for twelve weeks," McHutchison recalls. "So it was either wait and dose for twelve weeks or get some info on a small group of people. We wanted to get this done and get it done fast." Enrolling subjects for trials before Christmas is difficult, but McHutchison enlisted investigators at busy clinics in San Antonio, Texas, and Puerto Rico. Again the study was small, just two dozen patients.

A month after the Morgan meeting, Vertex announced the results. When its drug was added to current common therapy, the virus became undetectable in the blood of all twelve patients after four weeks, compared with fewer than one in three of those taking only peg-riba. That morning, the *Times* published a story on Vertex and Boger that quoted McHutchison: "We've never seen data like this before, where everybody's negative so early on in treatment." Doctors are advised to wait twelve to twenty-four weeks after finishing treatment to declare a cure, because of viral breakthroughs—relapses. An absolute rule of clinical trials is that the results in large studies are always worse than in smaller ones. Still, "twelve of twelve" became a benchmark, a talisman, a signal that the standard of care could soon be shattered. It was a calling card for Vertex, which now looked to have the dominant HCV drug in the field.

Pushing the data hard, Vertex began suggesting at investor meetings that VX-950 might clear HCV in twelve weeks by itself without significant side effects. Stories of miracle drugs are legion in the drug and biotech industries, but this one had started to have a ring of truth about it. Investor buzz, already feverish, spiked.

Of course, there was also a competitive logic for urging shorter-duration trials. Schering-Plough, which also had a protease inhibitor, had started Phase II trials in the fall, a few months ahead of Vertex. Its drug didn't produce as sharp a drop in viral activity after two weeks, and Schering was testing its molecule for six to twelve months, like the common therapies. Boger and Alam were gambling that they could leap ahead with twelve-week trials, but the FDA continued to resist. McHutchison, working to repair the relationship between the company and the agency, started to accompany Alam when he visited the FDA's sprawling campus in the Maryland suburb of Silver Spring to plead his arguments.

"The meetings were contentious," McHutchison says, "because the FDA was saying let's be safe and let's go slowly, and Vertex and John were doing it in the classic Vertex way, which is to push in all directions. There were negotiations over sample size, and how many people you could put at risk, not knowing. Nobody knew what you were going to do. The fear was you would create a resistance in these patients, and they would be harmed for life, just like with HIV."

❖

As head of the cystic fibrosis program, transplanted from San Diego to Cambridge, and with a charity to account to instead of a corporate partner, Eric Olson grappled with how to move VX-770 into clinical development across uncharted ground. The collaboration with the CFF had been severely tested by the decision to go first into clinic with a potentiator aimed at helping just 4 percent of patients. "That was a problem for them: 'How can we do this? How can we sell this? Should we support this?' " Olson recalls. Vertex, with other funds from the foundation, continued to develop its hits for correctors; but because he was going ahead first with a potentiator, Olson needed to convince both parties to bear eventually hundreds of millions of dollars in clinical development and buildout costs that neither had anticipated or could easily afford. "I was getting continuing pressure from Vicki and finance: 'Can you get some more money to fund this?' " he says.

Before Sato retired, he'd persuaded her that he could ask the foundation to fund a development program only if Vertex also agreed to invest. "The model was to fund research, and it wasn't really clear on what basis, other than our needing more money, they could do this," Olson says. "We couldn't really say to them, 'And oh, by the way, we're committed to this, but not committed enough to spend our own money.' When I was putting the development plan together for 770, it was clear that we were gonna have a problem at some point. We were going to do our proof of concept in twenty to thirty patients, whatever it was, and if it worked, our next study was going to be our pivotal study. So we needed to be ready at the end of proof of concept to go to Phase III: tox studies, final formulation, final synthetic route. All these things needed to be done before we even knew whether the drug worked. In a normal program, you would wait for POC before you invested."

Olson and Ken discussed Vertex's options, and Ken suggested that the company propose two alternatives to Beall: a step-by-step lower-risk process, and a higher-risk accelerated plan where activities were run in parallel. As Van Goor had warned, the development costs for a drug for patients numbering in the low thousands were no less than for one that targeted tens of millions of people, especially during the early and middle stages—before it was proven to be effective. At the same time, speed to market can mean everything to those who may suffer and die without it. The faster you go, the more you need to spend up front, and the higher the risk. Olson estimated the cost to the foundation of about $20 million.

During the fall, he and senior business development director Phil Tinmouth had traveled to the CFF headquarters in Bethesda, Maryland, to present the plan to Beall, chief medical officer Dr. Preston Campbell III, and several others. "They bring us into this conference room,"

Olson recalls. "Bob is already furious. In his mind, the agreement had always been 'We'll fund it up to there, then you guys are gonna fund the rest.' And here we are saying, 'You know, we need some help.'"

The conference table was bare except for a single item: a book, *The Billion-Dollar Molecule*, a dramatic nonfiction account of the early days of Vertex. After Beall had enlisted the Gates Foundation to support the search for drugs targeting CFTR, Bill Gates Sr. had urged him to read it, and ever since, Beall had referred to it and recommended it when the question of Vertex's commitment came up—especially among the many skeptical CF researchers who still doubted the approach and resented the company as a favored novice.

"That was his prop," Olson says. "He was saying, basically, 'I trusted you guys.' He ranted for an hour. We listened and listened. I said, 'Bob, it's fine, we can wait till we get proof of concept. I'm not asking you for money. I'm just saying here's an opportunity.' I remember he used the phrase 'I thought I bought the Cadillac. Now you're telling me I bought the Chevy.'

"Then I said, 'Well, Vertex is willing to put up half. Are you?' What could he say? He'd just had his board meeting, and his budget was set. At the end of it, he pulled me over and said, 'Let's start making the slides for the board. If what you're telling me is right, we have no choice. We have to do this.'"

In March Vertex announced that it had entered into a new collaboration with Cystic Fibrosis Foundation Therapeutics (CFFT), the nonprofit drug discovery and development affiliate of the CFF, to accelerate clinical development of VX-770. CFFT agreed to pay Vertex $13.3 million over two years so that the company could begin testing the drug in patients by the end of 2006 and move quickly from there to Phase II studies. Vertex retained worldwide commercial rights.

Olson supported moving into patients as quickly as possible, but he also anticipated a problem. With so few patients having the G551-D gating mutation, Vertex had been unable to find a suitable lung from which to extract HBEs: that is, as convincing as Van Goor's videos were about the overall approach, their beating cilia were from cells treated with a *corrector*, in a genotype for which VX-770 wasn't designed to work. The team had no comparable evidence that the drug would work in the very genotype for which it was developed. Boger clearly understood the issue, but Olson wasn't sure whether many others in Cambridge appreciated how big a risk the company was taking by committing to clinical trials by the end of the year.

As Van Goor and the scientists stepped up the search to produce cell lines isolated from a scarce and precious resource—the lungs of one of the world's three thousand or so G551-D patients, Olson reckoned with another new type of hurdle arising from personalized medicine: the ever more complex challenge of what it means to "prove" a biological concept before the FDA will allow you to test it in sick people.

❖

At the May meeting, the board relieved Boger of the job of chairman, which he'd held since Benno Schmidt retired nine years earlier. The directors named longtime member Dr. Charles Sanders, seventy-four, a former chairman of Glaxo and vice chairman of Squibb, to replace him. Since the wave of business and accounting scandals that rocked Wall Street during Bush's first term, business ethicists and shareholder groups had crusaded to strengthen boards of directors, which in the go-go years had increasingly become rubber stamps. Many corporations were voluntarily moving away from the unitary executive model—just, it's worth noting, as the White House embraced it. Boger didn't care about the change but believes he should have.

"That was a mistake," he says. "When Charlie took over as chair, that was presented to me as where good corporate governance was going— that there would never again be in the history of the world, except by historical accident of companies that hadn't yet got to this advanced level, another CEO-chairman, so it's about time for us to lead the way and split these roles." The succession to Sanders cut him off from direct responsibility for communicating with other directors, isolating him from their discussions. He stopped knowing what they were thinking.

As Vertex advanced and grew, it was all but inevitable that the company would need to acquire what Murcko called "adult leadership." Sanders was a paragon, a gregarious native Texan with a large presence like Schmidt, who'd led the war on cancer and later helped found and capitalize dozens of biotech companies. Also like Schmidt, Sanders had strong institutional and political ties and a jumbo Rolodex. A distinguished cardiologist, he was a former general director of Massachusetts General Hospital, a past chairman of the New York Academy of Sciences, and chairman of the Foundation for the National Institutes of Health. He also served on the boards of Genentech and Merrill Lynch. After retiring from Glaxo, he registered as a Democrat and ran for the party's nomination to unseat North Carolina senator Jesse Helms, spending his own money to buy name recognition but losing ultimately to the former mayor of Charlotte.

It was equally certain that with "adults" in charge, the company's remaining original torchbearers would leave or recede from running things—"put down their torches," as Boger says. Vertex had kept its soul, its corporate DNA, during the long, hard years leading up to the swoosh in large part on the strength of Boger's hub-and-spoke, open-door management style, and he still made it a point to have lunch with all new Cambridge employees within their first three months—a practice that kept him in touch with what was happening down the ranks. Fewer had left than he had feared. But Mueller gradually had tightened his control in drug discovery, displacing and unsettling the old guard.

Mueller wanted one person in charge in San Diego, and he chose Negulescu. Tung agreed to remain as head of discovery at Aurora, but his wife, who wanted to return east, considered it a slight and a demotion, and in the fall, he left Vertex. (Shortly after he quit, he reconnected with Aldrich, who'd become a leading Boston biotech investor and venture capitalist, and together they started a new drug company, Concert Pharmaceuticals, with Tung as CEO and Aldrich as chairman.) Mueller wanted new blood in Cambridge and replaced Thomson with the former head of the Novartis collaboration, biologist Mark Namchuk, a laid-back Canadian who set out to undo some of the reporting strictures that had thickened through the years of having to account to corporate partners for every one-quarter full-time employee.

Thomson opted not to put down his torch. Managing Vertex's mushrooming collaborations, he had witnessed over the years the necessity of strategic networks and was eager to replace the prevailing model by

developing effective new ways to treat neglected diseases—those that
no drug company will spend its own money to investigate not because,
as with CF, there are so few patients but because the ample patient populations live in countries too poor to buy drugs. Seeing an opportunity
to leverage Vertex's expertise in small-molecule discovery against major
scourges such as tuberculosis by building alliances with academic researchers and local partners in the regions of the globe where hundreds
of millions are infected, Thomson discovered another challenge worth
his industry, one that he believed he and Vertex were well positioned to
meet.

Murcko also found a large role to play in shaping Vertex's strategic
direction without having major managerial responsibilities, operating
more within the interstices of Mueller's new order. On the ever-shifting
organizational chart, he remained chief technology officer, a description
he chafed at because it invoked wizards whipping up new gadgets. But
his larger purpose was to champion "disruptive innovation": ideas that
change things radically by unexpected means.

"My job," Murcko says, "was to identify new technology—very
broadly defined—but also to evaluate it and bring it into the organization in a way that would provide value. So it isn't about playing with cool
toys. It's not a sandbox. A lot of people, when they hear the title of chief
technology officer within pharma or biotech, they assume it's the guys
who go play with lasers. It's the guys with beards and suspenders playing
with cool toys in the basement. What Peter and I were trying to do is figure out what are critical problems within the whole organization where
any kind of new approaches, new technologies can be brought to bear
in ways that would shed new light on a problem, make it easier to make
certain kinds of decisions, make something go faster—have some material impact. It's not limited to research. It covers the whole organization."

Equally vital was Murcko's history as Boger's sidekick and his ability
to read Boger's actions. In a place where everything was debated deeply
and head-snapping decisions came down too often with insufficient
explanation for why they were made, puzzled scientists across the sites
looked to Murcko for understanding and perspective. After Vertex's own
stock swoosh had started in 2000—the nosedive after the biotech bub-

ble popped, the crash of the p38 kinase inhibitor, the layoffs, the limbo of pralnacasan—even longtime Vertexians frequently got anxious. They'd been millionaires during the run-up, and now the company seemed to be sucking wind, or throwing everything behind a molecule that Lilly rejected. Many had trouble filtering out the internal volatility and external noise. They would wander into Murcko's office early in the morning or late at night for counseling. "We got a little ahead of ourselves," he advised them.

"The underlying mission of the company hadn't changed, the goals hadn't changed, the leadership hadn't changed," he recalls. "What I would tell these people is, 'So exactly what are you worrying about? Yeah, it's unsettling, I get it. I don't like it either. It sucks. But what are you going to do? Do you or don't you believe that we're still on track to something really, really useful? If you've come to the conclusion that we aren't going to get there, then go away, leave. Some of us still think this is a game worth playing.' "

Boger's losing Vertex's chairmanship was of little consequence in the labs and conference rooms. The lions had been thrown some meat so that they wouldn't breach the fence. The adult who mattered to the scientists was Mueller, not Sanders. Vertex was vaulting, the *Boston Globe* wrote in June, "out of its research driven niche into a prominent new role carrying the banner for the idea that new science can turn into real profits." If anything, the three and a half down years from 9/11 to "twelve of twelve" had been astoundingly successful, producing VX-950 and VX-770, which the FDA now granted fast-track status. The company's market value had soared as high as $5 billion in March before falling back by nearly a third. Murcko, like Boger, couldn't fathom exactly what people were worried about.

❖

It had been a decade since most Americans first heard the term "drug cocktail." Combination therapy had turned AIDS into a survivable scourge; you only had to look at Earvin "Magic" Johnson, the former LA Lakers basketball star-turned-entrepreneur brimming with vitality, to see the change. Ever since the first patients starting choking down as many as forty pills per day, the challenge to drugmakers, clinicians, and the FDA was to determine which drugs to combine in which specific subgroups of patients and for how long.

Though the microbes that cause AIDS and hepatitis C share many similarities—both are RNA viruses that hijack human cells to make more of themselves—they're vastly different as disease agents and drug targets. HIV inserts itself into host genetic material in the nucleus; HCV doesn't. Thus, except in a handful of reported cases, HIV can't be eradicated without killing the host, but HCV can. And so while the goal of treating the AIDS virus is to lower the viral load to the point where it stops infecting new cells—requiring a lifetime of vigilance—the body can be rid of the virus that causes hepatitis C, the patient cured.

On the other hand, HCV produces trillions of new viruses daily, a thousand times more than HIV, mutating at a much higher rate. And because HIV lasts only a few minutes once it's exposed to air, it doesn't spread as easily as HCV, which can last up to sixteen hours—explaining, for instance, why tattooing carries a risk of transmission for HCV but not HIV. Most crucial for drugmakers, HIV resistance arises only under drug pressure, while every patient infected with HCV carries multiple subspecies—a "population of mutants," Boger calls them—that are resistant before the drug is even there. "To control and wipe out HCV, you had to get every last bugger," he notes, which is why a cocktail is absolutely needed. Both in the United States and worldwide, four times as many people are infected with HCV than with HIV.

In July the FDA approved the first complete treatment for AIDS taken once a day as a single pill. It combined three drugs already on the market made by two companies, Bristol-Myers Squibb and Gilead Sciences. A fixed-dose cocktail called Atripla—a pink, film-coated tablet with "123" stamped on one side—it was to first-generation protease inhibitors like Agenerase what the iPod was to the Sony Discman. Not only was it an elegant, future-inspiring marvel of miniaturization and packaging but also a triumph of intuitiveness. Not a single drug but an all-in-one therapy, Atripla had all the hallmarks of a game-changing product.

Gilead sold only two of the molecules itself; the other it licensed in. When it first tried to combine the three compounds, "we had glue on our hands," an engineer recalled. It took about a year to find a formulation that would produce the same exposures as when the drugs were taken separately, and Gilead had tested five different formulations in volunteers. A midsized Northern California company, it had bootstrapped itself to become the sales leader in AIDS. With a higher valuation now than Sears and rich in cash, it was on a buying spree, aiming to branch out from its historic antiviral franchise into circulatory and respiratory diseases. Of special note to Kwong and her group were Atripla's targets: none of the compounds inhibited HIV protease. Like AZT, all three blocked reverse transcriptase, the enzyme that enabled the virus to make DNA copies from RNA, and one was a nucleoside analog—a nuc.

Kwong hoped to determine how the hepatitis C virus progressed in the bloodstream of patients on VX-950, and she had hired a young researcher named Tara Kieffer to build a clinical virology group. Like Kwong, Kieffer was a true believer, her heart deep in her work. In first grade, she fell in love with the structure of the double helix during a visit to her father's biology lab at Montgomery College in Maryland, made a drawing of it, and had kept it on her wall ever since. While working on her doctorate at Johns Hopkins University, she devised and performed complex experiments to track the development of resistance to HIV drugs; her publications, along with others, helped scientists to conclude that the virus couldn't be eradicated with one drug by itself.

Looking at the viral load curves from the second fourteen-day study with VX-950 alone and with peg-riba, Kieffer and her group assessed the overall problem with treating HCV: how to devise a better drug cocktail. They sequenced both the typical form of the virus most common in nature and also numerous mutations—both before and after treatment—and then examined the patient profiles. Kieffer tracked their progress in the presence of the drug. She found that it was powerfully inhibiting the most common form of the virus but that there were variants with specific mutations in the protease that by degrees were less sensitive to the drug. Only with peg-riba were these least sensitive variants suppressed. "We learned that the primary role of VX-950 is to get rid of the wild-type virus, which is the completely sensitive virus, and some low-level resistance," she recalls. "Then we were really relying on the interferon-ribavirin component to get rid of the higher-level resistance.

We were able to look at these profiles and determine how long our drug is needed to do that job."

Beyond boosting Alam's case for shorter-duration treatment, Kieffer's work gave scientists their first understanding of how HCV responds to direct-acting antiviral medicines. Clinical and commercial interest in her research was keen, and she was asked to give a plenary talk at the Liver Meeting in November at the Hynes. Meanwhile, Vertex signed a broad international collaboration with two Johnson & Johnson subsidiaries to develop and sell VX-950 outside North America and Japan, securing ample financing for its late-stage development and commercialization. Terms of the deal called for J&J to pay the company $165 million up front in cash, with an additional $380 million in potential milestone payments and tiered sales royalties averaging about 25 percent. Wall Street analysts, estimating that the European Union had twice as many patients as the United States and Canada, cheered the collaboration. More than one projected peak US annual sales of up to $3 billion by 2013 and overseas figures cresting at $2.4 billion two years later.

Smith pounced. After three years of pressing Boger to get the company in shape while insisting to Wall Street that Vertex had not just a breakthrough medicine on its hands but also a true blockbuster, the headwinds had stalled. Like Aldrich, he knew the time to sell new shares to investors was not when you had to but when you could. Smith had converted the remaining debt, and the balance sheet now showed about $500 million in cash. Vertex would lose more than $200 million this year and projected losses of more than $300 million in 2007. With Ken and his legal team filing quickly with the SEC, the company issued nine million new shares, the offering netting about $300 million. Boiling down the math, what Smith realized he could do—now that J&J was footing the bill for Vertex to get through approval to market with its newly renamed HCV protease inhibitor telaprevir (VX-950)—was to fund all the rest of the company's own R&D for another year simply by minting more stock, which eager fund managers and institutional investors gobbled up in nine days. Perhaps no other biotech had ever been in such a position.

"When we went out and spoke to investors, we were very bullish," Smith recalls. "It wasn't inappropriate—we were basing it on the data—

but we were very optimistic. We were very aggressive with what we might achieve with clinical results: the potential and the opportunity. I make no apologies for that. We needed to create a big story with critical mass that caused a Wall Street following. This was a big story. Hep C is a big disease. Nobody had ever seen a result like ours. You're kind of like the Pied Piper. They start to follow along. There's also this viral marketing aspect among the investors themselves. There's a lot of momentum; let's call it greed as well: 'I want to be a part of this.' And then they start to live vicariously through you."

From the beginning, science, people, and money had been the vital reagents in Boger's experiment in building a better company. Throughout the fall, for the first time, a critical mass seemed to be reached. Kieffer stole the show at the Liver Meeting, where a crowd of several thousand packed the main conference hall to hear her explain how telaprevir worked, how it encountered resistance, and how it could be used effectively in a cocktail with existing drugs to cure even genotype 1 patients. Boger, anticipating Vertex's transformation from a scientific and clinical leader to a commercial one, again turned his attention to a non-obvious foundational issue that felt to him far more urgent even than the singular mission of getting telaprevir to patients.

❖

Bob Kauffman pressed hard to convince the FDA that in placebo-controlled midstage studies, Vertex could truncate the standard treatment time while also improving cure rates. Long-term safety concerns arose. Telaprevir was highly potent. The longer you took it, the higher the risk to your body, but if you went off it too soon, it could unleash viral resistance. Kauffman partly won over regulators with virology data and modeling. "In the end, the FDA agreed to a multiprong study that allowed us to study shorter duration with some patients," he recalls. "We were really obsessed about twelve weeks at the time, and the FDA was very concerned. We believed in our modeling. But they did let us do a small number of patients with twelve weeks, and in Europe they allowed us to do more."

The race to market was decidedly on, not just with Schering-Plough and Boehringer, but with dozens of other companies, including several

that, like Gilead, had started trials with nucs and other polymerase inhibitors that in preclinical assays looked as potent as telaprevir. "We were
pushing very hard for an aggressive development program," Kauffman
says. "We knew we were in a competitive area, there were other people
working in the field, and it was starting to heat up. We really wanted to
both advance our program quickly but also really give patients something that was good at the end—not just higher SVR rates, but higher
SVR rates with shorter-duration treatment. All of our advisors kept telling us that would be fabulous.

"We had a real impetus to stay in the lead. It was frustrating sometimes when it took a few months to negotiate a protocol with the regulatory agencies. We were pushing to get stuff done. But you have to see
this from the regulators' point of view. This was a new area, new field. Nobody had ever done this before. Nobody had even shown that you could
get an SVR with a direct-acting antiviral until we did it. So they were
somewhat skeptical, and concerned. They wanted things tied up well."

Within two months of starting Phase II, while Kieffer prepared her
slides for AASLD and Smith captivated Wall Street with tantalizing
visions of a monotherapy for the leading cause of liver cancer and transplantation, a few patients at different study sites began developing severe
rashes. Investigators reported them to Vertex and the FDA. Kauffman,
keenly aware that dozens of drugs cause two forms of life-threatening
skin toxicities in rare cases, and that a clear correlation can halt a drug
from being approved, grew increasingly worried.

"Things had been swimming along: fabulous swoosh, great data,
start Phase II. Then this," he says. "There was one case, another one,
and another one, and we all started looking at each other. You know, the
first one, you just go, 'What was that?' After three of them, you go, 'Oh.
This could be telaprevir.' After five of them, you're pretty sure it really is
the drug."

❖

During World War II, George Merck made sure that his drug company
was central to the war effort. Pharmaceuticals were in low regard—
deservedly, since very few of them did anything useful—and profiting
from medical research was anathema; an idea whose time had not yet

come. Merck, approaching age fifty, chaired the government's biological weapons program. He also helped drive—and was a prime beneficiary of—Washington's efforts to marshal the nation's research labs in defeating the fascist powers.

President Franklin Roosevelt's chief science advisor, Vannevar Bush, chose Merck's top consultant and scientific architect, Dr. Alfred Newton Richards, to chair the Committee on Medical Research—the biomedical equivalent of the Manhattan Project. On August 7, 1941, four months before Pearl Harbor, the CMR heard intelligence reports that the Germans had isolated the active substance from the cortex of the adrenal gland—cortisone—and were feeding it to their pilots to help them withstand the rigors of aerial combat. The committee made cortisone production its highest priority. That evening, after returning by train to Philadelphia, Richards received two visiting researchers from England, Howard Florey and Ernst Chain, who had experimented with a highly scarce mold extract—penicillin—to treat burns and infections, harvesting it in desperation from patients' urine and reinjecting it until they ran out. By the end of the war, under Richards's direction, Merck led other drugmakers and federal labs in making enough of the antibiotic to supply the nation.

Driven to expand his business and assimilate himself and his company at ever higher levels of American life, Merck built deftly on his wartime public service. Bush, forward-seeing architect of the nation's postwar scientific order, joined the company's board, soon becoming chairman. In the labs, chemists working under the fiery, protean head of research Max Tishler—Boger's mentor—labored another decade to develop a synthesis to make cortisone commercially available. George Merck became a spokesman for the industry's idealistic nature, his company exalted for its scientific leadership and social progressivism. In August 1952 he appeared on the cover of *Time* over the caption "Medicine is for people, not for profits."

Boger had a similar trajectory in mind for himself and Vertex, but the War on Terror hadn't produced comparable opportunities for drugmakers to pitch in. The only Axis was the name-only "axis of evil," and the wars in Afghanistan and Iraq offered few opportunities for ambitious, public-spirited company builders to assume national leadership roles against a major threat. Boger seized on an alternative route. Envisioning a leading position for Vertex locally, within the global health industry and with government regulators and payers, he set about building a highly visible web of outside responsibilities, connections, and interests. "Externalities," he called them. To keep his employees up on what he was doing, he started to blog about his activities internally, for Vertex-only consumption.

Boger became the first biotech executive in nearly a decade to chair the Massachusetts High Technology Council, and he joined the board of the Biotechnology Industry Organization (BIO), the national biotech trade organization. He sat on the board of fellows at Harvard Medical School and became a major fund-raiser for the Greater Boston Food Bank's capital campaign. Thinking globally but acting locally, he became a close advisor of Governor Deval Patrick, representing the state's broad scientific, academic, and business interests, and advising Patrick on how to position Massachusetts for the future. "Unlike many of his peers— who feel that since they represent the future, politicians should come to them—he sees the value of involving himself deeply in Beacon Hill affairs," *Boston* magazine noted, ranking him among the city's power elite.

He also heartily adopted a few of the trappings of the role, the VIP/inside-the-ropes attention, access, and other entitlements that major CEOs enjoy. In January, a week after returning from the Morgan conference where he outlined Vertex's plans for the trials of telaprevir and VX-770, he and his wife, Amy, traveled with the New England Patriots to the AFC championship football game against Indianapolis, sitting three rows behind the team's cerebral head coach, Bill Belichick, during the flight. In March they flew to Mountain View, California, for the XPrize Foundation's Radical Benefit for Humanity at Google's sprawling head-quarters, special guests of one of the foundation's board members. With no structure taller than a few floors and "curiously conservative land-scaping," he observed in a ten-page, single-spaced blog post, the campus looked "like an unnaturally clean summer camp for geeks."

He was seated at a table with Charles Lindbergh's grandson Erik, who four years earlier had piloted a replica of his grandfather's *Spirit of St. Louis* on the seventy-fifth anniversary of the first solo flight across the

Atlantic—a grueling physical and mental feat. Erik Lindbergh had been diagnosed in his early thirties with rapidly progressing rheumatoid arthritis and had been forced to stop flying before he started to take Enbrel, which halted the disease and restored his life, and for which he was so grateful that he'd become a celebrity spokesperson for the product. "And so, I casually asked him," Boger wrote, " 'How important to you would be a once-a-day pill that did about the same thing?' Unhesitatingly, he jumped forward. 'That would be great. I hate Enbrel. I'd be curled up in a home without it, but I hate it. The injections: They burn. The new less-frequent injections burn a lot worse than the older ones. The infections. I hate it.' We talked about Vertex's decade-plus-long commitment to arthritis. Vertex has a new fan."

The *Boger Blog* recounted in colorful detail the glittering events of the evening: the A-list sightings (Larry Page, Sergey Brin, Richard Branson, Jerry Brown, Arianna Huffington, and Robin Williams, who performed after dinner); the extravagant table setups and posh, all-organic cuisine by Google's in-house chefs. It showcased Boger's far-flung interests, erudition, cultural commentary, and own geek "cred"—unscrolling in a rush of asides. About a lackluster finale of "Hallelujah" by singer Rufus Wainwright, for instance, he wrote, "Get the Leonard Cohen version, on his *Various Positions* album. It's chillingly superb recitative, a mature creation to Wainwright's empty simulacrum. (The pop Jeff Buckley version, on the album *Grace,* is second derivative and even more self-indulgent than Wainwright's.)"

During the silent auction, Boger hoped to win a zero-gravity flight with Stephen Hawking. A lifelong space travel enthusiast who views Hawking as the greatest physicist of our time, he was enthralled by the idea of flying weightless alongside Hawking in NASA's "Vomit Comet"—a specially modified Boeing 707 with a padded cabin where, for twenty to thirty seconds, as the plane flew sweeping parabolas in an invisible, protected corridor above the Atlantic, they would float free. Hawking is as famous for his motor neuron disease, paralysis, electronically modified voice, specially equipped wheelchair, and expansive wit as he is for his theories. Boger confessed to feeling bereft when he at first learned that he'd lost out to another attendee who outbid him, a mix-up that wasn't resolved until the end of the night. He blogged, parenthetically: "*Star Trek* fans have already seen Dr. Hawking in space, playing himself (the only person ever to play himself in any *Star Trek* episode) in *Star Trek: The Next Generation* episode #252, 'Descent—Part I,' first aired 21 June 1993 (Stardate 46982.1). As every fan knows, Hawking, as himself, plays poker with the android Data, in a holodeck simulation including Albert Einstein and Isaac Newton. Actors played the latter two characters. Best episode ever."

It would be hard to overstate Boger's joy and fulfillment on a Tuesday in April, when he flew to Florida to meet up with two dozen other passengers to take the trip on "G-Force One." Earlier in the month, McHutchison had given a late-breaker presentation in Barcelona, Spain, at the European Liver Meeting on the first shortened-duration telaprevir trial. The results suggested that twelve weeks of telaprevir-based therapy with peg-riba enabled some patients to clear the virus. The company also now had a corrector for CFTR—VX-809—on track for clinical development, the key ingredient in the drug cocktail that could be the elixir of life for the great majority of those with cystic fibrosis. Boger was soaring. "Jet Blue to Orlando," he wrote. "This was no Mickey Mouse tour. I didn't go to Tomorrow Land. Felt a little Goofy, but that was understandable. I was headed for weightlessness. Weightless. With Steven Hawking. Weightless."

Hawking was accompanied by four handlers and two physicians who laid him on the padded floor among the other fliers, guided and monitored him through eight plunges, and concluded that he was in "tremendous condition"—heart rate, blood pressure, oxygen levels all normal and perfect. "Space, here I come," Hawking pronounced ecstatically after the flight. Boger felt much the same way.

"Zero gravity is unlike anything you have experienced," he blogged. "It's not at all like cresting a hill in a fast car. It's not at all like dropping rapidly in an elevator shaft. It's not at all like jumping out of a plane. It's not at all like scuba diving in blue open water. It is not at all like any of these. Imagine a dimmer switch for a room light. You glide it slowly down, and the light slowly dims. Raise it back, and the light comes back on, slowly, evenly. There are no jumps, no bumps, no stomach shifting, no wind, no resistance. The light just goes down, and then it comes back up. That's what zero gravity feels like."

It was in this frame of mind that Boger consolidated his external positioning at the May BIO convention in Boston. At the annual membership meeting, he was elected chairman of the board—in effect, the industry's face in Washington. At the same time, his efforts on Beacon Hill bore fruit. Governor Patrick—with Boger beaming at his side, rolling his right thumb and forefinger together like a safecracker—touted to thousands of conventioneers recent gains in the Massachusetts economy that helped make the state a world leader in biomedicine. With one in seven jobs in the health care sector, and with more than $2 billion annually flooding into its research hospitals and university labs from the NIH, the state was proof of the power of government-subsidized science and assembled intelligence as a platform for prosperity.

"Within this small state," Patrick said, "we have an extraordinary confluence of research universities, teaching hospitals, brainpower, venture capital, and a long tradition of entrepreneurialism that has helped define this economy as being fueled by innovation. We are quite simply the largest life sciences supercluster on the planet. And that is a thing to be very proud of."

Patrick announced the goal of a $1 billion Life Sciences Initiative: a package of seed funds, training programs, subsidies, and tax breaks to attract more private investment and jobs. Boger, a key promoter, was already lobbying legislators for its passage. Rolling up his sleeves, influencing those he needed to make things happen, he saw no distinction between his public activities and his role as CEO. "I don't see boundaries all that clearly," he remarked.

Inside Vertex, Boger's blog had the paradoxical effect of bringing his external life to everyone's laptop just as they saw less and less of him in meetings and at beer hour. Many were thrilled by his rise, it becoming a foretaste of Vertex's future. His fun and enjoyment were in some way an extension of them, a vindication. Yet in the labs and cubicles, some also had the feeling that *they'd* like to be mingling with the Patriots instead of grinding away at midnight projects. They didn't see him driving the executive team or trying to put together the remaining pieces of an organization that within a year hoped to submit a new drug approval (NDA) application to the FDA. Boger was trying to do all that too, but he didn't think hiring the right people was the key to sustainability; culture, and spirited citizenship, were. Boger couldn't explain himself and made no effort to. He recalls:

> Ironically, as I got more visibility about what I was doing, people got more nervous about what I was doing. What do they think a CEO does? Go ask the CEO of any real company how much time they spend walking the factory floor. As little as it takes to take the picture for the annual report. That's just not what a CEO does. And if he does, you're growth bound. That's just wrong.
>
> If we were going to be a major company, I needed to be the chairman of BIO. I was convinced that we were on our way, so I wanted to make sure that I was building the other piece of Vertex's externality, which couldn't just arrive the day the first drug arrives. So suddenly we were gonna have a community relations department and start paying attention to the place where we live and pay attention to government? You can't turn it on like a light switch. You've gotta have authenticity. We needed to start building an external presence. That's not a function that you can outsource. That's crucial. It was very calculated.

❖

Garrison cochaired many Vision into Practice (VIP) teams with Murcko. Now inside Vertex, he no longer was conducting an archaeology dig but a building project, and the key was to enlist talented and exemplary people. On management, Garrison favored the notion of "positive deviance"—the idea that there are people whose uncommon but successful behaviors and strategies enable them to provide better solutions than their peers. Identify those people, make them models; there's your varsity. Boger had told him, "It's not about the executive team, it's about everyone," and so he put out a companywide email asking, "Are you a deviant, or do you know one?" He got 360 responses, a deep pool from which he recruited the leaders of the VIP teams across the sites and functions. "Mark," he says, "is definitely a positive deviant."

Phase II of the vision process, EMBED, altered how management valued Vertex's key asset: its people. Like all companies, Vertex had goals and evaluated its employees on how well they met them. But Boger saw the values as a mandate to change the company's compensation structure. How do you recognize collaboration? Originality? Daring? Bravery? Intensity? Passion? Over the next six months, with discussion at every level of the company, Vertex linked the values to performance evaluation, thus hardwiring them to salary and incentives. Your bonus now relied as much on the way you met your goals—with what attitude and spirit—as on whether you met them. "This was huge," Garrison says. "Now you could *not* achieve all your goals—which is what I believe all innovative companies should strive for, because if you get all your goals, they're too easy. However, if your performance on the values was exemplary, you could get the highest possible compensation. By the same token, you could get all your goals but be in trouble. It really put teeth into it. We went around and made a lot of presentations about that."

Surely the greatest risk to Vertex, now that it looked to be on its way, was that it would stop taking the sort of risks that had propelled it all along, that it would cease rewarding those who were willing to fail for a bold idea, against bad odds, in order to come up with something useful—a better way. Boger believed that linking the values to compensation meant that the people who succeeded best in the company and set its tone in the future would bring to their jobs the same passion and sacrifice "beyond explanation" as Thomson, Murcko, and the other early torchbearers. Of everything he had set in motion so far, EMBED made him proudest, since he was confident it would guarantee—more than anything else—that the dominant Vertex phenotype would replicate for many years ahead.

Trish Hurter was another positive deviant. With some "serious hiring," she had built her formulation group up to thirty-five, going to dinner three nights a week with candidates, her lightning metabolism the only thing keeping her from becoming, she says, "a blimp." An early adopter of the Vertex values and a true believer in Garrison's mission, she had pushed the group to develop its own BHAG—Big Hairy Audacious Goal—a guiding vision of what it wanted to achieve and become. The result was an expression of Hurter's own aspirations: "Everything that can be miniaturized, predicted, or modeled has been," she explains. "It means there are no mysteries. One of the things people say about formulation is that it's an art. That drives me ballistically crazy. To me, it's science. It's very difficult, complicated science, but it's science, not freaking art. Nothing is black magic, like the witches and the cauldron."

In December Mueller had made Hurter head of pharmaceutical development, putting her in charge of everything pertaining to the physical nature of a drug substance and drug product, the manner in which both are made, and the regime by which the manufacturing process is shown to be in control. CMC, as the overall function is known, faces extreme scrutiny by the FDA and other regulators, and Hurter's department, now seventy people, would more than double before telaprevir could be launched.

She took everyone off-site to develop a "pharm-dev" BHAG: "everything we want to be twenty years from now," she told them. Garrison catalyzed the session. Phase III of the Vertex Values Process, SUSTAIN, was now under way, with the goal of putting in place a culture that was, as he says, "forever." There had been much discussion about how to foster original ideas and help them rise through the company without being strangled by middle management. How do you originate an idea? And if you're a supervisor and somebody brings you an idea, how do you respond? (Originator: do your homework; do feasibility studies; gather support; get people on board with you. Supervisor: be an honest sounding board; help the originator build momentum; adjust the workload to allow exploration of the new idea.) The VIP teams had recommended proposals that Garrison workshopped with task groups like Hurter's.

Breaking into customary small groups and using flip charts, Hurter's team divided its goals into four categories: science; team and business processes; personal character and well-being; and prestige and reputation. What the members arrived at was an aspirational consensus: they wanted Vertex to be a standard-bearer for great research-driven drug development and manufacturing, lauded for its industry-leading precision and efficiency, done by researchers who worked hard but had fun and high integrity: "basically scientists leading full lives," Hurter says.

She and a small group distilled the daylong discussion into a paragraph, including specific language about delivering drug molecules expertly to the precise sites in the body where they are needed, and about producing the exact desired substance, whether in milligrams or metric tons, through the most advanced continuing processes.

Garrison pressed her, as ever, to be briefer, to zero in on the singular message that would encompass "what it looks like; the top of the mountain." He said the paragraph was a fine example of a so-called "vivid description," but it was not a BHAG, which wraps all that you want to say into a slogan. Hurter's group again went off-site. They talked about what they were trying to achieve; what could be embossed on a company coffee mug or T-shirt and also have the power of truth. In its earliest days, Vertex held a slogan competition for its first company T-shirt. It was won by Aldrich ("We don't leave success to chance"), but the scientists, more grounded in the grueling realities of science, generally preferred "I'd rather be lucky than good." A right and true message isn't easy.

The pharm-dev BHAG that Hurter and her group decided upon matched the larger reputational vision of the company: "Made by Vertex: the gold standard for drugs delivered right on target." Here in a phrase was the same proud confidence that for fifty years had echoed in the words "Made by Merck." It didn't come from an inspirational founder or from marketing or from the ET or a consulting firm, but from the production line.

❖

Hurter took it and ran. Her group, with support from Murcko, began chipping away at the larger goal, doing more PK modeling to try to improve its formulations. It expanded its use of flow simulators that predicted molecular behavior in hoppers mixing thousands of kilograms of powders and compaction simulators to predict the precise physical results when Vertex was pressing pills by the millions. Working toward a final formulation of telaprevir, Hurter wanted desperately to engineer a smaller tablet and get rid of the food effect: its bioavailability, as with many drugs, dropped off on an empty stomach. For a year, she threw significant resources into doing extensive studies on five formulations, none ultimately better than in Phase II.

She worried that Mueller would be disappointed. "I'd spent huge amounts of FTE time, my time, to get this tablet to be where I wanted it to be, and at the end of the day, it didn't meet the goals I'd set." Mueller backed her up, telling her, "What are you talking about? We have a Phase III, commercial, scalable tablet that matches the Phase II. This is excellent work. Perfect! Great news! You gave it your best." He did what Boger hoped he would do: encourage Hurter to exceed herself, even if in the end she fell short. It was all in your state of mind.

"When people take a risk and it doesn't pay off, you've got to be supportive," Hurter says. "That's the biggest difference between Merck and here. At Merck, if you took a risk and it paid off, you were a hero. If you took a risk and it dumped on you, you were a disaster, and for the next fifteen years, your performance review reflected that. They were very unforgiving about anybody who did something that didn't pan out. Basically people become very cautious."

As a student of companies and CEOs, Garrison felt as if he'd landed in Oz, crashing through from black and white to Technicolor. "The company had me," he admits. Almost everywhere he looked, he found a hunger to invent and improve that organizations talk forever about instilling in their people, a systemization of the process of innovation that the country and the world were crying out for. And what it was, he believed, was simply *attention* to the nobler drives of a certain ambitious character type. Garrison:

> I was the creative director of an advertising agency and the CEO of an ad agency, and I love to tell my former colleagues that Vertex is the most creative company I've ever worked for. And it is, by light years. Why are they that way? Because that's the way Josh wanted it.
>
> It's what you do. The thing that Josh understands is, it's about the people. So he would do things that most guys don't do. He would have a lunch with every new employee. He never said the same thing twice, but he always said the same thing. The key part was: "You're here to save the universe. Drug development stinks, we've got to be smarter, we've got to do it differently, you've got

to have courage." He would tell stories. He would find ways to tee this up.

That's how you do it. You have to put the energy into the reason for being of the company, instead of putting the energy into the financial performance of the company. Josh built the culture. He paid attention to the culture. He was fascinated with the culture. *It's the culture, stupid.* That's the difference between great companies and not-great companies.

A degree of caution was in order. Merck, lauded for its values and vision, was one of the companies that Jim Collins, inventor of the BHAG, featured in his 1995 bestseller *Built to Last.* Only a dozen years had passed, half a generation, and now when great companies with promising futures were discussed, Merck wasn't mentioned. Nor was Sears, another of Collins's prized examples of companies that seemed to get the balance of mission and cultural identity just right. Both could well come back, but not soon, and not easily. The "secret sauce," as Garrison called it, had curdled. In presentations to the company and to the board, he identified the cause, which he called CLCD—"Corporate Life Cycle Disease"—the self-deception, corruption, and conservatism that come with time, experience, and Wall Street's tyrannical expectation of year-to-year growth, which is particularly trying in pharmaceuticals.

As much as Boger was creating Vertex to last, the company was still in adolescence—still hemorrhaging $60,000 an hour—with a long way to go to launch. Whether it was a great company or even a good one had not been tested in the market. Only time would tell if, in its deepest structures, it had become the enterprise he and other Vertexians were wagering their careers trying to build.

❖

As strained and remote as Boger's relations got with his board, his ties to his top executives suffered from the opposite problem: the inevitable paternalism and fractured alignment of hub-and-spoke leadership. A majority of directors now distrusted him, feeling misled. They grumbled that he withheld data and switched subjects when questions came up that he didn't want to discuss; that he was purposefully oblique. Emmens felt a troubling "disconnect" in meetings—a palpable mutual distaste. But the ET members—Alam, Mueller, Smith, Garrison, Ken, and the others— owed Boger their positions, admired him, and were most effective only when they could enlist him on their side. They too met as a body, but if any of them had a problem, that person went in to see Boger one-on-one, sitting across from him as he continued to work at his computer, multi-tasking furiously, yet outwardly as attentive and still as a stork. On the Vertex ET, the "we" in " 'We' wins" was much more "I and thou," more "imperial we" than an expression of unity and teamwork.

In late June Vertex announced two new senior management appoint-ments. Kurt Graves, hired as chief commercial officer and head of busi-ness development, came from Novartis, where he had launched a record nine drugs in eighteen months as chief marketing officer. "He had fan-tastic credentials," Boger recalls. "He was the Novartis commercial guy who sat in on FDA phone conferences." Still in his forties, Graves had started his career at Merck, where Roy Vagelos picked him to lead a new unit to market Prilosec, telling him, "Go build the twenty-first-century pharma company." With a first-class strategic, business, and analytical mind, he also had the requisite deep appreciation for R&D. Several board members considered him ripe to be groomed as Boger's succes-sor, one of them going so far as to anoint him the "Golden Boy" in his interview—before he was offered the job. "Thanks for taking away all my leverage," Boger recalls thinking.

The lawyer Amit Sachdev joined to direct the company's govern-ment affairs and public policy efforts. He arrived from BIO, where he managed the health section and the governing board. Boger didn't want just an in-house lobbyist with strong Beltway ties, though Sachdev was surely that: a former deputy commissioner at the FDA and before that majority counsel to the Committee on Energy and Commerce in the US House, where he was responsible for bioterrorism, food safety, and en-vironmental issues after 9/11. Boger needed a strategic-minded activist who could wake up Washington about hepatitis C and, by extension, the urgent need for telaprevir and other new drugs. Unlike those of AIDS, the dimensions and costs of the HCV epidemic remained largely beyond the ken of politicians and bureaucrats, who failed to recognize the size of the threat to public health and the nation's unpreparedness to pay for and fix it. A kinetic figure, casual but intense, Sachdev could backslap, but first he and Boger knew he had to educate.

The next week, the board held a retreat. Boger led a three-hour discussion with charts and slides to determine the criteria for the next CEO. Ever since the subject of succession first came up three years earlier and Boger advanced Emmens as his "get-hit-by-a-bus guy," the directors had supported his basic proposal: they should be on the lookout for someone with "a deeply experienced commercial sensibility married to a sincere appreciation for, if not mastery of, the R&D side." Now they fleshed out the description in greater detail: a great speaker; someone who could project equally well the image of a fast-paced scientific company to Wall Street and a solid, progressive business entity to scientists.

"You weren't going to find a lot of these lying around," Boger recalls. "The board arrived at a profile which seemed to be unmeetable, but it was sincerely put together. But there wasn't any search firm. I wanted their input on it—if we see the right person, then we should move, rather than a clock clicks, and it's time to go do it. I said, 'We're not ready for the transition, but we'd better be ready intellectually for the transition if the right person comes along.' Most of the board thought Kurt needed years of seasoning but that he had the right DNA and the right background. Kurt fit the profile except that he hadn't been tested in a growth situation."

Boger thought he might step down someday, but he saw no more than the same routine urgency for a thought-out, scalable leadership plan that had been there all along. Few modern CEOs lasted a decade, still fewer almost twenty years, as he had, and Boger knew that Emmens, for instance, the only serial CEO among the directors, thought that five years was the optimal tenure: long enough to take a company to the next stage or turn it around but not so long that you begin to think, like a professor or a judicial appointee, that the job should be yours as long as you want to do it. Taking seriously the dispensability of everyone in charge, including himself, Boger directed the head of human resources, Lisa Kelly, to develop succession plans for the entire ET. In parallel with Garrison, Kelly began to identify those within Vertex who seemed best equipped to move up quickly in the organization if anyone in senior management left suddenly.

❖

Boger and Sachdev huddled with small groups in Cambridge and Washington over the conundrum of having a blockbuster molecule for an underrated and undervalued disease. With diabetes or incontinence or depression or the other ailments depicted in the ubiquitous ads that now, because of demographics, saturated the nightly newscasts, there were few mysteries about the size of the problem or the market. The same was true with horrors such as CF and AIDS, where activists drove home the message. But hepatitis C was a true orphan disease—not by the FDA's numerical definition but in the sense that it was dispossessed, without a home.

Drugmakers don't invent diseases, but they do define how governments and people think about an illness by offering hope through treatment. The history of hepatitis C had none of the hallmarks of an emergency, an issue that would readily enlist a politician or an agency to get involved. Unlike AIDS, it rose from the shadows, a furtive infection resulting from shared blood that, in most cases, showed up only decades later, when it was too late to do anything about it. As it spread rapidly during the 1960s, 1970s, and 1980s—before the virus was discovered and the blood supply could be secured—it became stigmatized, a social disease. As disease historian Jacalyn Duffin writes in her book *Lovers and Livers: Disease Concepts in History*, sufferers "took on special identities or 'types' that became part of the disease concept.

"Infectious hepatitis affected the poverty stricken, the lax, the institutionalized and the unclean; it was a disease of beatniks, 'street people,' and 'hippies.' Those likely to develop serum hepatitis were a mélange of the 'guilty'—self mutilators and needle-sharing drug users—and of the 'innocent'—recipients of blood or the heroic nurses and surgeons who had cared for them."

The marketing of peg-riba elevated awareness but elicited little sympathy. Schering's Pegintron and Roche's Pegasys were effectively the same drug, but Schering initially dominated, largely by sending doctors unsolicited checks for $10,000 to prescribe its product and paying others

enormous fees to run clinical trials that weren't much more than marketing gimmicks. It sponsored unblinded, sloppily monitored, head-to-head
trials with Pegasys for which the doses of ribavirin were altered to favor
Pegintron's cure rate. Doctors who strayed or even spoke favorably about
Pegasys "risked being barred from the Schering-Plough money stream,"
according to the *Times*.

The payoffs stopped, but since the 2006 Medicare drug benefit took
effect, Washington now paid for almost half of all medicines sold in the
United States, and government interest in the pharmaceutical market
had spiked. Federal and state prosecutors across the country were aggressively targeting the drug industry, investigating alleged payoffs, Medicare
and Medicaid fraud, and off-label marketing. Earlier in the year, Schering
agreed to pay $435 million to settle charges including that it lied to the
government about drug prices and promoting Intron A and another drug
for cancers for which they weren't approved.

Sachdev arranged for Boger to meet with Dr. Steven Galson, a top
public health physician who directed the FDA's Center for Drug Evaluation and Research (CDER). The problem Vertex needed to solve was
what it would take to bring the issue of hepatitis C to the forefront. The
disease was a time bomb, affecting a tight cluster of largely poor baby
boomers whose medical care was paid disproportionately by Medicaid
and Medicare. The older and sicker they got, the more the government
would pay to take care of them. Within a decade, the costs would be
enormous, dictated by what treatments were available. Yet because only
one in four people infected with HCV knew they had it, the greatest
savings would come by identifying them *before* they needed treatment.
"That was the premise we took forward," Sachdev recalls. "The system
was designed to push the problem along, not solve it. How can we in-
centivize people to say, 'You know what? We should solve this problem
while it's solvable.' Steve told us, 'You're not ready to trumpet to the
world that everyone should be screened. In all the noise of health care,
what's the urgency to solve your problem? Why do you have to solve this
problem now?' "

Here was the necessity of Boger's outside game writ large. Vertex, a
company that almost no one in or out of government knew, held the lead

against a major public health threat. Once telaprevir made it to market, decisions would be made across the country and at every level of government to decide whether to pay for it and how much to pay. This was not a disease where prescribers and payers could be bought off, not that Boger saw any need—or value—in operating that way. To get its drug to patients, Sachdev believed, Vertex needed to entice Washington and the states to look at their contributions, then convince them: "You actually can put your money where your mouth is in terms of cost effectiveness."

In October Bush named Galson, a retired rear admiral in the public health service, acting US surgeon general. Adopting the company's preference for data, Sachdev began to assemble Vertex's case. He commissioned a health care consultant, Milliman, to conduct an actuarial study of the consequences of HCV and the costs of a baby boomer epidemic of liver disease, to prove the disastrous cost impact on government and private payers over the next twenty years if better alternatives to peg-riba weren't advanced quickly. Vertex needed to reintroduce hepatitis C to politicians who might not have an HCV epidemic in their districts but might appreciate the economics. In political terms, there wasn't much time. "The message was not for the whole world," Sachdev recalls. "I designed this for the government."

❖

As Vertex developed telaprevir, telaprevir remade Vertex. The molecule needed a powerful company behind it, and the people who joined to get the drug out had to work fast, improvise, pull together the pieces they were responsible for, think on their feet, and depend on networks being assembled at the same time by new people who were in the same boat as they were. In Cambridge, the company couldn't hire fast enough. "We have had to build half of a company that we didn't have to have before," Lisa Kelly told the *Globe*. A well-oiled machine it was not; instead, it was an assembly of like-minded smart people who felt day to day as if they were building a boat while trying to sail it, in rough seas.

Mueller hired a quality engineer and operations manager named John Condon from Millennium to assemble the supply chain and oversee manufacturing. With no time for a lengthy search, Mueller was as usual Teutonic and to the point, offering Condon the job after an evening of dinner and drinks at a restaurant. Mueller sketched out the overarching goal and the company's urgency with little guidance and few details: a globally integrated, FDA-approved process for getting telaprevir tablets to market fully up and running within two years. "Peter said, 'So are you interested?' I said, 'Yeah,' " Condon recalls.

"He said, 'Here's what we'll do. You'll come in. We'll meet with the executive team, we'll make you an offer that's not embarrassing, and then you'll start.' "

The obvious challenge was "trying to figure out how much we had to make and when," Condon says. He did some preliminary calculations. Although Hurter's team was still working on the final formulation and had just produced a 60-kilo batch for the Phase III trials, projections from the commercial team on dosing and numbers of patients told him that by 2009, when Vertex hoped to launch telaprevir, it would need 50 metric tons of drug per year. Pharma companies like Merck, which until recently had owned everything down to the railcars delivering raw ingredients to their factories, had grown up as towers of vertical integration and could afford to put $200 million into the ground to make one drug. Investing Vertex's scant resources in a plant and equipment was out of the question.

Condon is lithe, grey haired, and athletic, an enthusiastic bike commuter with an up-for-anything, get-it-done vibe. On lab doors all around Vertex, there were VIP photos of him whaling beatifically on air guitar. He envisioned a "virtual supply chain" none of which Vertex would own: a flat global collaboration not just fully outsourced but also highly integrated, a "just-in-time" inventory system where supplies got from the Pacific Rim to Europe to North America precisely and only when they were needed. Hiring a small staff for the buildout, he began to pull together the pieces of a cross-border network that could churn out, in sections, the five major constituents of the drug substance; ship them halfway around the globe; synthesize them; ship them again overseas; and then formulate, manufacture, package, and deliver tablets of telaprevir to hospitals and pharmacies. "We were not going to build a manufacturing plant," Condon says. "We would leverage the external world."

Among the key raw ingredients—Condon's group called them the

Fab Five—two were especially complex and tricky to synthesize, and for the trials, Hurter's team had contracted with two Chinese chemical companies to make them. This was an increasingly common, fiscally urgent, and politically controversial arrangement: Western drugmakers depending on the exploding Chinese industrial sector to produce pharmaceutical raw materials with an eye always on the larger prize, the exploding Chinese market. In 2006 Condon negotiated agreements with the two firms. Some ET and board members worried about the reliability of Chinese fine chemicals, as well as political interference and whether their manufacturing processes might not pass muster with the FDA. Condon overcame the pushback by instituting quality audits and collaborating with the companies in designing and building large reactors that were highly product specific.

Making the alliance attractive for the Chinese, Vertex placed significant orders for the materials in smaller lots that could be made with existing reactors, even though telaprevir still had to survive Phase III pivotal trials, and there were no guarantees that the substances would ever be needed. Spending months in China building relationships, Condon devised an overlap plan to ensure that telaprevir was available at launch. "In case there were manufacturing delays, we didn't want to have all the product resting on the success of building two plants in China, bringing them up to speed, and then scaling up the chemical processes from fifty-kilo batches to five-hundred-kilo batches," he says. "Between the two of them, we committed about thirty million dollars. That went a ways to encouraging them to take the risk."

John Thomson also traveled more and more to Beijing and Shanghai, building a company presence there, leveraging relationships of other sorts. As head of strategic networks, Thomson was looking far ahead. What he saw were vast East Asian populations urbanizing rapidly, their governments unprepared to pay for Western medicine but woefully in need of new drugs and eager to learn how to discover and develop them on their own. This emerging pharmaceutical megamarket faced tremendous challenges as it moved inevitably to different payment systems, but some experts were more and more convinced that it would be hugely attractive sooner rather than later. The Mitsubishi deal precluded Vertex from selling telaprevir in China and the region, so Thomson had to devise a noncommercial strategy that drew on other needs and interests, creating other angles of approach.

Drugmakers were keenly interested in China not just as a reawakened nation of more than 1.3 billion potential consumers, building frantically in advance of next year's Summer Olympics amid worries about human rights protests and athletes competing in smog-shrouded Beijing. They coveted it also as a place to access naïve—previously untreated—patient populations for their clinical trials, which were drying up in the United States and Europe. Thomson, surveying the country's technological prowess, concluded that Vertex could make fast inroads by pursuing parallel approaches. One was philanthropic, by working with Chinese and American public health officials, research hospitals, and drug companies to assemble a nonprofit network to develop new treatments against tuberculosis. The other involved partnering with local companies to codevelop a new Vertex antibiotic with a novel dual mechanism that had limited commercial potential in the West but could be highly useful against new infections arising across the region. Thomson:

> With it being cheaper to develop molecules in China, and with their yearning to get their hands on advanced Western assets to show that they're in the innovation sector, there are a lot of not-directly-commercial motives for people wanting to develop new innovative drugs in China. And therefore, this is an opportunity that gives us upside: new market, friends in high places, accessing naïve patients, working out which pharma companies we can work with and which we can't, developing a drug that might otherwise sit on the shelf. Above all, the world *desperately* needs new antibiotics. We've just got a situation where it's really difficult for economics to drive the development of narrow-spectrum antibiotics because they're compared against the broad-spectrum antibiotics. So this is an opportunity that scratches a lot of different itches.

It was Alam more than any of the others who supplied the timetable by which all of the company building revolved. Still on point with the

FDA, he pressed it to formally recognize Vertex's work on Phase II and discuss its plans for Phase III. Two large midstage trials—PROVE 1 and PROVE 2—showed interim SVR rates of telaprevir-based combination therapy of 65 percent and 61 percent, respectively, after twenty-four weeks, compared with about 40 percent for forty-eight weeks with peg-riba alone. "That gave us a lot of hope that we really did have a drug," Kauffman recalls. The most common side effects were fatigue, headache, nausea, and rash, sometimes severe, but dropout rates were low. Kurt Graves, as head of commercial, believed that with such overwhelming improvement, Vertex shouldn't have to do Phase III trials at all. When Alam started introducing Graves in teleconferences and agency meetings, reviewer Russ Fleischer, a clinical analyst with deep experience with antivirals, considered it startlingly inappropriate—too pushy by half. "We played into their concept of us," Alam agrees. "We were trying to leap ahead. I wanted to get this drug to patients, but I'm sure at the FDA it was perceived as all about advancing the company and making money."

Alam frayed the relationship further by failing to control his impatience with Fleischer, who had trained as a physician's assistant, had a master's degree in public health, and, not incidentally, was charged with providing guidance to drugmakers—the same job that Alam's father had done for decades while Alam was growing up. Alam recognized the problem but seemed powerless to act any more than passingly, reluctantly collegial toward Fleischer. "I absolutely was the wrong person; the smart-guy physician," he says. "His attitude was, 'Fuck you, I'm not going to do what you want me to do.' Both my scientific ability to connect the dots and leap ahead and my personal impatience and urgency absolutely did not fit with how the FDA wanted to move."

In pharmaceuticals, time is money in an expanded sense: every delay ramifies across the organization, ratcheting up the pressure, especially in companies where everything depends on getting approval for one vanguard product. Frustration and impatience—the drivers behind much discovery and innovation—seldom help. The more Vertex was consumed by its urgency to push through to market, the more slowly and more carefully the FDA seemed determined to see it go.

In October, a few days before Vertex announced the PROVE findings,

Schering-Plough disclosed preliminary, but positive, efficacy and safety data from a Phase II study of its HCV protease inhibitor, boceprevir. While Vertex had been pressing full speed ahead, Schering was moving at a slower pace, but the data showed that 70 percent of patients taking boceprevir and peg-riba went undetectable at twelve weeks—not a cure, but a strong indication of viral response. The company reported the drug was well tolerated—no rash, albeit significant anemia. Still hoping to push telaprevir into Phase III studies in early 2008, Alam and others at Vertex believed they remained eighteen months ahead of Schering's drug. But for the first time since the demise of the original Boehringer molecule, they had a direct competitor: one with deep pockets, and a longtime commercial leader in hepatitis C. The announcement sent shares of Vertex down by 15 percent.

Boger welcomed the challenge. Ever since seeing the structure of Schering's compound a year earlier, he knew that boceprevir would be potent and effective. He believed, however, that the features Vertex had engineered into its drug made it superior, and that its trial designs would lead to a simpler and more straightforward treatment regimen. What heartened him most about Schering's comeback was that he now had a partner crusading on behalf of the disease, another company with a large stake in publicizing hope for hepatitis C. By now Boger relied not just on his partners to propel him toward his goals but also on his rivals.

❖

Back during the early 1990s and their many death marches across East Asia, Boger and Rich Aldrich often stood in the aisle of a United 747, across from the lavatories, where they could stretch their legs, discussing long-term business strategy. Asia was the future. Almost two decades later, Boston, which in the 1830s became America's second-busiest port merchandising Chinese tea, porcelain, and silk, still had no direct flights to the region. The day after Thanksgiving, Boger flew to Washington Dulles International Airport to board a United flight to Beijing, part of a forty-three-person state trade delegation with Governor Patrick. The comforts and décor were vintage. Boger blogged: "I've got almost 600,000 United air miles—not just funny money miles but real 5,280 foot air miles, and I've seen a lot of United 747s. This plane looked like it had been playing host as a 'please touch' air museum piece since 1991. I glanced around for the eight-track tape player. Oh well."

At the opening ceremony at Tsinghua University in Beijing, Patrick gave what he called his "China stump speech": a paean to China's history as a world leader for thousands of years, the historic trade relations between Massachusetts and China, the state's leadership in research-driven industry, and a growing partnership. "We see great opportunity in building anew upon the blend of China's economic dynamism and matchless historical perspective and Massachusetts's tradition of innovation and joyful engagement with the future," Patrick said. "Common business and international ventures lead to common ground and, in turn, to a common future. In a world weary of war, scarred by so much suffering and struggle, and in search of reason to hope, we have come in essence to make the world a little smaller, our circle of friends a little wider, and the prospects for mutual prosperity a little brighter."

Boger picked up on the theme in his own remarks. He told the audience about Vertex's widening Chinese collaborations, including agreements with Beijing's Institute of Medicinal Biotechnology and two of the city's major teaching hospitals in tuberculosis, Thomson's joint-program development for an antibiotic, and Condon's agreement with Shanghai-based WuXi (*Woo-she*) PharmaTech to manufacture two of the Fab Five. "Everything here is more," Boger wrote. As he apprised the coming tide of pharmaceutical players reaching the Chinese market, establishing franchises, alliances, relationships, and beachheads, he believed as ever that the most effective companies wouldn't be the biggest but the fleetest, the most emotionally involved, with a long-term vision that could help both sides.

"In this new international network model in life sciences, China will play a major role," Boger said. "And Massachusetts is a natural partner for China. Both of us have the opportunity to grow up our industries in the new model, without substantial inertia from the old, twentieth-century model. Both of our business cultures are fast moving, innovative, and entrepreneurial. Both of us are playing to win on the world stage. So we at Vertex are happy and confident to say that our success in hepatitis C *depends on*, and in tuberculosis *depends on,* network collaborations in China. And vice versa. That's a formula for success in the twenty-first century."

The trade mission had broad goals and a crammed schedule, but Boger managed to insert a visit to WuXi, a seven-year-old pharmaceutical, biotech, and medical device R&D outsourcing company that did contract research for nine of the ten biggest drugmakers. Its founder, Ge Li, was a returnee: a prominent member of the rising class of business and political leaders who'd left the country in adolescence to study abroad and were known as *haiguis*: sea turtles. (Most of Shanghai's emerging political leaders were *haiguis*.) Ge had a doctorate in organic chemistry from Columbia University. In Shanghai, Patrick and the delegation met with the mayor. Boger chaired a life sciences forum. After a quick turnaround at the St. Regis hotel, there was a reception at a trendy, alpine-style restaurant in the former French concession. Boger blogged:

Afterward, close to midnight, I ride back across town with the governor in the back of his Audi. He wants to talk about the next morning's visit to WuXi PharmaTech. This is to be the only life sciences company we visit on the whole trip. The governor particularly wants to know what I think the tone of his remarks should be. When he talks about health care, he says, he can tend to get emotional, focusing on the human needs and the human benefits and the human emotions rather than "the business aspects." I laugh. "That will be just fine," I say. "You'll find, I think, WuXi PharmaTech is run by some kindred spirits. Most of us in the innovative drug business are idealists. It's too hard a challenge to be in this business for just coldly rational reasons. There are a lot easier ways to make a living. Be yourself. They'll love it."

# CHAPTER 8

FEBRUARY 11, 2008

It was an hour after dark by the time Boger, Smith, Graves, and Alam took their places in the windowless conference room for the year-end earnings call with Wall Street analysts. They sat at a large round table, a speakerphone positioned between them. Vertex's new senior director for strategic communications, Michael Partridge, directed the call. On a whiteboard he printed the names of the dozen analysts listening in, prepping for questions. Self-effacing and soft spoken, with a nuanced view of the business of science and a deft knack for the care and handling both of ET members and fund managers, Partridge had worked his way up in the company, joining a decade earlier to write press releases after a stint in market research. Now he was in charge of the complicated Kabuki of providing information to those who regarded Vertex not as an experiment in New Pharma but as a stock: VRTX.

All publicly traded companies must disclose material information to all investors at the same time under the 2000 SEC ruling Regulation Fair Disclosure, better known as Reg FD. Hence the ritual of the quarterly call and the after-hours timing. Whatever trading advice the analysts would provide for their clients, management's presentation would be the same to each of them—operational results, earnings outlooks—delivered while the stock exchanges were closed and thereby theoretically leveling the playing field between large institutional funds and individuals. Reg FD governed Partridge's every communication. Opening the call, he delivered the usual cautions about the risks and uncertainties in any "forward-looking statements"—projections of future performance.

Wall Street was in a deep funk, fearing the coming year. The housing bubble had run its course. The credit crisis was no longer just a subprime mortgage problem. It was sucking in all borrowers. As home prices fell and banks tightened lending, even people with good credit histories were falling behind on their mortgages, car loans, and credit cards. Steep losses across the financial industry warned of a crash. The once-triumphalist Bush White House team hoped to leave Washington before it struck. Over the past weekend, Illinois senator Barack Obama swept Democratic presidential primaries in Washington, Nebraska, and Louisiana, while conservative Christian Mike Huckabee won in Kansas, embarrassing Republican front-runner Senator John McCain.

Graves took the lead addressing the analysts, focusing on Vertex's pipeline, portfolio strategy, and key priorities for the coming year. "First and foremost," he said, "we aim to execute Phase III policy to fully leverage the differentiated profile and first-to-market opportunities with telaprevir and key telaprevir life-cycle programs." In other words, the company had designed its large-scale pivotal trials to broaden its FDA label and intended to exploit its pole position aggressively.

"Second," Graves continued, "now that telaprevir is in the final stage of development, we aim to take definitive steps to build a multidrug HCV portfolio around telaprevir, our second-generation protease inhibitors, and other target external assets to build a leading HCV franchise . . . Telaprevir has set very high hurdles on the two most important attributes of efficacy and duration of care. However, if something can be designed to compete with it or even define a new combination treatment approach, we intend to be the company that brings that to the marketplace through our knowledge and expertise in the area."

Asserting that the company meant to keep and expand on its lead in hepatitis C was necessary to reassure analysts and investors that it was adequately positioning itself against the arrival of new agents. Graves went on to detail plans to study telaprevir across all types of patients, especially those who failed previous treatment: "null responders," or so-called nulls. Vertex hoped to prove in a big way early on that by increasing cure rates for the hardest-to-treat patients, telaprevir would be the drug of choice for those high-prescribing doctors who influence what others in the field think. Turning to other molecules for other diseases, Graves briefly mentioned the ongoing trials with VX-770 and VX-809 for cystic fibrosis before alighting on a major new project: a returnee from the Novartis collaboration.

The post-Gleevec stampede after targeted cancer cures and the fever-dream of an oral Enbrel had concentrated tremendous attention on drugs that block Janus kinases, or JAKs. No one had brought a JAK inhibitor to market, but Pfizer led the field. The big question was selectivity. Pfizer's compound, which it was testing against rheumatoid arthritis, targeted JAK3, a key signaling molecule that helps trigger immune cells to attack, but it also was considerably active against JAK1 and JAK2, which had other basic biological functions throughout the body. It was not a clean drug.

Graves announced that Vertex expected by midyear to start clinical trials of a highly specific oral JAK3 inhibitor for treating a gamut of immune-mediated inflammatory diseases, including RA, psoriasis, organ transplant rejection, and several others. Not coincidentally, it was the same constellation of diseases that Boger had gone after with Vertex's first project in 1989, when the basic biology of cell signaling remained a black box, and that had propelled the rise and fall of the p38 kinase inhibitor and pralnacasan. If Vertex had a grail, this was it: an all-in-one, safe, effective immunosuppressive pill that wouldn't raise infection risks and would capture the $10-billion-plus market now belonging to Enbrel and the other follow-on injectable biologics. "A JAK of all trades," as the journal Nature Reviews put it.

Graves handed the briefing over to Smith, whose job was to explain how Vertex, with insufficient revenues from royalty payments and research collaborations, would pay for everything. Earlier in the day, the company announced two concurrent public offerings of notes and shares aimed at bringing in about $400 million. The stock rose on the news. By now Smith's strategy of selling the telaprevir story to fund all of Vertex's efforts except telaprevir had widened the company's prospects, although it still confounded most analysts. Nearly all of them, accustomed to biotech companies putting all their resources behind one drug, and biased staunchly against dilution, continued to value the company strictly on anticipated revenues from hepatitis C.

"People started to get more comfortable on the data, and so it all begins, we raise the money," Smith recalls. "But the interesting thing is you're raising the money based on the HCV data, but HCV is being paid by J&J, and the money we're raising is being applied to basic research and cystic fibrosis. We could have consolidated down behind telaprevir, allowed the J&J partnership to fund the business, and issued fewer shares. But you'd be a one-drug company. We took more risk to keep the business wider. That's why we're so different. I'm sure people in our labs will say it was scientific brilliance. I like to say I was riding that scientific brilliance, and we went and raised money to give the company the runway and the ability to go and invest in its science."

The gloom in the economy scarcely intruded. Boger, in his brief remarks, echoed Smith's theme. His concept of what Vertex was worth and his time frames had always differed sharply from Wall Street's. Weary of arguing, he sounded matter-of fact, even taciturn.

"We are increasing the fundamental value in our business," he began. "We believe the progress we are making with telaprevir, as well as across our organization, is positioning us to succeed at our corporate and commercial goals . . . We talked about several milestones for you on the call today. We know it is up to us to deliver in order to establish a successful business for many years to come, and we are focused on delivering . . . There are a lot of data being generated from clinical trials across our pipeline. It should be an exciting year."

❖

Wall Street, where story stocks can float above the churn, imbibed the offerings in a week. Four weeks later, the investment bank Bear Stearns, the most admired securities firm in *Fortune*'s "most admired companies" survey, known for its tough, overnight-is-long-term trading culture, collapsed. The first major casualty of the financial crisis, it spiraled down from being healthy to being practically insolvent over a long weekend before it was rescued from impending bankruptcy by a deal with JPMorgan Chase & Company. In Washington, Treasury Secretary Henry Paulson conducted a hastily arranged conference call with Wall Street titans. "He didn't want rapacious trading tactics to further wound a gravely injured Bear," the *Journal* reported, "so he decided to put it to the firms straight: I expect you to behave yourselves."

Eric Olson and the CF team in Cambridge and San Diego remained largely unnoticed by the analysts. VX-770 went unheralded and underestimated even within Vertex—"the well-adjusted second child," according to Virginia Carnahan, a longtime company manager who in January became vice president of strategic marketing of new products. As in families where parents obsessed with videotaping firstborns almost forget to photograph their younger siblings, biotechs tend to favor their lead programs, celebrating these programs' accomplishments and convulsing over perceived flaws. For two or three critical years, CF had enjoyed the advantage of flying well under the radar. Olson: "We had a very small army of very committed people who weren't thinking about HCV."

Clinical development had advanced smoothly, with few glitches and a strong safety profile. In the fall, the company started a midstage study with VX-770 in twenty patients with the G551D gating mutation. Patients took the drug for fourteen days. Dosages were varied. Expectations were deliberately set low. The company's chief marker of activity was sweat chloride; if a patient's perspiration was less salty, it would show that the potentiator was keeping the molecular gates in CFTR open longer and allowing the protein to pump chloride ions across the membranes of epithelial cells just under the skin. Whether that translated also into improved lung or pancreatic function because of a reduction in mucus went beyond the modest goals of the trial. The medical community, as ever, remained deeply skeptical that the compound would work in patients.

At the annual North American Cystic Fibrosis Conference in Anaheim, California, in November, a nurse from one of the trial sites in Iowa had approached Olson and blurted, "I know I'm not supposed to tell you this, but one of the patients in the trial says he hasn't felt this good in five or six years." Olson recalls: "I didn't want to believe it. I said, 'Listen, I've seen this before. People get on drug. There's placebo.' " The brain's role in improving health during trials is a known clinical effect—often the only effect—of most experimental medicines, and Olson tempered his expectations. In San Diego, most of the attention was on VX-809, the corrector, still the greater hope for the majority of patients.

In late March Vertex received an interim analysis of the study results. After two weeks, the investigators saw an average improvement in lung function of 10 percent in patients receiving the highest dose, compared to no improvement in those receiving placebo. A few of the patients reported breathing better within days of starting on the drug. The lung-function data were supported by improvements in sweat chloride and another biomarker, nasal potential difference, a measure of ion flow in the nasal lining. In the executive suite at Aurora, site head Paul Negulescu huddled expectantly with project leader Peter Grootenhuis to hear the findings. On the phone was Dr. Claudia Ordonez, a Vertex physician, who, along with Kauffman, monitored the trial.

"We hadn't heard anything in months," Negulescu recalls. "The deal for us was we had to show a pretty dramatic effect in a very small number of patients to get excited about this. So Claudia comes on the phone. The data are streaming on the computer, and we're listening to Claudia's voice. Before any of the data came in, I remember the first thing she said—and she's a very conservative person, not prone to exaggeration. She said, 'Well, guys, this was an absolute home run.' We looked at each other. We had no idea what a home run looked like in CF. But we had hit everything we hoped and more.

"It answered so many questions. It said CFTR is fixable in people and that it's been sitting there like a switch that's off, waiting to be turned back on. That was one of the other worries: you're born with it defective, maybe your whole system develops improperly, such that even if you fixed it later, it wouldn't matter, because all the biology has changed. But no, it turns out the biology is there, just dormant, waiting to be awakened."

Mueller, driving Vertex hard to deliver a second drug, remembers being "incredibly stunned" by the data. Far and away the company's most seasoned drug hunter, he had seen dozens of proof-of-concept findings over his career, but none remotely as dramatic and powerful as this. Even the Vertex swoosh paled. The implications for patients, for Vertex, for the possibilities of taking on larger problems and harder diseases—even for the way certain diseases could now be viewed—were epochal, he thought. Mueller recalls:

> I think this is the first time ever that any company has had a small molecule that is able to correct—in a subpopulation, at least—the gene defect those people have. It was proof that Vertex is really doing transformational medicine. Cool as HCV is, from a reputation point of view, this drove it home. CF is a lot more complicated in nature than hitting a virus over the head. It shows the commitment and also the capabilities that we have as an organization, to go after problems that nobody else really has ever done. They always stepped away: it's too complicated, not doable. The seven, eight years of hard science that we did finally paid off.
>
> And that gives you a feel that there is an evolution going on in Vertex that is beyond rational drug design. It's the new-century approach where biology drives the effort, and it's not just the chemistry. I believe that the twenty-first century has a mandate, and the mandate is that you make an advance in health, not just a drug. This is what you get with transformational medicine.

It took a few days for the euphoria to recede and the larger reality to hit full-on. Sabine Hadida, the chemistry leader, didn't sleep for a week, she said. Team members were alternately elated and terrified that they might fail the other 96 percent of patients, that they might never produce a corrector as effective or, apparently, as safe as 770. Olson flew to Washington to tell Beall and the foundation the results before Vertex announced them publicly.

"Bob and I and a few others were on this BIO panel in DC," Olson says. "We'd gotten the data. They knew the data were coming sometime. Ken wanted me to tell them in person what we had. I said, 'Bob. Can we go back to Bethesda after the panel? I've got some things we need to talk about.' I think he kind of got a sense, because instead of going back to his office, they had a room at the Hyatt, which was just a few blocks away, and he suggested we go there. So we go into this room. I had a sealed envelope. They opened it. I said, 'Here are our first data.' They were just stunned. Bob was so ecstatic."

For Beall, the results vindicated a decade of all-out heterodoxy and entrepreneurial risk taking: his determination to push venture philanthropy into for-profit research; his refusal to yield to established opinion that CF drugs could treat only the symptoms of the disease but not its cause; his insistence on going after CFTR; his recruitment of Aurora; his faith in Olson and the company; and, especially, his willingness to foot part of the bill for Vertex's clinical development. If the determination of an individual to make a big difference can be measured by volume, the significantly improved FEV1—forced expiratory volume in one second—that measured the exhalations of patients taking VX-770 was Beall's deliverance. Calling the findings an "unprecedented milestone," he said in a press statement: "The results from this trial are among the most important announcements in the history of the foundation. Although we are still early in the process, we have increased confidence that we can treat the basic defect of CF, and these data show that we are on the right path to cure this disease."

❖

At the end of April, John McHutchison presented the data from the two PROVE trials at the European Association for the Study of the Liver (EASL) meeting in Milan, Italy. The cure rate rose to 68 percent in one arm of the second study, when telaprevir was given for twenty-four weeks. A tolerable number dropped out due to rash: 7 percent. McHutchison also reported on a Phase I trial evaluating the oral availability of a nuc, a polymerase inhibitor developed by Pharmasset, a ten-year-old New Jersey company with early-development programs in antivirals and cancer. McHutchison told the audience that after four weeks, the Pharmasset compound showed an ability to knock down the virus "similar to HCV protease inhibitors, and [it] has an encouraging short-term clinical safety profile."

Mueller, like Boger, thought nucs made bad drugs, and so far, that had been borne out by the data. All types of other compounds had failed routinely against the virus: nucs, non-nuc polymerase blockers,

IMPDH inhibitors, humanlike antibodies, gene regulators, therapeutic vaccines. A much higher percentage of follow-on protease inhibitors were still proceeding in the clinic, strongly suggesting that as with AIDS, the protease was the most inviting target. Ann Kwong, keenly aware of Gilead's conquest of the HIV market, was less dismissive of Pharmasset's approach. She began to press Mueller to consider nucs more seriously in planning for second- and third-generation cocktails, but Mueller was unimpressed.

With skin rash turning out to be a manageable but not insignificant problem, Kauffman was working with the FDA to shorten telaprevir's treatment time still further, eventually winning the go-ahead for an eight-week treatment arm. Enrollment in the two late-stage trials was filling quickly as word got out about the Phase II data. Meanwhile, the company was rapidly advancing expanded trials of VX-770 and VX-809, also expected to start by summer.

Alam, after eleven years, put down his torch. He had helped elevate Vertex from a research organization to a clinical leader with break-throughs against two major diseases, but his relations with the FDA had become a liability, and he didn't have the "tactical rigor" Boger knew he would need to manage the expanding wheel of Phase III trials and the filing of new drug applications with the regulatory agencies. "I love John to death," Boger says. In the usual Vertex fashion, personal affection and loyalty were not enough when measured against where the company needed to be in five years. "It came as a shock," Kauffman remembers. "John came into a staff meeting one morning with a sort of Cheshire cat smile on his face and said, 'I have news for you.' 'What's that?' He said, 'I'm leaving.' They had been conducting a search, it turns out, that I didn't know anything about. I had not been told about this at all, nor had anyone else in the room."

News of Alam's departure was wrapped inside the announcement that Vertex had hired Dr. Freda Lewis-Hall as executive vice president for medicines development. Formerly of Bristol-Myers Squibb, Lewis-Hall would lead the company's buildout of its regulatory affairs and medical affairs teams in addition to running all clinical and nonclinical development, but her real urgency was to improve relations with the FDA's Fleischer. Tracking him down at medical meetings, engaging him in conversation, she became the company's "Russ whisperer."

Since Sato retired, more and more frequent turnovers in senior management had rattled investors, perplexed analysts, and confused employees. Boger believed it was the price paid for forever blowing out functions to create the future. In relentless succession, the company had built a research organization, a development organization, and was now rushing to create a commercial organization to launch two drugs. The ET staggered from move to move. First, Tony Coles had left when Boger hired Victor Hartmann. Then Hartmann left when Kurt Graves joined as the seeming heir apparent. When Boger started at Merck in the mid-1970s, it was assumed that he would retire there, but in biotech, if people didn't change jobs every five years, they felt they weren't advancing fast enough. Companies adapted to the abrupt comings and goings like a sloop catching a side gust, entering a state of perpetual disruption where head-snapping departures became the norm. "Thrills and chills," Ken calls it.

❖

A month later, Emmens retired as CEO of Shire, although he remained the company's chairman. When he took over in 2003, Shire's flagship product had been the attention-deficit drug Adderall XR, a reformulated amphetamine under patent threat from generic drugmakers. Soon after, its most promising late-stage-development prospect was rejected for FDA approval. With most of his business in the United States, Emmens moved Shire from England to the Philadelphia area, settled the legal standoff, acquired smaller companies with innovative technologies and franchises in rare diseases, and encouraged them to develop their own cultures and programs. Now Shire ranked near the top of specialty drug firms worldwide.

Emmens, fifty-seven, anticipated getting back to his garage to work on his fleet of restored sports cars, flying around the Caribbean, keeping up with his directorships, visiting his adult children, maybe getting a new dog to train. He thought Vertex's revolving door belied a larger problem—Boger's hub-and-spoke leadership model—and he told the other directors in his firm, accentless monotone that he didn't think it was a scalable design. Development of telaprevir was ongoing. The company had its first FDA submission coming up. It faced scores of inspections, both of its supply chain and clinical activities. It had no commercial organization. And it expected to launch a drug into a highly specialized market in eighteen months to two years. Without thorough integration and alignment among its senior executives—not to mention the faith of the board—Emmens doubted that Boger could succeed.

"Did I think it was time to put somebody else in? Yeah," he recalls. "I said, 'If we don't do that now, leave Josh in. I know what it's supposed to sound like. You can have the greatest sense of urgency in the world, but unless that team works together, you're not gonna make it.' "

In July, Charlie Sanders was named to head a special committee of Genentech's board of directors to assess a surprise takeover bid from Roche, thrusting him into strenuous high-stakes negotiations. In 1990 Roche had bought 55 percent of Genentech, which went on independently to become the most productive R&D organization in pharmaceuticals. With some shares still trading publicly, the company had been able to attract entrepreneurial-minded researchers with stock options and promises of autonomy. Now Roche wanted to own everything, and there was deep unease at Genentech that losing its independence would destroy the science-focused culture that had led, over a five-year period, to the rollout of Roche's three biggest-selling drugs: the cancer breakthroughs Rituxan, Herceptin, and Avastin.

Hiring Goldman Sachs as advisor, Sanders and the special committee evaluated the bid. They rejected it, saying it "substantially undervalues the company." As spokesman, Sanders said the board would consider a higher offer, and he tried to dispel the notion that Roche's takeout proposal was unwelcome. "We look forward," he told the *Times*, "to the company maintaining its successful relationship with Roche, regardless of the ownership structure."

A case-hardened truth in industry is that the only sure way to avoid being taken over by a bigger company, or "taken out," is to become too costly. Vertex, with a market cap of $5 billion, a harrowing burn rate, and no immediate prospect of profits, seemed safe for the moment. Ken and his team had negotiated a "standstill" agreement with J&J: a contract barring one partner from making an unsolicited offer for the other. The company's antitakeover provisions, including a poison pill, had been in place since the public offering in 1991. Taking another look to see that they hadn't forgotten anything, the brothers agreed they were as well defended as possible. "Like trying to eat a blowfish," Josh Boger said.

In mid-September, two days before Vertex planned to announce another stock offering, Wall Street shuddered again. This time the disruption engulfed the global economy. The prominent securities firm Lehman Brothers hurtled toward liquidation after it failed to find a buyer because Washington, unlike with Bear Stearns, refused to step in. Merrill Lynch agreed to sell itself to Bank of America to skirt the same abyss. Less than a week after the government took control over the troubled, government-sponsored mortgage finance companies Fannie Mae and Freddie Mac, the insurance giant American International Group (AIG) sought a $40 billion federal lifeline, without which it said it might have only days to survive.

On Monday Smith told Boger, "We need a constant supply of capital. If the system goes down we've got a real issue." With Vertex heading into the completion of pivotal trials of telaprevir, expected losses for the year of around $400 million, and less than $500 million in surplus cash, Smith "needed to top off the tank." Vertex decided to go ahead with the offering, announcing the sale of more than eight million new shares as Wall Street teetered. Goldman Sachs, one of the two remaining investment banks, acted as the sole book runner for the sale.

Since the last time Smith had raised money, in February, Vertex's share price had ridden a whirlwind, driven down more than 50 percent by a few prominent analysts touting fears over telaprevir-related rash and concerns over competition from Schering, before regaining everything and more in the wake of the company's scientific publicity blitz after the PROVE and VX-770 results. Doubling down on telaprevir, Smith had sold the rights to Vertex's royalty stream for Agenerase and Lexiva for $160 million in cash—the company was no longer in the business of treating AIDS. Heading into the teeth of the worst capital market in seventy-five years, Smith, with his usual brio, approached the fund managers he hoped to shepherd together to complete the offering.

The following week, Paulson and the White House submitted a $700 billion bank bailout to Congress, which was now considering it. McCain, declaring it was time to "put politics aside," said he would temporarily stop campaigning so that he could return to Washington to help forge a deal. In the maneuvering, he suggested postponing the upcoming presidential debate. Obama rejected the delay, saying, "It is my belief that this is exactly the time when the American people need to hear from the person who, in approximately forty days, will be responsible for dealing with this mess. It is going to be part of the president's job to deal with more than one thing at once."

Smith's urgency reflected the nation's financial instability and Vertex's make-or-break race to market with telaprevir. He compared the company's situation to a road trip into barren, unknown country. He thought, "This might be the last gas stop before our destination." When one fund manager from LA protested the price of the offering in the face of the larger crash, Smith cut him off. "I've got to call a few other investors," he snapped. On the same day that Vertex completed the sale, netting $225 million at a share price of $28, the Dow plummeted 650 points.

Defying gravity had long been Vertex's unofficial scientific credo. Now it became its financial signature as well. Starting the third week after Black Monday, the Dow shed 2,400 points—22.1 percent—in eight harrowing trading days. Trillions of dollars of wealth were wiped out as volatility raged. Despite Washington's apocalyptic rhetoric and roiling discomfort, Wall Street got its bailout. By Election Day, VRTX shares were up almost 30 percent on the year, making it the top-performing stock in the NASDAQ 100.

❖

Boger realized that Emmens wouldn't stay retired long. After thirty-five years in pharma, Emmens was receiving approaches from headhunters and venture capitalists alike, and he was entertaining them all. If he returned to business, Vertex would surely lose him as a director. He had recently written and published a business primer in the form of a fabled quest, coauthored with an illustrator, an old friend. Titled *Zenobia*, after novelist Italo Calvino's mythical city on stilts, it tells the story of a young heroine who gets a job at a former industry giant bedeviled by paralyzing hierarchies and who triumphs over "yes-men, cynics, hedgers, and other corporate killjoys." He believed passionately in the power of individuals to rescue organizations by inspiring them, and having just proven his thesis again at Shire, he wasn't through. "Anyone who hasn't read Matt's book doesn't understand Matt," Boger says.

Sanders was preoccupied by the Genentech takeout, which had "taken a big toll on him," and he was "fading in strength," Boger believed. No longer intimate with the pulse of the board or the directors' thinking, Boger sensed a growing vacuum, a free-floating anxiety: "background noise," he recalls, especially about Graves's ability to pull together and manage a commercial team. "The board fell in love with Kurt and then fell out of love with Kurt and blamed me for it." Despite Vertex's all-around positive results, there grew a sense of crisis. Members, taking seriously the succession issue, worried that Vertex needed more commercial leadership—now. Sanders was unable or unwilling to rein in their apprehensions.

"Remember, there's lots happening in Charlie's life," Boger recalls. "A company where he had spent a lot of time on the board was getting engulfed, and their mission was going to get derailed. . . . A lot of Genentech people were very upset with Charlie for not taking a harder line. I don't blame Charlie at all, but it did go on for a long time. There was a lot of back-and-forth. It took a lot of Charlie's energy and time. He was worn out."

In late fall, Sanders invited Boger to breakfast at the Harvard Club of Boston in the Back Bay. "We're thinking of making the transition now to Matt because Matt's available," he said.

"Is he interested?" Boger asked.

"Yeah, he's interested."

Boger weighed the news as he did any new data point. He was discerning and dispassionate. "So at first I was thinking to myself, that's a pretty good idea," he says. "I wasn't happy with the way that Charlie presented it. I wasn't happy that there had apparently been—by the way, illegal—board meetings without me. They had called all the directors, except me, together to have a discussion about this, so by the time Charlie told me, it had already been decided. I'm a director, so how can they exclude me from those discussions? Would I have opposed it? I don't think so, because I think it was a pretty good idea."

Boger thought of Vertex as *his* company. He had imagined it, realized it. Had he named it Boger Pharmaceuticals—had he given it the family name as, pre-biotech, generations of founders had done when they entered what was then called the "ethical drug" business—his stamp on it could not have been more personal. In seconds he knew it had been completely taken over by others without his even being asked what he thought. "Too many roads had been crossed before I even had a chance to have an opinion," Boger says. "Charlie was not playing a strong role there, and I was cut out of the discussions. It was so obvious that Matt had effectively managed the whole process so that he was completely in the driver's seat, and so that anything he was asking for they were saying yes to, even things that they had told me—like being chairman—were off the table. I just didn't understand this."

Emmens recognized the depth of Boger's attachment—and his continued irreplaceable value—to Vertex. He appreciated Boger's position, as well as his own. "When I sat down with Josh," he recalls, "I said, 'If you don't want me to do this, I won't do it. I'm not a scientist, and I know that's lacking, but I think I can come in and do what needs to be done to get commercial in place.' Plus, we had to finish out development. 'I think I can get a team of people to make this work. I have the kind of skills to get just what we need. I'm not gonna make the science any better, but the rest of the stuff I've done.' If he said, 'No, go away,' I would've."

Boger had a choice: to lead Vertex at least through the launch of telaprevir and maybe beyond, or leave the company. Many on the board thought there was no place for a charismatic founder except forced retirement. Emmens needed Boger involved and insisted, over their protests, that he be kept on as a director. For Boger, it was a no-brainer. As always, he was thinking leagues ahead to what the company could become. There was no question that Emmens was the better candidate to do what Vertex needed to do in its next phase of transition in becoming a profitable, global drugmaker, and that now was the right time to make the switch. Boger reflects:

This is where people don't believe me when I say this: it's *never* been about me. But I had studied various examples of succession. I did know that there are fifty examples of people staying too long for every one who left too soon. So the mistake that people make is staying too long.

I also knew from talking with people who had made these transitions, one side or the other, that an absolutely terrible time to make a transition is just after you've had a big success. Because then the guy coming in has got nothing to build his loyalty on. What's he gonna do? It's another four to five years until the next big success, and meanwhile he's just a caretaker of your success legacy. So you can't really ask a new leader to come in and just take the trophy. It's not fair. And people who've done that have not had a good track record.

Boger opposed Emmens's becoming chairman. He thought Emmens could run Vertex and talk to Wall Street as CEO, and that he—Boger—could serve as non-executive chairman, managing the board and representing the company to the rest of the world. He thought he could continue, basically, to do what he had been doing for the past three years, using his public persona and worldwide connections to prepare the company to step into a larger footprint, without having to deal with executive reports or raise money or manage anything.

"I thought they had a beautiful opportunity," he says. "It was going to be terribly delicate for the rest of the company to have me suddenly vanish—needlessly, since we had the culture thing up, and it was the most important thing in the company. That was putting all that at risk. For what? . . . I was worried that whatever Matt did was going to alienate a significant fraction of the torchbearers in the company, enough so that they were either going to leave or put their torches out. And that he was never gonna recover from that; that he was gonna lose the sustainability momentum that I had built up over the past five years. That's what I was worried about."

Emmens prevailed. He knew he had to have complete control. Indeed, he'd seen up close what happened to Boger when *he* yielded the chair. Boger agreed to stay on through a three-month transition period beginning in February, when the announcement of his retirement would be timed to celebrate Boger's twenty years at the helm. As Vertex closed for two weeks at Christmas, he had his resentments, but he held them in check. That he'd made the right decision was affirmed by Emmens's doing just what he'd have done if the roles had been reversed. "I thought he was asking too much, and I didn't see why he was being given everything that he was asking for," Boger says. "After me saying for two years I was being paid like we were a start-up but we were going to be a big company, suddenly they buy it, and his salary is triple. Why did he get the argument? I was making the argument before. There were things like that that annoyed me. But leaving too soon was by far the better choice than leaving too late."

# CHAPTER 9

## JANUARY 12, 2009

Boger had chased investors for two decades. He'd endured depressing periods before. At his first JPMorgan H&Q cavalcade in San Francisco eighteen years ago, it seemed Wall Street had simply abandoned the sector. "The mood of that conference was that there will never, ever be another biotech IPO again in the history of the world," he would recall. Four months later, he took Vertex public during a bubble that hyperinflated after Amgen won a five-year patent fight to market epoetin alpha—Epogen, or EPO—a protein drug that stimulates the production of red blood cells. Notably, it was the promise of more billion-dollar drugs and the first Gulf War in 1991 that led Wall Street to finally rebound from the crash of 1987.

The mood at the Morgan conference now was equally grim. "If people don't want to buy bonds in General Electric," a CEO grumbled, "what's going to make them want to invest in an early-stage biotech company?" It was a fair question. According to BIO, 45 percent of the country's 370 publicly traded biotech companies were operating with less than a year's worth of cash; of those, two-thirds had less than six months' worth on hand. Nineteen companies had recently pulled IPOs, and eight went bankrupt. In the post-Darwinian, post-meltdown economy, a new order had taken hold. "Too big to fail" was surpassing "survival of the fittest" as a measure of selection in business.

As chairman of BIO, Boger was characteristically stalwart and upbeat. What fueled the industry—what had always fueled it—was "the fundamental optimism that there are great things happening," he told the *Wall Street Journal*. He made no general predictions about the year. Not so with Vertex. He believed that the time to sell a company's sizzle was when it delivered on its steak, and he sought to take advantage of his last appearance as CEO, using his standing as a spokesman to drive home its message. Having been ousted in secret by his board, he nonetheless remained the public face of the company.

In his slide presentation, Boger likened Vertex to Genentech in the 1970s, when it developed the first genetically engineered biologics, and Gilead in the 1990s, when it advanced AIDS medicines that made it possible for patients to take fewer pills per day. "Ever the showman," the website Xconomy reported, "he accompanied those historic examples, respectively, with the Bee Gees hit 'Stayin Alive' and Nirvana's classic 'Smells Like Teen Spirit.' "

Veteran CNBC medical correspondent Mike Huckman caught up with Boger in a hallway. Bristol-Myers Squibb had just announced a global collaboration to find a cure for hepatitis C with ZymoGenetics, a Seattle company with a unique interferon that BMS hoped would fit with its emerging portfolio of small-molecule antivirals. The studio anchor from New York, introducing the segment, called the conference a "who's who of the health care field. It's where investors look for the next big thing, not just these days in biotech, but what might be the next big thing for innovation and technological development overall, to get us out of the crisis." Boger, looming over Huckman, who stood a head shorter than him, seemed entertained.

HUCKMAN: So, speaking of the partnership with Bristol and Zymo-Genetics, you're already partnered with Johnson & Johnson, is that correct?

BOGER: For Europe and ex-China and Japan. We have North American rights.

HUCKMAN: So in a year when many are predicting a record year in M&A in biopharma, could you take that to the next level? Could J&J buy you, or could somebody else take you out?

BOGER: Well, Mike, you ask me this question every year, and we're still here. We're moving forward on our business plan to become a major pharmaceutical company, and we think we have the product to do that.

HUCKMAN: You're in a duel now, primarily with Schering-Plough, which is also in the late-stage development of a hepatitis C drug. What makes yours better?

BOGER: I think that from the data that they've reported and the data we've reported that telaprevir is more potent. They reported that their drug is not effective in people who have not responded to previous therapy, and we've reported dramatically positive results in those nonresponding patients. Every naïve patient is a potential nonresponder. This is a cure, so I think patients will choose the best shot they have.

HUCKMAN: A Bernstein analyst recently said you guys could potentially charge seventy-five thousand dollars for a course of treatment of your drug. Is that accurate?

BOGER: Well, we don't set the price until the very last thing right before the drug goes to the marketplace. The basis of that report, I will endorse. There's an enormous value to curing the patient with hepatitis C. Infected hepatitis C patients really have a pretty grim outlook, for very expensive chronic liver failure and premature death.

Boger walked viewers through Vertex's immediate goals to complete the late-stage trials, finish the scientific analysis, and assemble all the data for a new drug application. Huckman, noting Vertex's industry-leading stock performance, asked, "What catalysts are there to drive the upside in 2009?"

"We're looking to 2010 to file an NDA," Boger said. "So we're in the last strokes, and I think investors are paying attention to that."

Huckman pivoted to face the camera. "Joshua Boger, the CEO of Vertex Pharmaceuticals, thanks again for talking with us," he signed off. "And coming up, the CFO of a company that is taking a stop-smoking pill and an alcoholism pill, mixing them together, to potentially make . . . a *diet* pill. Can it work?"

❖

A week after Inauguration Day, Pfizer agreed to acquire a rival, Wyeth, for $68 billion. The deal was the first big merger backed by Wall Street in several months, financed by the surviving investment banks with federal bailout money sent to recharge the economy. It challenged the drug industry to contemplate the potential for a combined "Pfyeth"— "a pharmaceutical behemoth . . . a rarity in the current financial tumult: a big acquisition that is not a desperate merger of two banks orchestrated by the government," the *Times* reported.

Not to say that Pfizer wasn't desperate, facing a void in many ways as ominous as the giant lenders'. Its blockbuster Lipitor, acquired when it bought Warner-Lambert Company earlier in the decade, generated $12.5 billion in sales, or a quarter of its total revenues. The drug would go off patent in 2011, to be replaced by cheaper generics. Along with other patent expirations, Pfizer faced a loss of more than 70 percent of its 2007 earnings by 2015. Despite an annual research budget approaching $7 billion, it had no equivalent earners in its pipeline. Its net income for the fourth quarter dropped 90 percent because of charges for resolving settlement claims in off-label marketing prosecutions.

Buying Wyeth gave the company revenue, a potential new portfolio in biologics, and an opportunity to restructure while Congress began to grapple with health care under President Obama. Facing patent cliffs like Lipitor's, all large pharmaceutical companies hunkered down against rising threats making it tougher to market their products. Congress was widely expected to restrict direct-to-consumer advertising, on top of stricter new restraints on how the industry sold to doctors. Universities and hospitals were starting to limit what their researchers could accept from companies for promoting their drugs.

On February 5 Vertex announced Boger's retirement and Emmens's succession. In Cambridge, as at the other sites, the news dropped without warning. Boger had been tight-lipped for months, betraying no hint that he was leaving. A few nights earlier, he'd taken a group of a dozen of those who'd started at the company in 1989 out for a ceremonial dinner, giving them all inscribed Swiss watches and regaling them until after midnight. None of them had a clue of what was coming. All day, more than a hundred scientists, many of whom Murcko didn't know, traipsed up to his book-lined office in Fort Washington II, loitering in his doorway, trying to get his bead.

He was as startled as any of them. Murcko had expected "it was a real possibility" that Boger would leave Vertex at some point, but not so soon. He listened to the researchers' questions, distilled their concerns, and then responded at five o'clock the next morning with an email blast that focused on three worries in particular:

- What do I know about Matt Emmens? Not much except by reputation, which is good: straight shooter, good listener, respects the science, lots of experience, has lived through all kinds of situations (i.e., "battle tested"), knows about commercialization. Peter likes him a lot. Obviously, Matt does not equal Josh, so in a sense it will be a "shock," but different doesn't mean worse. Matt won't be around until March, I believe, so it will take a few months to get to know him and see what he's really like; but in the meantime, no worries.

- Does this mean we're for sale? Again, just a personal opinion, but I would say no. We have two drugs to launch in the next few years. The potential value of that is huge, and the board understands that. And Emmens doesn't have the track record of one of those "dress-it-up-and-sell-it" hired guns. And Josh will still sit on the board . . .

- Is this the end of Vertex as a unique organization? This is the most important question of all. Well, it's kind of up to us now, isn't it? Or as a colleague rich in wisdom put it yesterday afternoon, "Do you still live with your parents?" This day had to come sooner or later. Do we really believe that the ONLY thing that made Vertex unique was Josh? . . . My response to the news (after shaking my head a few times to clear it) is exactly what Josh wrote: the mission continues. The best way to honor what he's built is to rededicate ourselves to that mission. That's what it means to say

Vertex transcends any one individual—you, me, even Josh. I'd like to think that five years from today, Vertex will be an even MORE amazing place—that we've kept improving our game—that we are tackling even harder problems. It's up to us now—as in fact it has *always* been up to us.

Garrison considered the board's decision a monstrous mistake. He didn't think the directors understood or appreciated what Boger had done in building Vertex and what lost opportunities lay ahead while Emmens tried to fill his shoes. "If you did a study," Garrison says, "Josh Boger is a complete anomaly. Nobody—nobody—starts a company, goes through angel [investors], goes through venture [capital], goes public, and then goes twenty years, never makes a dime, only loses money. Nobody. Only a guy with Josh's incredible ability to learn and listen to others. There's nobody like Joshua Boger. Ned Johnson at Fidelity—spectacular. But he's no Joshua Boger. Jim Ranier at Honeywell? Josh is in a class by himself. They got him at the wrong time. The war was won, and they shot him. The board thought, 'Now we're gonna go commercial; he hasn't got the experience.' But he never had the experience at any point along the cycle. They weren't worthy to fire him, in my opinion."

Most Wall Street analysts cheered the announcement. They had their own ideas about what Vertex could become, and Boger's disdain for their role was no secret. Geoff Porges especially felt that Boger was holding back the company from delivering on its promise. "I think this will be viewed positively," he told the Reuters news agency. "He is perceived to have been an obstacle to the sale of the company."

Investment analysts believed that the only way ahead for pharmaceuticals was aggressive consolidation, and Porges had strong opinions about how he'd like to see the future shake out. He thought Vertex would be an excellent fit for Gilead, which was finding little success in trying to buy its way toward new therapeutics for respiratory and heart diseases. In a note to investors, he also proposed that Vertex buy Pharmasset, which had a market value of $300 million, to add nucs to its stockpile. Porges didn't know Emmens, but his intelligence about Vertex was reliable enough for Smith to worry that he might have an inside source.

❖

The ninety-day overlapping transition was Boger's idea. It said to the company and the world that Vertex was on an orderly trajectory. Emmens took the opportunity to listen and learn. He had his own urgency: "Do we have enough time and money to get to commercial?" He also had major concerns about the psychology of the organization. Being unprofitable for so long afforded Vertex the luxury of adhering to its virtue while its stockholders tolerated almost $2 billion in accumulated losses. Patience was running thin. The pressure to perform was mounting. "My biggest worry was whether introducing commercial thinking would meet resistance. The day you become profitable changes the world," Emmens says. And so while Boger still ran the business, Emmens studied the people and operations of the Cambridge site—1,300 employees spilling across eight buildings and desperate for more space—putting his ear to the engine, so to speak.

"This culture does not suffer fools at all," he says. "When people come in, they notice, you get up to speed or you're not gonna make it here. Or, if you do it wrong, you're not going to have any respect, and you can't lead. So the hires are key. The stuff that I was doing wasn't here yet. The big deal was we didn't have time to make mistakes. If we had hired the wrong folks in this, we would've been in trouble. That was the key thing. I think I'm good at making teams, selecting people, feeling out emotional intelligence to see if people can work together well. You don't want to have a Ford and try to fit VW components in it."

Emmens's management philosophy was to get ET members to function voluntarily as a whole, something that was impossible under Boger or any other founder. He saw himself chiefly as a subtle facilitator, leading through stealth and modesty as well as strength, fostering honesty while breaking down defensiveness, prompting his senior staff to exceed themselves and do what needed doing not because they wanted to impress him but out of a sense of shared purpose. "It's not groupthink," he says, "it's a combined brain.

"You can't be the smartest guy in the room all the time. The smartest guy in the room is the guys in the room. How do you get that? I've spent my life really thinking about how it's better to ask a provocative question than it is to give your answer. If you give your answer quickly, and you're a real smart guy and the answer's right, it's relatively ineffectual. They don't learn from it, so you're not developing your people. And it's not washed through the giant brain of the combined people at the table."

Vertex announced another secondary stock offering, its third in twelve months. With ten million new shares at $32 a share, it netted $320 million and, as importantly, a new crop of long-term shareholders. Most development-stage companies depend miserably on biotech hedge funds—"people who aim to get rich trading volatile stocks, second to second, and make big bets, long or short, on whether an experimental drug will work," as the Xconomy writer Luke Timmerman describes them. Now that Vertex had begun the subtle process of "derisking"— eliminating one by one the things that cause companies to crash and burn—more buy-and-hold fund managers were increasingly attracted to VRTX. They took small positions, but Smith envisioned that they would want to raise their stakes as the recession deepened.

The M&A frenzy in pharmaceuticals, a rare bright spot for Wall Street, peaked in March. The business development fever sweeping across the industry made hepatitis C an especially promising opportunity. The basic equation—big companies were cash rich and pipeline poor; small ones had promising leads but desperately needed funding and could neither borrow nor go public—had produced a permanent state of consolidation. But now after "Pfyeth," it seemed all at once that every company (except the weakest) was a potential player while at the same time all (except the truly gargantuan) were set in play.

A small Canadian company outside Montreal, ViroChem, had two non-nucleoside (non-nuc) polymerase inhibitors in midstage development. Graves, competing against players with much deeper pockets, persuaded the drugmaker to come to terms by leveraging Vertex's lead position. ViroChem decided that it could get its drug to market sooner by latching itself onto Vertex rather than onto pharma. On March 3 the company announced it would buy ViroChem for $375 million in cash and stock. Its lead compound, VCH-222, had been tested in thirteen people.

A week later, Merck announced a deal to buy Schering-Plough for

$41.1 billion in cash and stock. Boger was not surprised that Merck, which had long avoided the acquisitions market—in large part because its people couldn't believe that anyone else could make something that Merck couldn't produce better on its own—was drawn into the competition. But the merger with Schering, which most analysts rated as better than the "Pfyeth" coupling but not by much, shocked him. It was not that Vertex would now, at last, and after so many disappointments, clash head-to-head with Merck, as telaprevir and boceprevir progressed to market. Nor was it the irony that the competition arose just as he was being pushed out. It was the brutal cultural disconnect. "Like the Four Seasons taking out distressed housing," he said.

Three days later, Roche, overcoming eight months of resistance and the threatened exodus of top managers, agreed to buy full ownership of Genentech for $46.8 billion. Emmens wasted no time. Upping Vertex's preparedness, he retained a new Wall Street "defense bank"—a relationship that effectively gave Vertex, if it got hit, professional muscle he felt he could trust in weighing its response.

❖

In the trenches, the roiling above was concussive, dizzying. To manage the business of getting telaprevir through the approval process, and to smooth the contentiousness with the FDA, Lewis-Hall hired a politic and pragmatic industry veteran, Jack Weet, to run regulatory affairs. "Our relationship with the FDA is terrible," she told him. "You need to fix it." Barely had Weet set foot in Cambridge than Lewis-Hall left to become chief medical officer at Pfizer. People internally were shaken by the announcement, their small-company insecurities aroused. What did it say, they asked one another, that the top medical person representing their first drug would depart after just nine months with the company?

The buildout of the development organization remained undone. The challenge from Merck and boceprevir loomed, a race to the finish despite Boger's insistence that Merck's drug, which caused many patients to become so anemic that they had to take EPO as an add-on, wasn't a serious threat, and, indeed, was unsafe and wouldn't be approved. Emmens knew it would take a year to find a new development chief who could also keep the clinical organization on track. He had no time for that.

He turned to Mueller, who had pushed drugs over the finish line before, and who possessed an almost superhuman ability to toggle between breathtaking imaginativeness and exacting execution. Mueller could expand on a blue-ocean vision for curing the crippling neuro-degenerative disorder Huntington's disease by designing a molecule to fix a mutant protein in the brain, and then, just as capably, comment on a minor technical issue related to the advanced spray-drying technology for manufacturing telaprevir. In May Vertex promoted him to head global research and development. Mueller took charge more or less of the entire science side of the business. Kauffman, who'd managed the nitty-gritty of the ever-widening clinical studies under Lewis-Hall, became chief medical officer, reporting to Mueller.

Mueller believed that telaprevir had made—and would make—Vertex the company it needed to be, but the hepatitis drug by itself would not be enough. The greater pressure was always to deliver on the next big promise. The progress in CF had proved that lightning could strike twice, but Mueller now thought that both those products could be just a warm-up: that hepatitis C, as a crucible, had made the company better equipped to address larger and larger problems than it otherwise might ever have become. Mueller:

> The one thing that made the company is that we were capable to move Josh's original vision that he had, which, in a nutshell is: you have to be on the forefront and the frontiers, constantly. We were able to move this frontier with us, across different disciplines. That is what made Vertex. It wouldn't have been good enough to be great in clinical. You have to be great in everything. That's why, in a way, telaprevir was an ideal molecule, because it put challenges on all angles. Wherever you looked with the frickin' thing, okay, it was just a nightmare. I had sleepless nights over many, many, many problems there. But that's what made Vertex Vertex, at the end of the day.
>
> So there is a feel, okay, the sky is maybe the limit. You see what hell really looks like. You see the heat goes up every step you walk. But I think this is the real fantastic outcome; this is what telaprevir did to us. Telaprevir made the company, but it made it because

we made telaprevir. But I think we have not to forget, this alone would not have done it. It is that telaprevir helped basically spark an environment that allowed us in parallel to do other complicated things—to have not just a one-trick pony. You have to have enough shots on the goal.

In Washington, Boger and BIO CEO Jim Greenwood made much the same case for the industry. Like all tech-industry lobbies, BIO had one focal issue: how to sustain profitable innovation and, more generally, America's *lead* in it. Greenwood, a former six-term Republican congressman from the Philadelphia suburbs, took the position that his members had a stake in every part of the health care economy—"the entire step-by-step process from NIH funding until the patient is cured," he said. Sizing up Obama's campaign promises and the new Democratic-controlled Congress, including a sixty-member supermajority in the Senate, Boger and Greenwood shuffled BIO's priorities to suit the new political reality: specifically, that the generic industry historically had better relations with the Democrats than the "brands," which usually favored Republicans. Greenwood worried especially about the industry's campaign for extended protections against follow-on biologics: cheaper versions of drugs like Enbrel and Avastin.

Boger viewed the overall health economy as bloated and wasteful but, in its basic structures, still optimally designed to find new cures. And so he supported, for instance, increasing funding for the FDA. "It's notable," he said in a BIO podcast about the new atmosphere inside the Beltway, "that an industry is asking for more resources for its regulator. But the reason we do that is that we all believe in the enormous value creation of the whole system. And we are standing up for the whole system, not just our narrow piece of it." He conflated the issues of innovation, drug pricing, and access to new drugs for patients, and he believed that the surest route to improving all three resided in the existing government-driven regime of extending patent exclusivity and price protection so that owners and investors can expect a return on a risk they otherwise never could afford.

He continued: "This balance of innovation and exclusivity is paying off royally for society without damaging the innovation on the small-molecule side. And it would be tragic even as that Solomonic balance is paying off to have not as good a balance on the biologic side. The ultimate solution to the health care cost crisis is to make people healthier, and that's gonna take innovation. If cost savings come at the price of poor health care, it's a false saving. And so we can't have a long-term solution to the cost crisis in health care without better health care, and part of that solution is technology."

In mid-May Boger flew to Atlanta for the BIO annual meeting. That morning, Vertex published the actuarial study that Sachdev had commissioned to show elected officials the size of the looming threat of hepatitis C. The authors wrote that without changes in how the 80 percent of patients who don't know that they have the virus are identified and managed, the annual cost of advanced liver disease in HCV patients would jump to $85 billion in the next twenty years. Medicare costs would soar 500 percent, from $5 billion to $30 billion. The authors found that most individuals living with HCV were born between 1946 and 1964, and a disproportionate number were African-Americans, who were almost twice as likely to have HCV as the general population. The $85 billion estimated cost included overall direct medical costs for patients with HCV infection but not other societal burdens such as lost productivity.

In the cacophony over health care costs, Sachdev knew he had a serious talking point: an epidemic about to explode at a moment in history when the bottom seemed to be falling out of everything. Government medical spending was spinning out of control. Here was an actual solution: identify the millions of people in the country with the virus and treat them *before they got sick,* thereby saving hundreds of billions of dollars in a future that was identifiably near-term on a graph of current spending obligations. People who knew they had HCV either were symptomatic or found out about it when they applied for life insurance and were sent for a blood test. Sachdev positioned Vertex to become a central mover behind a national screening program for baby boomers.

In Atlanta, Boger rocked to the B-52's, now in their fifties, at the opening reception at the Georgia Dome. As outgoing chairman, he titled his valedictory "Save the Planet." As an expression of the values he'd tried to impart to new employees for the past twenty years, the theme of salvation had become a useful shorthand, as when he'd convinced Garrison to join Vertex by telling him: "If I'm right, I'm gonna save a million lives a year. What else are you doing that's so damn important?" Now Boger was being literal. In his remarks, he asserted his usual strategic advice: that the only sustainable business plan is to produce innovation that is of high value to society. He also said that millions of people who could solve the challenges the world faces could well die from disease before they make their discoveries without the production of innovative therapies and widespread access to them. "Josh and I would always have those debates," Mueller recalls. "Josh believes that the world will get killed by asteroids, at the end of the day. My philosophy is that we'll be killed by viruses. We came to the conclusion, let's make sure that we basically cure people that have viruses so that we have some left to think about the asteroids."

Boger had always equated doing well with doing good. The Pharmaceutical Research and Manufacturers of America, the drug industry lobby known as PhRMA, also considered that its mission was to ensure patient access to high-quality medicine, but its reputation was for standing first and foremost against any attempts by government payers to cut prices. During the 2003 debate over the Medicare prescription drug plan, PhRMA had it written into the legislation that Medicare couldn't negotiate with drug companies. Now its lobbyists feared that Obama wanted to bring down drug prices and that he supported reimportation of lower-cost prescription medications.

Approaching summer recess, as the White House talked more about overhauling health care, the PhRMA operatives grew anxious. In early June, a few days after Boger cleaned out his office in Fort Washington (the company gave him as a parting gift the original door from the converted garage at 40 Allston Street where he started out), one of the lobbyists wrote to Obama's senior health care advisor. Although Obama was overseas, meeting with Saudi King Abdullah in Riyadh, the advisor wrote back that she and other top officials had "made the decision, based on how constructive you guys have been, to oppose importation." Two weeks later, before a packed East Room audience, Obama reversed Bush's policy on stem cell research. He said he would release federal dollars to fund significantly broader experimental efforts because "medical miracles do not happen simply by accident," and he vowed to make up for ground lost under Bush and the Republicans.

❖

"With Josh gone," Garrison recalls, "I was out of there." The company's culture was set. Whether it would remain now fell to Emmens and his team, and Emmens didn't want or need Garrison's help. Boger's social experiment had produced an organization that was committed to applying a different set of values to overtake markets and revolutionize institutions—including itself. It characterized what influential Harvard Business School professor of management Clayton Christensen calls disruptive innovation, though the preferred term within Vertex, since its challenges were ultimately scientific, was "disruptive technology."

No one embodied the mission more than Murcko. "I was the guy who was supposed to think about: As smart as Vertex is, what are the things we could be doing differently, where are the opportunities that we've been missing, what are the out-of-left-field ideas that we should be looking at more seriously?" he says.

In 2004, Murcko had proposed that the company completely revamp the way it looked at new projects. Mueller supported the creation of a sixty-person team to look for new leads not by rational drug design or high-throughput screening but by other methodologies: a team B, so to speak, an independent global unit with Murcko in charge that would scour the environment for alternative approaches and disease areas to broaden both Vertex's discovery system and its pipeline. Kwong, especially, was interested in pandemic flu viruses—not seasonal varieties but those killers that regularly sweep around the world, such as bird flu and SARS (severe acute respiratory syndrome).

Without knowing the protein targets, Kwong's group started feeding molecules to infected cells to see their effects, an approach called phenotypic screening, since the goal is to observe changes in physical properties and behavioral traits and deduce from them drug targets of interest. They soon identified several tantalizing compounds. Murcko:

One of the things I was really interested in, and Peter was as well, was the idea of using cellular screening for finding new projects. The trick with phenotypic screening is that you have to pick a screen that really does teach you something about the disease biology. We found that in the area of influenza, it was possible to construct such an assay.

The whole idea of getting the Cambridge site used to thinking about phenotypes was something that was violently opposed by many people. A whole long list of objections arose. It's a stupid idea. We don't know how to do it. It's too risky. You don't know what the target is. You can be blindsided later. The agencies won't ever let you file for the drug if you don't know what the target is. But we just did it anyway, with Peter's blessing.

As with the kinase collaboration with Novartis, Murcko's Boger-esque optimism and self-assurance failed to take fully into account the tendency within organizations, even forward-leaning ones, to resist disruption. Even as the new project group pushed ahead with flu, eventually discovering the molecular target that allowed the company to address it as a more traditional enzyme inhibition problem, the site heads complained that its centralized standards and processes were impeding the speed of discovery. Murcko countered that introducing check steps and validation steps earlier in the process would produce better data and drug candidates and that a decentralized system would be less effective overall. He lost the argument. In 2008 Mueller rolled out a reorganization that gave the sites control over which targets and diseases to pursue, effectively closing down team B and giving Murcko more time to think about where the company should be headed next.

After Boger left, Murcko pressed hardest on two main fronts where he and Mueller agreed that Vertex could stake out a lead. Cell therapy is the process of introducing living cells into tissue to repair function, a type of regenerative medicine that, from an experimental, basic biology, and drug-delivery standpoint, is the extreme opposite of designing small molecules. Murcko, seeing synergies, began to dig into dozens of cell therapy companies and develop key academic partnerships in the field. Mueller quickly became enthused by the prospects. Murcko also started building a team in systems biology, a speculative realm where complex computations are used to understand biological processes. "A lot of the work was dodgy," he says. But he believed the field could someday be useful and intended to stay ahead of the bandwagon.

With time to stare out the window, Murcko could do what Mueller couldn't: look beyond Vertex's pipeline to the medical landscape of the 2020s and 2030s. Mueller had up to twenty direct reports and a $600 million R&D budget. He received emails at a rate of one every ninety seconds, and was chronically double and triple booked. He worked sixteen-hour days, regularly bolting down to the courtyard for a cigarette. When he left, verging on exhaustion, for two weeks' vacation in August, Emmens worried about his capacity to shoulder the ever more grueling pressures that would face him when he returned: a year of daily, critical-path deadlines leading to the submission of the NDA, with Merck and Wall Street constantly looming.

Emmens's team building revolved around Mueller's outsized personality and needs. Before he left for vacation, Mueller complained to Emmens about Graves, who was pushing for John McHutchison to become chief medical officer. Emmens spent two hours interviewing McHutchison, deciding quickly not to hire him because he thought McHutchison's character and style wouldn't fit at Vertex, and also because he was certain that McHutchison and Mueller would clash. "It would have been World War Six," he says.

Just after Labor Day, Emmens told Graves he was making a change in commercial. The announcement, buried in a corporate update, included a perfunctory note from Emmens: "I thank Kurt for his efforts to build Vertex's commercial infrastructure over the past two years. Today, Vertex is well-positioned to advance telaprevir to commercialization, and I wish Kurt well in his future endeavors."

❖

After taking over regulatory affairs, and without the buffer that Lewis-Hall had briefly provided his predecessor, Jack Weet learned quickly how far he could push in all directions and still advance toward submitting an NDA by the end of 2010. Dry witted and low-key, he had to orchestrate a complex collaboration involving R&D, commercial, clinical, US and European regulators, and J&J's Tibotec Pharmaceuticals subsidiary, which was located in Antwerp, Belgium—all while mobilizing a fast-morphing team of medical staff, statisticians, IT specialists, lawyers, outside consultants, and independent experts. Formerly a vice president at Bausch & Lomb, the eye products company, Weet knew the terrain, though he was unacquainted with the ways of Vertex and the players in antivirals at the FDA who would decide whether and when telaprevir would get approved.

During the summer, Graves had urged him to press the FDA on safety issues related to EPO, which Merck was adding to its treatment to help many patients overcome the anemia that resulted from combining boceprevir with peg-riba. Sold by Amgen as Epogen and Aransesp and by J&J as Procrit, the most lucrative product ever developed by the biotech industry and the largest drug expense for Medicare, the anemia drug was under harsh attack by the FDA. It had added a "black box" warning—the strongest type—to the labels of the drugs, advising doctors to use the lowest possible dose to avoid blood transfusions, and was reevaluating previously approved advertising claims that the drugs raised energy levels and improved quality of life.

"We were aware that Schering-Plough was using EPO in their trials with boceprevir. We knew they had an anemia problem," Weet recalls. "We also knew there were safety issues related to EPO and that there had been safety warnings sent out by FDA relative to the use of EPO. So Kurt put a lot of pressure on me, saying, 'Write a letter to the FDA. Tell them they shouldn't be allowing these studies to continue because they're using a product that isn't approved for use in hepatitis.' "

Weet thought it was a good point. He composed the letter, signed by several Vertex executives. "We got this pretty nasty letter back saying, 'Mind your own business. This is under our auspices. Butt out,' " he recalls. Though the tone was stark and negative, Weet considered the response useful in making it clear inside the company that if Vertex wanted to succeed, it could no longer act in ways that smacked of its earlier arrogance or made it appear that its controlling motive was profits, even if that wasn't how it saw itself. It showed everyone where the boundaries were.

Not all contact with the FDA was toxic. Kwong, Tara Kieffer, and

Dr. Shelley George, who had taken over the clinical trials from Kauffman, worked closely with the FDA and outside researchers on the unfortunately named HCV DrAG, a working group that focused largely on resistance issues. Kieffer and George had had positive dealings with Russ Fleischer and his team. Approaching the upcoming Liver Meeting, Weet expected that Vertex would face the usual controversies. He saw an opportunity to calm the waters by arranging to bump into Fleischer. He recalls:

> My goal for that meeting was to track him down, not to stalk him. In previous AASLD meetings, he just went after the people who manned our booth like a pit bull. He was tearing them to pieces, criticizing them, just raking them over the coals, saying they were doing preapproval promotion and all this kind of stuff. And internally, people were asking me, "Please, hang around the booth, because if Russ comes, we'd like to have you sort of stand between us"—like another guard dog, for interference. I said, "I'm happy to do it."

Weet approached Fleischer delicately, but directly, as if he'd happened to wander over.

> I said, "Russ, I'm Jack, I'm the new guy here." He said, "Oh yeah, I've seen some letters that were signed by you. You're one of the ones that signed that letter about EPO, weren't you?" I said, "Look, I've been doing regulator stuff for twenty-five years. Never had a bad relationship with FDA. Don't expect to start now. My philosophy is I work with FDA, I take your counsel, I take your advice. I'm not here to tell you anything or cajole you in any way. If you have a problem with our booth, you call me. Here's my card. Don't hassle these guys. They're just doing their job." We had a nice conversation. A wall just came tumbling down.

❖

Emmens knew what kind of person he needed to lead the commercial organization, and it wasn't someone with a blinding or abrasive intelligence. In the mid-1990s, when he headed sales and marketing for Astra Merck, he led a commercial team that drove the most successful market expansion of any drug in history: Prilosec. For heartburn sufferers, the drug was truly a miracle. But the real amazement was what happened after the company started telling millions of TV viewers in newly legal ads to ask their doctors about "the purple pill"—also known affectionately, by those relieved of the hellish heat in their lower chests and upper abdomens, as "purple Jesus" and "purple crack." With a great drug and effective commercial operations, Astra Merck turned its franchise into the biggest-selling medicine of the decade.

Emmens was promoted to CEO in 1997, and he replaced himself in sales with Nancy Wysenski, a lieutenant of his who'd also started out at Merck. Wysenski began her career as a VA nurse, and she had a nurse's healthy directness. Outfitted in trim pantsuits, short hair streaked, with a blaze in her eye, she energized the sales reps with her excitement and passion and impressed her bosses with her thoroughness. When Emmens later went on to found North Carolina–based EMD for Merck KGaA, he invited Wysenski to join as vice president of marketing and sales.

Women CEOs in biopharma are rare except among the smallest businesses. When Emmens left EMD to take over Shire, Wysenski replaced him, presiding over the slow-motion collapse of a privately held company of 1,300 that couldn't be averted; this had little to do with her leadership and entirely to do with the fact that all but one of its eight drugs failed in trials. "The importance of having good R&D," she says now. Wysenski landed on her feet. Having twice built operations from scratch, she joined Endo Pharmaceuticals, outside Philadelphia, as chief operations officer, participating for the first time in investor calls and learning how to navigate Reg FD as a company officer.

In August Endo announced that Wysenski was resigning to attend to "family matters." (From an anonymous post on Cafepharma, the industry's online chatter site: " 'Family matters'??? What woman in a top corporate position is able to care for a family? If female reps can't, then one in upper management certainly cannot either.") Wysenski negotiated a separation agreement that included, in addition to accelerated vesting of her stock options, a $75,000 reverse-relocation allowance and $200,000 for the loss of equity on her Pennsylvania home.

Her return to Durham was brief. In December Vertex announced her appointment as executive vice president and chief commercial officer. Arriving in Cambridge before Christmas, Wysenski encountered elements of the Vertex culture that concerned her and that might have troubled her more deeply if not for her comfort and history with Emmens. She recalls:

When I got my first exposure to the board and the ET, what I found was an incredibly bright, high-intellect company that seemed to enjoy intellectual sparring. That was difficult for me, because that's just not my thing. So I would sit in meetings and say, "My goodness, are we trying to get this done, or are we just trying to explore conceptually how many ways we can twist it and look at it?" It didn't make sense to me. At times I'm sure that people felt that I was a little overwhelming—I've heard the word "hard-ass" used. But we had to get it done. The future of the company depended on it.

I was very fortunate reporting to Matt. I know him, trust him, totally. No BS. We can just talk very directly. The first thing he did before I ever arrived is put a place mark in the financial planning process that was more than enough money than I needed. When I looked at that, in my gut I knew I didn't have to worry. The other thing is, since Matt had come from the board, it paved the way for acceptance there. The thing that made it easier than if I had been at another company is I didn't have to prove myself to him. I *wanted* to prove myself to the board, and I wanted to win them over. But I also knew he would fly cover for me.

Six months after replacing Boger, Emmens had his "giant brain." The central lobes were Mueller, to drive telaprevir across the goal line; Smith, to finance the company; and Wysenski, to push the drug out the door and into the sales channels, to get it to patients. There were other important voices on the ET, notably Ken, repository of Vertex's received wisdom, business history, and, as the sole remaining Boger, collective consciousness. "Yoda," Garrison calls him. But in the small circle that Emmens drew around himself, fewer was more.

Smith started to be recognized outside Vertex as the architect of its financial story, and he enjoyed his rising status. At the Liver Meeting, where the company reported a spate of impressive data, he strolled the poster session trailed by a gaggle of analysts and reporters eager to hear his opinions of other companies' molecules. Vertex held its lead by showing that telaprevir could cure even the toughest-to-treat patients with hepatitis C and that it might work just as well in twice-daily doses as when administered three times a day. Results from a midstage study showed an SVR rate exceeding 80 percent. But forty or more other drugs were in clinical trials—trying, as one fund manager would say crudely but effectively, "to crawl up the company's ass." As Vertex's new data mounted throughout the fall, shares of VRTX surged, driven by an analysis by Cowen and Company estimating that telaprevir would generate $2.3 billion in US sales by 2013.

Smith was soaring, each of his financing schemes more ambitious than the last. In July he'd announced that Vertex, which held up to $250 million in milestone payments associated with the filing, approval, and launch of telaprevir in Europe, would sell the milestone payments for cash up front: the deal netted $155 million. After the Liver Meeting, the company announced yet another follow-on stock sale. Boosted by investor confidence that telaprevir was a better drug than boceprevir and that it would become a blockbuster soon after launch, the company banked another $450 million by selling another 11.5 million shares, bringing its cash reserves to about $1.3 billion, polishing Smith's reputation for having a golden touch, and strengthening Emmens's confidence that the company would have enough fuel to get to launch.

Wysenski returned to Durham for the holidays, to spend time with her family before moving to a waterside condo in Boston. She was convinced that telaprevir was a transformative drug but not at all certain that the economic assumptions underlying both the company's valuation on Wall Street and its sense of its own worth were supported by any real understanding of the commercial issues it was about to confront. Selling new prescription drugs isn't like selling anything else. The people who use your products aren't your customers. Your customers are exclusively doctors and hospitals. But your payers are third parties: governments, giant managed care companies, and insurers. If you haven't lined up managed care reimbursements, it doesn't matter how many prescriptions get written. The key is to identify patients *and* their payers and somehow reduce all the psychological, medical, and financial barriers that keep doctors from "writing" your product and patients from receiving it.

Vertex had exorbitant expectations about the number of patients telaprevir would reach. Its market research revealed that the highest-writing doctors were withholding thousands of patients from treatment until better alternatives came along—"warehousing" them. Sachdev had helped line up voices in Congress to support an expanded health advisory from the Centers for Disease Control and Prevention recommending age, rather than a history of high-risk behavior, as the reason baby boomers should be tested for HCV, which would dramatically enhance diagnosis rates. The company envisioned a flood tide of new patients. As Wysenski reviewed the company's projections, she became more and more alarmed.

I'm going through documents that I hadn't had time to read, and it took my breath away . . . There was a market expansion rate in that forecast, in terms of diagnosing HCV patients, that had never been achieved by anyone, ever, with maybe the exception of the French government after they had dealt with an infected blood supply that had gone out. The other market expansion plans that we could benchmark came nowhere close. We also went back and looked at what Schering-Plough and Roche had done during the Pegintron and Pegasys launches, and *they* fell flat.

I said, "We've got great science here, and we're gonna leverage that, and we're gonna do everything we can to build a customer-intimate organization so that we can marry up the knowledge of the customer with that great science. But I think we *really* need to pull back on this thought that we can change the world. Because a certain amount of that is out of your control. Right? We've got a great effort. We've got Amit in DC, and they're working on changing the guidelines with the CDC. We're doing everything right. But you still have to look at how the market has responded historically and then make sure that you're being prudent about what it is you think you can achieve." So we brought that rate down dramatically, to about half the growth rate that we originally had.

❖

On the day that Emmens flew with Smith to San Francisco for his first Morgan conference as Vertex's CEO and chairman, the momentum lurched unexpectedly in Massachusetts in the race for the vacant US Senate seat held for forty-six years by Ted Kennedy, who'd died in August. In its latest survey of likely voters, Public Policy Polling showed a little-known but well-funded Republican state senator, Scott Brown, leading Democratic attorney general Martha Coakley by 1 percentage point. A *Globe* poll still had Coakley ahead by 15 points, but the contest had drawn much closer than anyone had imagined, with cosmic implications for both sides. Brown emphasized that he would be the deciding vote against Obama's health care legislation, proposed by the president in the wake of Kennedy's death and championed by the White House as a fitting legacy for Kennedy's career-long crusade for health care reform. Coakley inexplicably had taken a week off for a Caribbean vacation in the midst of the two-month campaign, giving an impression that she was taking her ascent too much for granted. In US politics, it was a resoundingly bad time to appear smug.

Emmens recognized the risks and benefits of being compared with Boger, but he concentrated on the larger challenge: convincing Wall Street that Vertex was closing in on Boger's formula for a new model of success in pharmaceuticals: New Pharma. "Two thousand ten will be a defining year for Vertex as we seek to evolve into a fully capable biopharmaceutical company," Emmens told a packed audience. Dressed in a light grey suit and matching tie, his sharp gaze framed by rimless glasses, he appeared crisp—monochromatic, polished, and on point.

"Phase III data for telaprevir, our lead drug candidate for hepatitis C virus infection, will begin to emerge in the spring of 2010 to support the planned submission of a new drug application in the second half of this

year," Emmens continued. "We also recognize the need for continued innovation in the treatment of this disease, and we are preparing to initiate
the first clinical trial combining telaprevir with the investigational HCV
polymerase inhibitor VX-222 this quarter.

"Beyond HCV, Vertex is conducting midstage and late-stage development of two novel compounds aimed at addressing, for the first time,
the underlying mechanism of the orphan disease of cystic fibrosis. The
VX-770 Phase III registration program is advancing rapidly, and we expect to obtain Phase III data for VX-770 in early 2011. Additionally, we
also expect to obtain clinical data from a Phase II trial of VX-809 in the
coming weeks that could potentially support the evaluation of VX-770
and VX-809 as part of a combination regimen in patients with the most
common mutation of this disease.

"Supporting our vision to become a fully capable biopharmaceutical
company, Vertex is also planning multiple proof-of-concept clinical trials
in other diseases, such as rheumatoid arthritis and epilepsy."

Moving ahead with the JAK-3 inhibitor in RA was no surprise. Pfizer
had shown proof of concept; it was up to Vertex to show that it had a
better molecule. But Vertex's epilepsy drug was a case of a benched asset
unexpectedly reactivated in light of a new biological insight—a medicine, so to speak, before its time. Such stories are all too common and
herald the value of never giving up on a promising target.

It had been almost seven years since Vertex and Aventis had stopped
testing the company's ICE inhibitor pralnacasan because of ambiguous
liver signals in animals suggesting that it was too toxic to be a drug. The
company's fast follow-on, VX-765, proved to be clean in similar studies,
but without a corporate partner to pick up the cost of development, it
languished in the company's compound library: a massive robotic repository of four hundred thousand molecules in the basement of the San
Diego site. Then, in 2009 Italian researchers at the University of Verona
reported a new observation about epilepsy, a brain disorder where clusters of nerve cells signal abnormally, causing seizures. They found that
a faulty immune reaction exacerbated the progression of the disease by
making the blood-brain barrier more leaky, allowing harmful molecules
into the brain that affect neural activity and make additional seizures

more likely. All existing epilepsy drugs were anticonvulsives, which suppressed nerve cells from firing uncontrollably, but here was a rationale for blocking seizures with an anti-inflammatory. Preclinical results for VX-765 were striking, and Vertex, in collaboration with the Italians, was pressing ahead to be first to test the hypothesis in patients.

Emmens, like Boger a year earlier, did a hallway interview with CNBC's Mike Huckman. When anchor Larry Kudlow cued the segment by saying that the Morgan meeting's seven thousand attendees were a who's who of those making news in biotech and those who knew how to make money from it, Huckman said, "Good Morning, Larry, yeah, this year in that group of see and be seen is Mr. Sam Waksal, of course the Imclone founder and ex-con." It was a theatrical reminder of the scandal that also put style-maker Martha Stewart behind bars, after Waksal, a friend, told her that the FDA was about to turn down an Imclone drug for submission, and she sold her stock before it sank. Huckman turned to Emmens:

HUCKMAN: You're going to have, as I mentioned, major data rolling out this year on this drug that you have in late-stage development for hepatitis C. You plan to file for FDA approval of it by the second half of this year. But Merck is close on your heels in development with a similar drug. Is it gonna come down to whose data is better, whose drug works faster, whose drug is safer?

EMMENS: Well, I can't predict the future. But I can tell you that our data look real good. The drug has been pretty consistent in clinical trials. We plan to launch that drug sometime in the first half of next year. And it works really well for those patients who have failed the current therapy.

HUCKMAN: Mr. Emmens, as you well know, this is a place—this event—where deals historically have been born. Are you being approached by Big Pharma, by Big Biotech? Or does your partnership with Johnson & Johnson put handcuffs, if you will, on other suitors?

EMMENS: Mike, we get that question all the time. That's gonna be up to our shareholders. But I'll tell you one thing. We're here to build as much value as we can, first with the drug for hepatitis C, then with cystic fibrosis. We have twelve more preclinical projects. We have now six programs in the clinic. So the scale of our company is quite amazing.

HUCKMAN: Yeah, but certainly an acquisition would create shareholder value?

EMMENS: We'll have to see.

Boger thought the reason business reporters and analysts persisted in fanning the takeover-target story line was that they still didn't understand Vertex's business model and the risk-mitigation strategy embedded in its portfolio. Emmens thought it was because he and Wall Street were in totally different businesses. His was to make new medicines. Theirs was to make money speculating on changes in the price of VRTX due to short-term developments and chatter. Back in Cambridge, he faced challenges but no one had the sense that he was any less committed than Boger to keeping Vertex from being taken out.

❖

Scott Brown's startling Senate victory burst through the halls of Congress and across America. The political center of gravity lurched rightward. "Tonight the independent voice of Massachusetts has spoken," Brown told his cheering supporters. Kennedy had called overhauling the health care system "the cause of my life." Now the state would send a Republican to Washington to try to kill it. Losing their supermajority, Democrats scrambled to salvage Obama's bill. PhRMA and BIO both backed the measure, the big drug companies having made a deal with the White House to block any effort in Congress to extract cost savings from them beyond an agreed-upon $80 billion, and BIO having inserted a provision to create an accelerated pathway for the FDA to license so-called biosimilars that were interchangeable with branded biologics.

Wysenski had twelve months to build a national commercial organization equipped to go head-to-head against Merck. Absorbing Schering's clinical and marketing muscle in hepatitis C gave Merck an initial advantage in the market, where the Schering team already knew and had contracts with nearly all the key opinion leaders (KOLs) in liver diseases and gastroenterology, and had friendly relations and well-established channels across the reimbursement landscape. Conceptually, designing an organization requires figuring out what your goals and priorities are, and then reverse-engineering from that what it will take to be successful. With the sharply reduced market expansion forecast, Wysenski now had reasonable goals for the launch and beyond. With a small team, she started to devise a framework.

"There are three buckets," she explains. "A strategy drives people, process, and technology. That's what I assessed, and that's exactly what I took to the board. I said, 'I've looked at the people, and we've got *great* middle-level management, but I've got to hire around that. It's not that they're not good, it's just that they've never done the lead roles, so I've got to supplement that because we can't fail.' You couldn't worry about people being a little bruised, because you had to make this happen for the company."

Wysenski fleshed out her proposal to the board. There were five core technology systems that Vertex would need to be ready to go on Day One. She identified the five key positions and the talent she would need. She also spelled out the processes by which everything would come together—"the operations, the policies, the infrastructure, the organizational structure you create"—and, with the help of one of Murcko's hires, created a program management system to track what was getting done. As spring approached, with Emmens's blessing and the board's support, Wysenski dove into delivering on her plan to turn Vertex into an operating company.

"I immediately started aggressively recruiting on what I thought were the most important three critical jobs: sales, managed markets, head of marketing for HCV," she says. "I also met with the CF team and told them, 'Please don't take it personally, but it looks like from a distance you're doing a great job. Just keep doing it; I'll be back in about six months.' They were on autopilot. They had the core there."

As Emmens had anticipated, many people in the company were jolted by the abrupt change in style brought on by Wysenski's commercial thinking. Wysenski knew what she needed, and it wasn't advice from people who had little or no experience in her world. She created her own subculture, a specialized army within an army. She knew she was treading on the company's mores but thought the need to do what needed doing trumped social norms, especially given the culture of open, democratic argument that drove Vertex from its earliest days. Wysenski:

> In an operating company, that debating can leave people with a sense of chaos on occasion. Because you never really know, "What did we decide?" I just had to shut that down in commercial and say we're not going to have that. Let's be clear about what our job is, each of us, and how we can help the person next to us. We have to start specializing because that's the only way we're gonna be effective. There were some people who didn't like it because it meant their job wouldn't be as fun. But it worked, and we got it done.
>
> Along with that debate, I think there was a sense that commercial is so simple, we could just read about it. There I had to push back too. There wasn't much respect for the wisdom of experience on the commercial side. And maybe a sense that it was easy—eas*ier*. And therefore, if we just thought hard enough about it, we could probably figure it out. But that's not the case. You really do have to have people, and that's what I was looking for. People who had the experience. People who had been through it before. People who had fallen and skinned their knees and learned from that, who could come in and drive the commercial buildout and execution.

She would have preferred to hire a managed-markets person first, since in pharmaceuticals you can't sell your product to your customers, and your consumers can't get it, until your payers are willing to pay for it. You have to have access to managed care reimbursements before your reps can really sell. Instead, she hired her sales lead first. Joe Cozzolino had a proven track record: he was Gilead's senior director of sales for the East HIV Division that generated yearly sales of $3.5 billion. "Joe, technically, had everything it took," Wysenski says. "He was a great guy. Great, great sales leader."

Obama won his landmark health care overhaul—the most expansive social legislation in forty years—at the end of March. He signed the measure, the Patient Protection and Affordable Care Act, during a raucous ceremony in the East Room. "The bill I'm signing will set in motion reforms that generations of Americans have fought for and marched for and hungered to see," he said. "Today we are affirming that essential truth, a truth that every generation is called to rediscover for itself, that we are not a nation that scales back its aspirations." Republicans attacked the measure as an example of big government run amok. "This is a somber day for the American people," House Republican leader John Boehner said. "By signing this bill, President Obama is abandoning our founding principle that government governs best when it governs closest to the people."

❖

As with AIDS, the first approved HCV protease inhibitors would surely usher in a new era in treatment. But the larger prize remained a superior cocktail: ideally, one that would further shorten the duration of therapy while sparing patients peg-riba's grueling side effects. Since taking out ViroChem in order to acquire the non-nuc 222, Vertex had pressed to stay ahead of a phalanx of competitors advancing a variety of approaches. Bristol-Myers Squibb developed a protease inhibitor to be used in combination with a compound that blocked a fourth target, NS5A, a nonstructural protein with its finger in multiple pathways and a key modulator of viral reproduction. Roche and its collaborators Pharmasset and InterMune were attempting to combine a protease inhibitor and a nuc. So was Gilead.

In March Vertex announced it had begun enrolling one hundred patients, chiefly in the United States, to test its combination treatment. The telaprevir dose stayed the same for everyone. Some patients got VX-222 taken, at different doses, as a twice-a-day oral pill. Others received both medicines in addition to peg-riba. "No one has proven the point that you can get an SVR without pegylated interferon and ribavirin," Kauffman told Xconomy's Timmerman. "It's a leap that has to be made, and we want to be the first to make it." Whether two antivirals would be strong enough to kill all the resistant variants, or whether a third, or a fourth, would also be needed was a crucial scientific and commercial question, as Vertex, like every other company, gamed out the future of the disease.

After ten years with no new advances in treatment, patients who knew they were infected with HCV were growing desperate and, due largely to an explosion in online information and interest, more and more sophisticated about the process of bringing a drug to market. They were no longer voiceless in society. Social media gave hepatitis C patients a new forum, and everywhere the cry was for a cocktail that would save them and their loved ones from having to endure the standard of care. Kauffman's vision of an interferon-free treatment elicited a small avalanche of hope-filled readers' comments to Timmerman's post:

Michele: My husband did the pegintron/ribavirin shuffle three times. The first two times the side effects were miserable but his viral load dropped to undetectable for several months until it gradually returned. The third and last time he developed interferon psychosis, which leaves him with no current treatment option. A new combo therapy without interferon would be a life-saving dream.

Diane: I hope and pray that a new drug therapy minus interferon will be developed. Was on the combo therapy and although everyone on treatment gets hemolytic anemia, I developed autoimmune hemolytic anemia, meaning my own white cells were targeting my red cells, caused by interferon. If anyone knows about non-interferon based trials, please post. I've got a lot of living to do . . .

William: all I hear is about relapsing 6 to 18 months after treatment. are there any people out there that are still clear of the virus 5-plus years on? i never hear of any and am beginning to think it never goes away. great for the drug companies. am coming up on 24 weeks of standard treatment plus first 12 weeks were also with experimental drug. standard flu symptoms all the time, horrible eczema, and am feeling a little nuts and miserable to everyone. is it really worth it?

Robert: PUT ME ON TRIAL TO SEE IF IT REALY [SIC] WORKS

Accelerating Vertex's internal pulse were several converging dead-lines. While Kauffman and his clinical team pushed to finish the two pivotal studies, Condon circled the world assembling the last crucial pieces of his virtual supply chain, and Weet and his group started compil-ing, hyperlinking, and submitting the mountain of data the FDA would require for filing an NDA. In every important meeting and discussion—holding what Boger, Emmens, Smith, Murcko, Ken, and many others worried were too many reins—was Mueller, dressed in his creaseless black uniform, often with a pale-hued pressed sweater draped crisply over his shoulders, peering over half-rim glasses in Bavarian field-general mode. Meantime, with Emmens promising Wall Street that the company would submit its application by the end of the year, Wysenski scrambled with headhunters to fill the yawning gaps and vacancies in the commer-cial organization, and Cozzolino laid the groundwork for putting a sales force into the field.

"It was very, very intense from beginning to end," Kauffman says, "be-cause we were pushing timelines that most companies would not have tried to do. And the data were continuing to flow. There was an enor-mous amount of effort going on. Part of the pressure was self-induced. We at Vertex really wanted to prove ourselves. We wanted to do a good job and show that we had really made it. The other thing is, every day of delay is a day of sales, and you just want to get this thing done as quick as you possibly can."

At the end of May the first data on the ADVANCE study—clinical trial 108—were unblinded. With over a thousand patients, it was the biggest study ever done on people infected with HCV. About 75 percent of patients receiving the standard dose of telaprevir for twelve weeks in combination with peg-riba still had no detectable virus in their blood twenty-four weeks after treatment—essentially a cure—compared with 44 percent of those who received standard therapy alone. Vertex's stock jumped 12 percent in after-hours trading. Yaron Werber, an analyst for Citigroup, estimated that global sales of the drug could reach $3 billion a year within five years.

ADVANCE and another Phase III study, REALIZE, gave Vertex most of what it needed to submit the drug for approval, but a third large pivotal study called ILLUMINATE—trial 216—was still under way in Europe. Because of the difficulties Alam had encountered with the agency over submitting the company's Phase III protocols, there remained some ambiguity about whether the studies would be adequate. Jack Weet was anxious. Vertex had requested a target date of September 28 for a pre-NDA meeting. At that point, it would outline the content and format of the NDA. If there were problems with either study, the result could be ruinous. "The way regulatory works is: you submit an NDA, the FDA files it," Weet says. "They have sixty days. Within that sixty-day period, they can come back and say, 'You don't have enough data, or the data that you have aren't sufficient,' and they can refuse to file. If they refuse to file it, that can kill a company. That's what happened to Imclone. They refused to file the company's application. Our stock would go down to zero if we had an RTF (refusal to file)."

Now in regular and cooperative contact with Fleischer and dozens of others at the FDA, Weet and his team pressed to accelerate the process. Waiting to hear about the pre-NDA meeting, he asked if Vertex could submit the ADVANCE data ahead of ILLUMINATE, so that if there were questions, the company would have time to address them. Was the study robust enough? Were there enough patients? Were there enough African-Americans? Weet hoped a rolling submission on the two trials would relieve some of the pressure he was feeling from Mueller and keep Vertex a step ahead of Merck, which was finishing two pivotal studies of boceprevir and also had announced that it expected to submit an NDA by the end of the year. As deadlines tightened, so did nerves.

"It got to the point," Weet recalls, "where we decided we're all going over to Antwerp because we've got to get the briefing document done, and we can't even spare six hours' time difference for them to get data and send it to us. The team flew over. We literally sat down with Tibotec as they unlocked the database and unrolled what the data said. We didn't know until that week what it said, so we had our writers there, we had the clinicians there, everyone together hammering out the document and the key response data. We packaged it all up in a pre-NDA submission, sent it out at the end of the day there, so that it got here in time to get it into the gateway that same day. We literally didn't have six hours to spare. We were hanging on by our fingernails."

❖

Joe Cozzolino hired Alexander "Bo" Cumbo away from Gilead to build Vertex's US sales force for telaprevir. As fellow rising stars there, they both "knew what good looked like," as Wysenski liked to say. More to the point, Cumbo knew the KOLs, not just nationally but region by region, city by city.

Reared and educated in Alabama, he was soft spoken, mannerly, silver haired at age thirty-nine, compact, and driven, an evangelist for antiviral drugs. If he hadn't gone into pharmaceuticals, he might easily have excelled at coaching or religion. A couple of years after graduating from Auburn University in the early 1990s, Cumbo, single and unemployed, took a job in Jacksonville, Florida, selling vitamins and birth control pills. After interviewing at Merck, Roche, and Glaxo in 1997, he chose Glaxo because it had the biggest HIV program, and because he wanted the pace of change and the opportunity that a fast-moving disease area could offer. He covered a territory stretching from Jacksonville to Savannah and Macon, Georgia, to Mobile, Alabama, and back, and for the first time but not the last he worked himself into the hospital, needing IV fluids. Cumbo joined Gilead in 2000, when it was on the verge of launching its first AIDS drug, and he had held senior positions managing state and federal accounts and sales to prisons. Telaprevir would be his tenth drug launch.

On his first morning in Cambridge, Cumbo screened the accumulated résumés of those seeking jobs as Vertex sales reps. There were about four hundred of them. He found them all useless and sent them back to HR. What he was looking for were not just the best, most experienced managers, trainers, and reps in gastroenterology and infectious liver diseases, but people like him who were doing very well and were happy in their present situations but would find the allure of being part of the next big thing in the field ultimately overpowering.

To identify these people, Cumbo contacted the KOLs, who more than anyone else knew and could recommend the reps from the eight companies that dominated in antivirals. Getting top prospects to leave their jobs was something else. "Specialty reps are out on their own islands," he notes—isolated from their colleagues and the home office. Their connections to their companies channel wholly through the products they promote and the reporting chain. With the disadvantage of representing a company that still, after twenty-one years, had no products or profits and had burned more than $1.5 billion in the past three years alone, Cumbo found it "very, very challenging."

He says, "I felt like I was that college football coach recruiting players: 'Why don't you come to our school? I know we're not LSU or Auburn, but we're Southern Miss, and you can come over here, and we'll take care of you.' I had to actually talk to a couple of my representatives' wives to make them feel at ease, because they were scared. They have families, they have kids, and they don't want to come over here and the drug not get approved and fail."

A sales leader leads by selling the value of personal experience. Cumbo told them his own story, a pilgrimage that in its contours and parallels sharply resembled Boger's and many others' at the company, but with the notable distinction that he had *his* epiphany not long after he launched Agenerase. He was at Glaxo, and the company was selling $7 million a day in HIV meds—70 percent to 80 percent of the AIDS market at the time. Bristol-Myers was its biggest competitor, with Sustiva, the first once-a day-medicine for HIV.

Cumbo lunged, joining Gilead. The small company had seen its market share collapse after its first drug failed, but it now had a promising once-a-day treatment for HIV called tenofovir, a non-nuc polymerase inhibitor. "My marketing manager called me up and said, 'Why are you going over there? They're gonna get bought out. Or they're only gonna get ten percent market share. Most likely you'll end up on the streets if it doesn't come to market. It has renal issues, bone issues. You're making a huge mistake. We're Glaxo.' And I'm like, 'You know, I can just see where the market is gonna go. I can tell you that it's gonna go once-a-day.' The arrogance at GSK was very thick at the time. They were the dominant player; thought they were invincible."

Cumbo's parable dovetailed neatly with Boger's story line about

Vertex being the next challenger to the twin legacies of Genentech and Gilead. The fact that Gilead had displaced Glaxo in the marketplace, that it now was the behemoth in AIDS and scrambling to expand in antiviral therapies with an unproven play in HCV, added further resonance. The interviewees would hear, as part of the company's pitch, about the three Vertex values, but what they connected to and drew strength from was Cumbo's warmth, inspiration, reverence for history, and go-big-or-go-home commitment.

I'd say, "Look where Gilead is. You just don't know what's gonna happen, but you have to take a chance. The people who take a chance have a lot more to gain. Do you want to stay at that company that's doing eight billion and just be that average rep, or do you want to be the guy who comes over here, tries to create things, build things, and take down the biggest players? I mean, that's *fun*."

I'd say, "Previous success isn't future success, but at least you're part of the game." And being a part of the game is what drives you. It gets the blood going. That's why I can tell you that the people who join are all sort of cut from the same cloth. These are the people who want to run through a wall and change the entire pharma industry, by beating Merck.

In August Merck announced that it had completed its second pivotal Phase III study, tightening the race and ratcheting up the pressure inside Vertex. A new compressed timetable came into play. NDAs are filed confidentially. Once the FDA received either company's NDA, it would have six months to make a decision. If the agency decided it was warranted, it could convene a public hearing with a panel of outside experts, an advisory committee, or AdComm, which by a vote would recommend action. Whichever company submitted its NDA first would determine the AdComm date, but since AdComms convene only one or twice a year, it was possible, if Merck or Vertex submitted its application soon enough, it could set the other company back by six months—an eternity in a head-to-head marketing war.

Rumors reached Vertex that Merck had filed its NDA. In Vertex's

NDA war room, a secure conference room on the first floor of Fort Washington II across from Weet's department, anxieties spiked. Weet had contracted with a consulting company, ProEd Regulatory, to come in and facilitate its preparations for the advisory committee. The company had decided to treat the pre-NDA meeting as a dress rehearsal for the AdComm, yet the first preparation session had been a shock. ProEd's team drilled Vertex's people, and feelings were raw. "I remember Ann Kwong sent me a note: 'I'm glad Peter wasn't here. If he was, there would have been blood on the floor,' " Weet recalls. "It was awful. ProEd was throwing us questions as if they were FDA. We weren't able to answer them. We were fumbling over the responses. We didn't know who was gonna answer them. It was really chaotic."

It was Kauffman, who had been with the project for a decade, persisting on behalf of VX-950 after Eli Lilly dropped out, who emerged from the session as the team's leading voice. Kauffman dismissed the rumors that Merck had submitted its NDA. He knew the general timeline for its studies and believed "there was no way in the world they could file that early." Still, with everyone at Vertex working to exhaustion to meet a target filing date of late November, the rumors tested his equanimity. "We had a timeline, and as much as Peter was trying to push us, we just could not accelerate any further," he recalls. "A certain number of things needed to be done. In the end, we just couldn't read the documents fast enough. If you want to actually review them, there's a limit to how much you can read. That was a time of real anxiety, and I was a little pissed off by that point. Ten days' difference didn't bother me much, but for them to be ahead of us by that much after we had been ahead of them for all these years would have made me really mad."

Kauffman had been in the role of NDA lead before, at Syntex with the immunosuppressant CellCept. Weet, too, had experience in getting organizations in shape for the FDA review process. At his last company, NDA teams decamped to another city so that senior executives wouldn't panic or intrude and exert pressure to influence the drug's label. At Vertex, the preparations were shadowed by Tibotec, which would be applying for European approvals. Weet took the information that Merck might be ahead gravely. "We didn't know when or if they'd submitted," he recalls. "There was some intelligence that leaked out that they had actually done their submission in September. We thought. 'Crap, what happens if they submitted in September, and they have a February advisory committee? We've got to get our NDA in as soon as possible.' What was driving us was this idea that there might be this advisory committee sitting out there in February and we missed it, because we didn't get our submission in."

ProEd organized two mock sessions for the pre-NDA meetings, bringing in a former FDA official in charge of viral products and two senior KOLs, and setting up in a conference room at the Cambridge Hyatt Regency, a short walk from Vertex. They fired away at Kauffman, who either answered or deferred to the person in the group who knew the data better than he did. For Weet, the critical goal was to get the FDA to agree that Vertex had sufficient content to back up its claims, so there would be no refusal to file.

The Vertex group, exhausted and strained by internal politics that produced sore feelings but no smashed furniture or defections, flew together to Washington for the pre-NDA meeting in Silver Spring. Cumbo and his recruitment team had been on the road for six weeks. Vertex had established hub cities for conducting mass round-robin interviews. Candidates cycled through three interviews in succession, and if they looked good, they met with Cumbo, who averaged twelve interviews a day. After a burst of hiring away more than two dozen sales trainers and reps from Gilead, he got a menacing letter from the company's lawyer, then its head of commercial, warning him to quit poaching. Emmens received similar threats.

At the meeting in Silver Spring, Weet, Kauffman, Wysenski, Mueller, and the rest of the Vertex group encountered effectively for the first time their opposite contingent: the senior people in antivirals and numerous other agency sections who would govern the reviews of both telaprevir and boceprevir. Weet expected that the old contentiousness was well known, and that the improved relations between him and Fleischer would make the division heads and department chiefs curious about the company. Merck, despite the Vioxx recall, remained something close to the gold standard for smart regulatory dealings. The question of whether the old Merck or the former Schering team was now driving boceprevir was debated exhaustively at Vertex.

Twenty-five FDA officials crowded into a conference room to hear Kauffman deliver Vertex's data package for telaprevir and the rest of the evidence supporting the company's claims for its superiority over peg-riba alone, across all patient groups. The officials asked to see a few slides, but otherwise the discussion was more of a scientific dialogue than the skeptical, Supreme Court–style grilling that most of them had prepared for and expected. The regulators agreed to sufficient content for filing. Kwong, Kauffman, Weet, Wysenski, and the rest of the group were impressed by the depth and acuity of the questions, the delicate position faced by the officials in having to evaluate two competing products at the same time, and their strong willingness to work together with the company.

Regardless of the jockeying with Merck, the meeting was a critical test of Vertex's readiness. All signs were encouraging. "It had very much the look and feel of an advisory committee," Weet says. "It was showtime. Peter was literally dancing in the halls after the meeting. He was doing his happy dance in the lobby of the FDA."

❖

The National Conference on Correctional Health Care in Las Vegas in late October focused on the problem of AIDS and hepatitis C in the country's prisons and jails. More than 40 percent of inmates in California were reportedly infected with HCV. Men and women over fifty behind bars represented a major market for new drugs, but so far prison health officials had seen very little reason to invest scarce medical resources to help them. High HIV coinfection rates were swamping prison hospitals with a complex secondary epidemic. Cumbo, from his work at Gilead and his contacts in the field, was Vertex's obvious point person with the wardens, prison doctors, public health experts, government officials, and patient advocates assembled for three days on the Strip. He flew to Nevada for the weekend conference while his staff moved on to next week's hub.

Cumbo had been feeling exhausted, run down. "I woke up in the middle of the night and couldn't breathe," he recalls. "I called a doctor

I used to work with at Gilead. He was at the conference. I told him my symptoms and he said, 'You need to go to the hospital.' So I went downstairs with a colleague, Alex Alvarez, to get in the taxi. Alex turns to the taxi driver and says, 'We need to go to the hospital.' And the driver says, 'Which one?' And Alex says, 'Take us to the one where there's less gunshot wounds.'

"So we went to the place. It ended up I had walking pneumonia. They gave me IV antibiotics and fluids. I checked myself out of the hospital at ten in the morning because I had a twelve o'clock appointment. I went back to the conference and kept my appointment."

If the Vertex sales reps wanted to run through a wall for their leader, they had their man. The battle with Merck was in high gear, and every confrontation was heightened inside the company by the sense that it was not just a head-to-head drug launch in a hot disease area but a title match, a fateful duel. Boger could not have scripted the showdown more perfectly if he'd tried—with one large, annoying exception: boceprevir wasn't a product of Merck's R&D but of Schering's, and it wasn't yet clear which Merck contingent would show up at the AdComm and in the field when the drugs were approved. In a more perfect world, Merck's molecule would have emerged from its own labs and been promoted by Merck's vaunted field force, but Boger couldn't expect to have everything.

The Liver Meeting in Boston was packed to overflowing, with more doctors and investors and analysts and company reps in attendance than ever before. Veteran doctors and public health officials felt the sessions had the same sudden aura of abrupt optimism and excitement that gripped the AIDS world in Vancouver in 1996, the year HIV protease inhibitors became the backbone of combination therapy—although now with more of a car-show feel, as Vertex, Merck, Roche, Gilead, and numerous other companies staged sleek sales and/or information booths in the caffeine-juiced lower-level exhibition hall. The sense of promise also raised the same problems and questions: *What now?*

Doctors were concerned that the new treatments wouldn't be an easy regimen, no matter which drug they chose. Compliance, already an issue, could become worse, especially with the introduction of a fourth toxic drug, EPO. Companies, investors, and analysts looked at the competition and the progress toward an all-oral treatment and tried to read the future a decade out. Where will the market go next? What's the shelf life of any one product? What would the new drugs cost? Who would have access to them? What about the hundreds of millions of HCV-infected people around the world?

Standing-room-only crowds packed the rooms where Vertex and Merck announced new data. The vital issue was how to compare two products that hadn't been tested against each other in clinical trials—indeed, whose trial designs differed widely and, in some cases, deceptively. Doctors who wanted to know which drug to give their patients could be forgiven for being confused and annoyed by the lack of apples-to-apples controls, but neither company desired a head-to-head trial for the obvious reason: it might lose.

Most Wall Street analysts agreed that Vertex had the better medicine—more potent, faster acting, safer, easier to use. The race to submit its application first, or at least keep the submissions close enough to avoid missing an AdComm, beat on inside both companies. The rumors of Merck's alleged September NDA filing still jangled already overfiring nerves at Vertex, where everyone in Cambridge, even Emmens, was susceptible to the terrible scenario that even one missed deadline could, after nearly two decades of pathbreaking work, irreparably set back the scheduled launch of telaprevir, derailing the company.

Since the pre-NDA meeting, nearly everyone in clinical, regulatory, and regulatory operations—the team that would finally put Vertex's one million pages of documents into a proper format for electronic submission—had become riveted on the November 28 deadline. A fast time from the moment the last patient in the last trial visits the clinic for the last time until an application is ready and thoroughly vetted is six months: Vertex had given itself four. The team followed a game chart linking time and events, with top priority going to polishing the data from the big European study, 216, alongside a group from Tibotec, which planned to submit its application two weeks later and was ensconced in another building on Hamilton Street, across from the gritty do-it-yourself auto repair shop owned and run by National Public Radio's *Car Talk* guys, Tom and Ray Magliozzi. By the Liver Meeting, several groups began hammering out a label that followed a common technical document format, submissible on both sides of the Atlantic. Where interpretations varied, or when European and American regulators had different policies or standards, tense standoffs on wording went on for days—and, on one occasion, weeks—sorely fraying nerves and stretching timelines taut.

Internally, Vertex set an accelerated date of November 23—the Tuesday before Thanksgiving. By the end of the previous week, the documentation was finalized. It was studded with a million hyperlinks so that the reviewers could cross-check, within the allotted sixty days before filing, every statement of fact and data point. Mueller, Kauffman, and Weet each reviewed the application in considerable detail. "We got it all done and wrapped up, and Bob stood up and said, 'I just have to say that this is probably the most special moment in my entire career,'" Weet recalls. "You could tell it was very emotional for him personally. It was kind of emotional for the rest of us too. Number one, because it all of a sudden relieved the pressure. But also, because we could say, 'We really have a drug here.'"

Champagne flowed in the war room, although after seventeen years, those celebrating—except for Kauffman, Mueller, and a few others—were not the same people who had made telaprevir. The baton had started with Deb Peattie and Charlie Rice, and then soon passed to Thomson, Murcko, Tung, Sato, and the other early champions of the program. It dropped to the ground with the collapse of the Lilly deal, and then was scooped up by Alam, Kwong, Kieffer, Hurter, Condon, and untold others before ending up now with Kauffman, Mueller, Weet, and their people. The inspiration for the drug had passed from individuals and groups before any of them could have more than a passing claim on true authorship. A drug, unlike a social network or a mobile app, was not something that you could dream up in your dorm room and turn into a billion-dollar seller or a $100 billion company. It was an ultramarathon relay in 100-degree heat. Boger, who had enabled it all to occur, was, of course, elsewhere.

The regulatory operations people spent the weekend running programs around the clock to convert the NDA to its final form, squeezing it piecemeal, section by section, through the FDA network gateway. "They have all these validation steps they have to do," Weet explains. "They go through the table of contents. You have to hyperlink everything you say in the labeling to something in the clinical summary. Then you have to hyperlink everything you say in the clinical summary to something in the summary of the clinical studies. Then you have to link that to everything in the studies, to the tables and the appendixes. The reg ops guys had to work all those links to see if they worked. They found some broken ones."

On Tuesday Vertex announced that it had completed the submission of its NDA for telaprevir—formerly VX-950. In its application, it sought priority review for the drug, requesting that the FDA shorten the normal approval time from ten to six months on the grounds that the medicine represented a major advance in treatment of the virus and so patients should have access without delay. "The submission is a milestone in our more than fifteen-year effort to change the way hepatitis C is treated," Emmens declared. "We are committed to working closely with the FDA to make telaprevir available as quickly as possible to the millions of people with hepatitis C who need new medicines to increase their chances for a viral cure."

Merck, no surprise, had gone dark, choosing not to publicize its own submission. As people left Vertex for the Christmas holidays, no one inside the company knew whether Merck had already filed, or whether its rumored September filing would yet prove to be true.

# PART 3

# Showtime

# CHAPTER 10

---

## JANUARY 9, 2011

Registered attendance was up about 20 percent from a year earlier at the Morgan conference, meaning that 1,400 more people jostled for seats at the breakout sessions. At lunchtime, thousands packed into a constricted hallway leading to the Grand Ballroom, inching ahead to get to keynote presentations by Obama's health advisor Nancy-Ann DeParle and—as was becoming routine—Morgan's chairman, Jamie Dimon, who'd fared best among Wall Street's top figures in the two years since the financial collapse. "The effect is not unlike pushing to get through the gates at a professional football game, only without the face paint and the drunks," an attendee blogged. "In this case, the elixir is money."

Vertex was poised "to take [it] to the next level," though not by the standard that enthused CNBC's Huckman and nearly everyone else there: monetizing the company's science to make an eye-popping bundle for shareholders. Sticking firmly to Boger's vision, if not his bold presumptions and brash tropes, Emmens told a standing-room-only audience that as committed as the company was to curing liver disease, its major focus remained set on becoming the first new-model independent pharmaceutical company of the twenty-first century. He announced that Vertex planned to submit an NDA for VX-770 by the end of the year, talked up its CF combination trials, and for the first time speaking with investors mentioned the flu program. He distributed drinking straws to everyone and urged the crowd to breathe through them to get a feel for what living with CF is like. An audience member tweeted that the straws were for sipping the company's Kool-Aid.

Vertex was squarely "in the hunt," as Weet put it. Merck had announced the previous Friday that the FDA had accepted its new drug application for boceprevir. At Vertex, everyone automatically back-calculated sixty days and realized that Merck's submission date was on or about November 6. Merck seized bragging rights to beating Vertex to a crucial milestone in the race to market, a race that Vertex had led at each juncture. But Vertex was close enough behind to guarantee back-to-back AdComms, which in real terms mattered substantially more than a symbolic gain of two weeks. Kauffman did not have to get really mad.

Emmens was unfazed by the mismatch in size and resources. He told the Dow Jones Newswires that he didn't believe the companies were in a race or that being first to market would be a significant advantage. He said quality is better than quantity in fielding sales reps in a specialty market and that the average person on Vertex's sales team had fourteen years of prior experience, with eight years spent marketing antivirals. What worried him wasn't his people but the uncertainties of the market. No one knew how many patients would show up for treatment—the biggest variable. Sales and marketing could accomplish only so much. "There is room for both of us," he commented diplomatically.

A few days after Emmens, Smith, and Vertex's investor relations and press teams returned from San Francisco, plunging back into making good on the company's promise—after more than two decades—to show a regular profit starting in the second half of the year, Vertex received its filing approval. It contained a late-May deadline for an official decision on telaprevir. Delivered by fax, the FDA letter started the clock on a heated sprint to market. The company had four months to prepare to go toe-to-toe with the remaining industry giant most committed to an R&D-based formula for success—albeit one still struggling to find its post-Vioxx footing. It was a long-anticipated showdown that the business press began to relish more now that the dates were set. With priority review assured for both medicines, Weet, Kauffman, Condon, Cozzolino, and their teams mobilized quickly to beat the new hypercompressed timetable.

❖

In December Vertex had run a "tablet validation campaign" for telaprevir—a test run of its virtual supply chain. For five years, Condon's small staff and groups from business development and legal affairs had hopscotched across continents, assembling the elements. After leaving the manufacturing sites in China, the molecular components known as the Fab Five were flown to northern England, where Vertex had partnered with an Indian company to build out designated capacity to synthesize up to fifty tons per year of the inhibitor. From there, the active ingredient was flown to Portugal, where another pharmaceutical manufacturer had built at its own an expense a three-story reactor to mix that compound with the polymer that Trish Hurter and her group had chosen to keep it from crystallizing.

The drug molecules were dissolved in solvent and sprayed through high-speed nozzles into two five-thousand-liter tanks. Inside the chambers, heated with nitrogen, a whirling column of gas, a cyclone, sucked off the solvent while the amorphous dispersion, a blend of the polymer and 950, fell to the bottom like snow from snow guns, a fine powder, highly stable. Spray drying, first used to make powdered milk, had become an effective technology for rendering highly insoluble drugs bioavailable.

The resulting powder, not a drug product per se but an intermediate, was shipped across the Atlantic Ocean to Cincinnati to be mixed with other chemical agents—some to help it adhere to itself to deliver precise dosages and enhance exposure, others to ensure that the tablets didn't stick to the machines that churned them out. The ratio of drug to excipients—the other chemicals in pills that make a molecule druggable—was 70/30, a striking feat of process chemistry that Boger enjoyed referencing whenever others brought up Lilly's prediction that VX-950 could never be manufactured. From there the tablets were taken by truck to a packaging company in Illinois and readied to go into blister packs and boxes. "It was all happening as we were building it," Condon recalls. "The WuXi plant started really cranking in 2010. What we were trying to do was time it so that we had the active ingredient sitting in front of the spray dryer, and that we could start to make the spray-dry dispersion as close as we could to the time we had to ramp the tablet production, to have enough tablets produced to feed the launch."

Condon believed he could wait until April to scale up full production of the drug, preserving shelf life. Throughout Vertex, the launch was paramount, but before the launch loomed the AdComm. The company advising Vertex, ProEd, arranged two mock sessions where outside experts would be brought in to play the part of commission members. In each of Weet's four previous launches, the head of regulatory affairs introduced the study presentation, which was usually conducted by a high-level physician—often the company's chief medical officer. He believed that ProEd agreed with him, and he started planning for the first "mock," in mid-February. Kauffman, too, started thinking harder about how the story of telaprevir should be told and who should tell it.

On the commercial side, Wysenski had hired as head of managed markets a graduate of the Virginia Military Institute and a former Pfizer executive, Jeff Henderson; and as head of HCV marketing, a rising star at Merck named Paul Daruwala. Huddling with Cozzolino, the three of them built out an operations plan that they would roll out in late January, when the field force would meet for the first time in Miami.

For a company that planned to post a loss of almost $750 million for the year, the FDA letter wasn't money in the bank but it represented substantial collateral. On January 24, less than a week after it was issued, Emmens signed a letter of intent to relocate Vertex's headquarters from Cambridge, where it had run out of room to grow and the political environment was less favorable, to a long-struggling harborside complex in Boston, Fan Pier. The company planned to take about 1.1 million square feet, filling a pair of eighteen-story buildings as the first major tenant in what Mayor Thomas Menino had been trying to rebrand as an "Innovation District" by offering tax breaks and other inducements to companies to locate there. Beyond allowing Vertex to consolidate its 1,300 local workers, now scattered across ten buildings, and giving city and state politicians something to crow about, the agreement made headlines as the largest commercial lease in Boston history and the anchor of possibly the largest privately developed project in the country since the onset of the recession. Of course, it was contingent on FDA approval of telaprevir.

❖

Cumbo felt beat, but relieved. He had done what he had set out to do, hire an elite sales force. "I feel good, I'm working for Joe, I've got my team in place, I've got a good team," he thought. Enjoying his first free weekend in six months, he flew home to Tampa before the sales meeting, picked up his wife and children, and drove down to Marathon Key to buy a vacation home. He promised to spend much more time with them after the launch.

After arriving in Miami a day later than Cozzolino and the sales team, he roamed the first-day training sessions. Monday afternoon, as the sessions wound down for the day and everyone prepared for an off-site banquet, Cozzolino gave him a dollar bill, a pat on the back for a job well done. It was meant as a joke. Back in his room, he got a call on his personal phone from Wysenski. "I've got to meet with you in fifteen minutes" she said. Cumbo, alarmed, texted Cozzolino: "Joe, I've got an emergency. Nancy just called me and wants to meet with me. What's going on? Am I in trouble?" Cozzolino reassured him, "Everything's all right. You'll be fine."

When Cumbo got downstairs, Wysenski pulled him into a private room. She said Cozzolino had just resigned, and she needed Cumbo to take over as the interim VP of sales. Cumbo reeled. He told her about the dollar bill, hoping for some misunderstanding. "No," she said, "he's no longer here."

The managerial revolving door—the Vertex vortex, if you will—swept Cumbo in at warp speed. One minute his world felt right, made sense; the next, his knees went out. "This is January. You're four months from launch. The whole launch was in peril at that point. Everything. Nancy goes, 'I'm gonna make an announcement first thing in the morning, eight o'clock. And I'm gonna call you onstage. And I need you to give a speech and tell the team why everyone needs to stick with you.' I said, 'I don't know if I can do this.' And she says, 'Well, I need you to. And oh, by the way, you can't tell anyone.'"

The lavish half-acre rooftop lounge at the Perry South Beach Hotel girdled a 110-foot outdoor pool with music piped underwater. Waves lapped eighteen floors below, and in the distance, the causeways crossing Biscayne Bay and the waterside towers of the downtown skyline glittered—a mirage of prosperity, as the city still reeled from the real estate collapse and a record number of skyscrapers were vacant. "See-throughs." There was a shark tank down in the sleek lobby.

Cumbo mingled at the bar and around tented chaise pods, each with its own flat-screen TV, in a crowd that included Emmens, most of the ET, and several new vice presidents. (Boger, chairing in "retirement" a half dozen boards, including the trustees of Wesleyan University and Harvard Medical School, couldn't make the meeting but had been "telepresent" earlier, telling the team members via satellite that they were "fabulous" and half-joking that if he were younger, "I'd be right out there with you.") People kept asking where Cozzolino was. "I acted like I had no idea," Cumbo says. "I didn't sleep at all. I tried to close my eyes, but I was sick to my stomach. I'm like, 'What am I gonna say to these people?' "

Boger's efforts to build Vertex into an organization capable of transforming lives ultimately relied on self-selection. Cumbo, who'd joined Vertex less than a year earlier and hadn't talked with Cozzolino or anyone else in the company, found himself overnight having to explain on his own to its newest, most vulnerable employees—all people he'd hired—why they shouldn't be alarmed by the abrupt, unexplained departure of their sales lead four months before launch.

The next morning, Wysenski announced that Cozzolino had left the company. There was a gasp. People were crying. Then she said Cumbo would start immediately as interim vice president, and everyone clapped. Cumbo had no speech, nothing prepared at all. Afterward, he wouldn't remember walking to the stage or what he said.

"From what people can recall, I talked about how I started at GSK in HIV and how I watched how people went from dying to living, how protease inhibitors changed their entire lives. I said, 'I'm here not because of Joe. I'm here because I want to be here, and I want to be with you. This is bigger than one person.' I said, 'We're gonna make history, and we're gonna make history together. I'm here, and I'm gonna support you, and I'm gonna lead you through this.' I was practically in tears. I know the audience was in tears. They gave me a standing ovation for five minutes, so long that I actually got off the stage shaking, and I walked out of the room, and everyone kept clapping."

❖

The first mock was held on February 10, off-site. It might have been odd at another company for the head of corporate communications to attend an early prep session for a regulatory hearing, but Megan Pace was responsible for the company's messaging and optics, and the AdComm and launch were once-in-a-corporate-life-cycle opportunities for Vertex to define itself. So far, as Boger would be first to agree, Vertex had not really attended to public relations—not since it wheat-pasted AIDS broadsides at night in the 1990s. Michael Partridge ran *investor* relations, but that was talking with people who might want to trade shares of VRTX, not branding a biopharma for the rest of the world. Sachdev spoke with Washington and the states, but not with people who associate drug companies with their products and behavior. Vertex would soon be introduced to the general public, and Pace, who'd joined from Genentech, where she'd been spokeswoman through both its blockbuster phase and the Roche buyout, was deeply involved with all aspects of the rollout. Blond and willowy, she was thirty-five, a brisk native of Orlando, Florida.

With the room staged to resemble an AdComm, the consultants sharpened their questions. Weet delivered the introduction. Mueller, asked by a reporter around this time what role, if any, he expected to play, said he thought that as the company's chief scientist and its chief medical figure, he and Kauffman should represent Vertex, but he met resistance within the room. "ProEd has done probably a hundred of these," Weet recalls. "They were very clear in their view as to who should be the spokesperson. You don't necessarily want to have the top person in the company, because if all of a sudden that person gets vaporized by the advisory committee, that's not good for public image. So they have very clear feelings: it should be someone senior, but not too senior.

"Their view was that the head of regulatory should give the introduction," he says. "So in the first mock, I gave the introduction, and Bob gave the clinical overview. I don't know exactly what happened, but I know Megan had said in her experience it was not the head of regulatory but always a clinical person, usually the CMO. We agreed to disagree with that."

The daylong session was by all accounts stressful, disconcerting, tense. In Weet's view, this was not because Vertex was far from ready to stage telaprevir's case but because it allowed people across the company to see how much work still remained to prepare for being interrogated by two dozen experts who had the time and inclination to advise the government on whether to approve a new drug. The FDA had scheduled a two-day AdComm for April 27–28, with Merck presenting its case first by virtue of submitting its application first. Vertex had ten weeks to prepare. Weet thought that this was more than enough time and that the mock did its job pointing out what needed fixing, but there was collateral damage in added tension and anxiety.

"It's one of the reasons you want to keep as many people away from these as possible; the first rehearsal is gonna show a lot of holes," he explains. "Well, but this was a spectacle. Everybody wanted to be there. Paul Daruwala was there. [VP of Marketing] Pam Stephenson was there. Peter was there, and people were saying, 'This is a disaster. We're not ready.' Of course we're not ready! My last advisory committee, we actually had our mocks in Cleveland. We were located in Rochester, New York. We had our mocks in Cleveland, just to keep people away. I mean, when we went off-site, we really went off-site. It was great. It was a disaster, but that's what it was for, and we improved it and got it right."

One notable effect of doing everything for the first time with the idea that it can and should be done better than anyone has done it before is to put a premium on the ability of veterans joining from industry leaders such as Merck, Gilead, and Genentech to push and prod selectively. Pace prevailed, in part with received wisdom, in part because she had the faith not just of Emmens but also of Boger and Ken, both of whom believed that, having been at Genentech, she "got it": the larger challenge of building a pharmaceutical vision into a beacon. Kauffman became telaprevir's undisputed spokesman and, by extension, its main external champion.

❖

New Merck CEO Ken Frazier didn't follow the usual trajectory to get to the top in Big Pharma. His mother died when he was twelve, leaving him in the care of his father, a janitor. Working hard in school and inspired by Thurgood Marshall, he went into the law, excelling as a student at Harvard and practicing in Philadelphia, where he gained a reputation as an aggressive litigator. He won a new trial for an Alabama man who'd spent nearly twenty years on death row for murder; the prisoner was later acquitted. Defending Merck against liability suits, he came to the company's attention, joining in 1992 and becoming general counsel in 1999.

Faced with thousands of plaintiffs' suits alleging that Merck knew about the health risks with Vioxx years before it pulled the drug from the market, Frazier devised a daring strategy. Rather than make the cases, as lawyers say, "go away" by settling quickly and quietly, Merck decided to fight each case to a verdict. The company won eleven of sixteen lawsuits at trial before agreeing to a $4.85 billion settlement fund. The final legal bill was under $8 billion, compared with $18 billion estimated by analysts at Merrill Lynch.

Value, in drug-making terms, has many metrics, and it didn't go unnoticed on Wall Street that Frazier's hard-nosed legal maneuvering created more shareholder value for Merck than its vaunted labs as the company sped toward a patent cliff with its biggest seller, Singulair, for asthma. Frazier rose rapidly to head Merck's human health business and then became president. On January 1 he took over as CEO, the first African-American to lead a major drug firm. Merck was the world's second-largest drugmaker measured by sales.

Less than two weeks later, Frazier got word that a safety panel had shut down a thirteen-thousand-patient late-stage study of an experimental blood thinner, vorapaxar. Analysts considered the drug, acquired as part of the Schering takeout, a jewel of the pipeline, with some annual sales estimates reaching $5 billion by 2015. Frazier and Merck's board faced a reckoning. Like Pfizer, Merck had hoped to buy its way out of going over the cliff along with its expired lead product by taking out another major drugmaker, imagining that the other company's R&D would fill the holes in its pipeline. The crash of vorapaxar and Schering's two other main hopes left only boceprevir to justify the acquisition.

Meantime, Pfizer, in its year-end-earnings conference with analysts—one day before Merck's and, coincidentally, Vertex's—touted a one-third cut in R&D spending and a big stock buyback. Its market value had sunk below what it had paid to take out several competitors over the past decade as it tried to build a broad portfolio of new products that never materialized and that Pfizer's new CEO, Ian Read, now said was no longer its goal. Read announced a plan to overhaul the company's research operations to focus on the most profitable programs. In addition, he said the company would increasingly outsource business and disclosed that it would shut its research hub in Sandwich, England, where Pfizer had developed five of its top twenty drugs, including Viagra. Read told the analysts, "At some point, your shareholders and stakeholders demand you have a return on investment in research."

Frazier and Merck were under stiff pressure from Wall Street to follow suit: shrink R&D, sell off divisions, goose the stock price by taking cash and giving it to investors instead of risking it on trying to discover new breakthroughs. Frazier doubled down. Torn between Wall Street's demand for profits and the long-term funding needs of drug innovation, Merck returned to first principles: if not "Drugs for people, not for profit," the idea that value in drug making ultimately results from dramatic new medical advances. Frazier told the analysts that the company planned to spend as much as $8.5 billion on research in 2011—putting it on a par with Microsoft and Pfizer, the country's long-term leaders—and he withdrew Merck's long-term profit forecast. Predictably, investors responded to Pfizer's plan by driving up the stock the most in six months, while Merck shares fell almost 4 percent. "I am not blind to what investors want us to do," Frazier told the conference call. "They want us to invest in prudent ways, in ways that actually drive return on investment and productivity. But as a company, we believe that the only sustainable strategy in the health care environment we're in is real innovation that makes a difference to patients and payers."

Here was the grand contradiction that the pharmaceutical industry—and society—had spawned during the golden age before 9/11. Millions of Americans had come to rely on the steady high dividends of drug stocks in their portfolios, making them complicit in Wall Street's relentless demand for escalating, year-after-year profit; the corresponding proliferation of marketing-driven me-too drugs; and the resulting precipitous decline in productivity. In the Great Recession, dividend-paying stocks were vanishing, while important new drugs were harder and harder to find. The point was: drug discovery, the least certain and longest term of enterprises, had helped prop up two decades of unparalleled prosperity at the cost of its own future. It was to Frazier's considerable credit that he would defy Wall Street and risk Merck's stock performance during his own tenure by hewing to George Merck's original vision.

Which, of course, inspired Boger's vision too, and Emmens's. In *his* remarks to analysts, Emmens made no apologies for Vertex's accelerating burn rate and galloping losses. With its research and development costs up 25 percent in the last year alone, the company had lost $180 million, or 90 cents per share, in the last three months of 2010. Revenue had nearly doubled: from an anemic $34 million to $66 million. Shares of Vertex fell 17 cents to close at $38.80, and the company, with a global workforce nearing two thousand as it added hundreds of jobs in less than a year, had a market capitalization of nearly $8 billion. Wall Street forgave Vertex's ballooning R&D budget, which more and more analysts were coming to regard in their forecasts as less of an enhancement of a company's worth and more of a "value destroyer," based on its rosy calculations about telaprevir and the consensus that Vertex remained a choice takeover target. Goldman Sachs biotech analyst Terrence Flynn initiated a "buy" rating, anticipating that telaprevir—TVR, he called it in his note—would capture 80 percent of sales and would so dominate the market that its revenue curve would extend several more years. "We expect the TVR launch to exceed high expectations, as the Street is underestimating the warehoused treatment-failure population awaiting new treatments and [the] pricing power of TVR," Flynn wrote, adding: "We believe we are more bullish than the Street on the sustainability of the TVR revenue stream beyond 2014, following the entry of next-generation competitors."

❖

Smith likened Vertex's disclosure meetings to "opening the envelope," a helpful analogy evoking a simpler time when life-altering news arrived— telegrams, say, containing the fate of loved ones—under a protective seal that you had to slice open in a rush of anticipation; or else the Oscars, a glossy ritual of suspense ending in crushing disappointment for most, elation for the fortunate few. In fact, what is distributed around the table is a printed booklet of a PowerPoint, a facsimile of a slide deck narrated dispassionately by a clinician as the others flip through it, absorbing the data.

The Vertex swoosh, "twelve of twelve," the Phase II findings showing improvement in CF patients taking VX-770—each had been a key internal milestone, a turning point inducing a heady blend of gratification and terror. In mid-February, heading into the long President's Day weekend, the results of a pivotal Phase III study of 161 patients with cystic fibrosis who were randomly assigned to get a VX-770 pill or a placebo—STRIVE, it was called—outdid them all. With Ken presiding as always, Kauffman presented the findings. The patients in both groups all had at least one copy of the G551D gating mutation.

The study found that those taking the drug had a 17 percent relative improvement in their FEV1—the amount of air they could blow out in one second—compared with those on placebo. The effect lasted the entire forty-eight weeks of the study. Side effects included headache, respiratory infections, nasal congestion, rash, and dizziness, but they were well tolerated. Smith recalls doing a double take when Kauffman presented the safety data. Remarkably, more patients from the placebo group dropped out of the trial than those on the drug.

The effect for patients exceeded anyone's imagining. Before starting on Vertex's drug, many couldn't climb a flight of stairs without erupting into a coughing fit; within days of starting treatment, some reported, they were able to breathe better than they had in years. It was as if a light switch had been flipped on, a radical upsurge in energy supercharged by a jolt of hope that they hadn't imagined was possible and for which they were totally unprepared. Because CFTR is also broken up in the digestive tract, CF patients generally struggle to maintain their weight, but patients on VX-770 put on weight readily and kept it on. Weight gain wasn't a primary end point of the study, but it was a hugely promising signal nonetheless that the pill worked throughout the body and could further transform the lives of patients by improving absorption, lessening the need for supplemental pancreatic enzymes, and making breakfast less of a painful battle of trying to choke down a few more calories.

As bacteria and detritus were swept increasingly from their lungs and airways, the patients on the drug had significantly fewer "pulmonary exacerbations": debilitating setbacks of worsening sputum and cough, shortness of breath, chest pain, appetite loss, weight loss, and lung function decline often requiring long hospital stays. The unambiguous take-home message from the trial was that for the 4 percent of patients most likely to benefit from a potentiator, VX-770 was a dream molecule, something that gave them back their lives by fixing the underlying cause of what was wrong with them. Science fiction made fact.

Far more than the Vertex swoosh, the STRIVE results opened up new horizons and raised daring possibilities about the future both of the company and of medicine, forcing Vertex yet again to lunge ahead and grow swiftly to keep pace. Everyone around the table instantly felt the impact: here was another blockbuster to build out under, the second in less than a year, an unprecedented repeat performance driving toward another launch, most likely within the next twelve months. Organizationally and financially, the company now had to prepare—while still burning record amounts of cash and preoccupied with building out operations—to emerge in 2012 as a full-service, multiproduct, global biopharmaceutical company.

The launch of telaprevir, the singular goal around which the company had organized itself for six years, now morphed into something both larger and smaller. It remained the essential driver of Vertex's near-term future, yet was diminished—if not dethroned—by the onrush of the next critical phase in the company's evolution. Though the impending launch was all most people in the company cared or thought about, what now compelled senior management was not just to cure as many patients as possible and crush Merck but also to bring in enough revenue from telaprevir to propel Vertex high enough and fast enough to reach a kind of escape velocity—putting it on a trajectory toward long-term sustainability. "The first stage of a Saturn V rocket," Emmens began to call it.

Bernstein's Geoff Porges noted the shift. He, like a number of other analysts, believed that 770 would be rapidly approved, and he forecast, despite the small number of people who would take it, sales of $500 million or more a year. Porges knew the company could price the potentiator at more than $250,000 per patient per year, and that governments and health insurers would pay that much because they believed the benefits justified the costs and because the number of patients was minuscule compared with other diseases. (Some recent biologics for patients with rare genetic disorders were priced as high as $400,000 a year.) He also recognized the magnitude of the findings: that Vertex, miles ahead of any competitors, would redouble its commitment to discovering a cocktail that worked for the other 96 percent of patients and to striking hard against other genetic diseases while the whole area of personalized medicine heated up. After writing a note exhorting investors to buy VRTX and setting a price target of $80, he explained:

Hep C we're going from a 35 percent cure rate to a 70 percent cure rate. That's really impressive. Tens of thousands of patients are gonna benefit from that. It's gonna save a lot of grief and cost in the future. That's great. But cystic fibrosis. If you take cystic fibrosis from a universally fatal and life-shortening disease into a chronic illness—you're on a lifetime of medication but you have a normal life expectancy, and that's now feasible—that's *breathtaking*.

I think the body language inside the company has transformed from where it was six to nine months ago—because of CF. They viewed themselves the way the Street still views them today, that is, as a company that's gonna have this short burst of fabulous profitability and maybe has a few other singles and doubles in its back pocket. The company inside now views themselves as having this short burst, and then having just this huge thing beyond that short burst. And the huge thing is this transformation of the most common inherited disease in our society.

Vertex shares rose 15 percent on the STRIVE data. Amid the ebullience in San Diego and Cambridge, there remained a nagging sense of incompletion, especially among those working in CF, who could not feel that they had done their job until the great majority of patients had access to effective medicine. With two correctors, VX-809 and VX-661, advancing in the clinic, hopes at both sites remained high, but nowhere near as high as on Wall Street. "We got more reflective," Olson recalls. "We thought, 'Yeah, this is fine, the four percent. It's almost like we've got the seed corn now. Now we know that plant has germinated, but to get more seeds, the whole thing's gotta grow. We've shown we can grow corn from that seed, but we've got to grow a lot more ears.' "

More than twenty years earlier, when Francis Collins and his collaborators located the gene that causes cystic fibrosis—and then a decade after that, when the human genome was decoded, holding out the promise of a new epoch in medicine—no one could foresee the larger societal dilemma presented by an ultraspecific, ultralucrative superorphan small-molecule drug for a narrow genotype of patients. An estimated six thousand of the seven thousand known human diseases are rare, affecting as many as thirty million people total. Yet with industry giants like Pfizer shedding unprofitable programs and struggling to survive, the number of new medicines filed with the FDA dwindling in a year from thirty to twenty-two, and average drug revenues dropping, Porges's bullishness introduced a new moral calculation. *Forbes*'s biobusiness chronicler Matthew Herper, writing about the STRIVE findings under the headline "A Big and Dangerous Day for Personalized Medicine," asked:

> If every drug is only going to work in a few thousand people, what then? One tough reality is that the new methods of studying genetic variation, which look deeply at the biology of relatively small numbers of people, are at odds with the way we've been thinking about drugs for years, studying common medicines like cholesterol drugs in thousands or tens of thousands of patients . . . What we're set up for is a collision between genetics and large-scale studies, between innovation and cost, and between a vision of medicine that is struggling and one that is trying to create itself. It could spark lots of great new medicines for everyone. It could also be a giant mess.

❖

Keith Johnson *knew* he was not on placebo. He was forty-two years old, an information storage specialist for a company in Manhattan, married and a father of two preschool-age girls. Since people with cystic fibrosis rarely live to middle age, he was an elder, an anomaly. Eighteen months ago, he developed a respiratory flu and his FEV1, which is measured by blowing into a plastic tube, dropped to 39 percent of normal—below the cutoff for severe airway obstruction and the point at which patients qualify for lung transplant lists. When Johnson was diagnosed in college with CF, he'd been told his life expectancy was just a few years. His older sister, Jennifer, also born with the disease, died at age thirty-six after receiving a double lung transplant. Even as he beat back his congestion with intravenous antibiotics and was able to raise his FEV1 into the mid-40s, he knew his breathing would continue to decline by 1 percent to 2 percent a year. He worried about how much more he could take.

What motivated Johnson to recover from his lung infection through late 2009 and early 2010—the fiercest months, as it were, of the health care war in Congress, when Republicans seized on the subject of end-of-life care and conjured a specter of "government death panels"— was the chance to participate in a placebo-controlled study of VX-770 in adults with two copies of the delta-F508 mutation, the largest subgroup of people with the disease. The 120-patient trial was the first to test whether the drug, by itself, might help the much wider majority of sufferers who don't have a gating problem but instead whose CFTR misfolds so that sufficient protein never gets to the cell surface. The biological rationale was to fix whatever scarce active protein the patient had.

Johnson's doctor told him she thought the drug "was a game changer," he recalls, but he had learned not to raise his hopes. A guarded but intellectually voracious Northern California native, he is a skeptic by nature and training—stolid, medium built, with short gelled black hair, intense hazel eyes, and graying sideburns. Life with CF was endlessly grueling: up to two hours a day of breathing treatments alone. Every morning, he sat for an hour at his computer, taking pills and inhaling drugs and saline solution through a nebulizer. Then he tried to do some work while jacketed by a powerful pneumatic vest that rattled his lungs until he coughed up three and a half ounces of mucus. A die-hard existentialist, he exuded a grim, weary resignation about his ordeal.

"This disease does not take a day off," he says. "It's very persistent. So as an *Art of War* type of thing, that's your adversary. You have to match or exceed its effort if you want to win, and that can be mentally and physically very draining. If you're going to live with CF and get what you can out of life, you have to be realistic about things. If you wake up every day disappointed that you didn't run a marathon, I don't think you're in the right place mentally."

Johnson and his family rented a town house near the Hudson River in Westchester County. He played in a pickup weekend soccer game, usually becoming too winded to run during the second half, walking the field, bent, tugging at his shorts, gasping. He loves golf, struggling a couple of dozen times per year around Hudson Hills, the premier local public course. On the day in February 2010 that Johnson started taking two unmarked tablets daily, along with his usual treatment regimen, as part of the randomized sixteen-week study, he "blew" 52 percent. A patient's FEV1 can easily vary from day to day; Johnson's baseline for the trial hovered in the low 50s.

The second week in March, on a warm afternoon, Johnson played Hudson Hills with a friend. The eighteenth hole is a 403-yard par four. Climbing to the green after slicing his drive, when normally he would be "huffing and puffing," he had little trouble breathing. "I felt amazing," he says. "I felt cured. I hadn't allowed myself to think that anything like this was possible. I could not believe that someone had figured this out." A few weeks later, he blew 58 percent. When Johnson returned to the adult unit of the CF center at Beth Israel Medical Center in Manhattan, one of several study sites, at the end of June, he was tested again. His FEV1 measured 60 percent.

"I knew it when I blew it," he recalls. "I could just feel. Within ten days, I would say that my life turned upside down for the positive. Not only did I feel that this was working, but that this opens up a whole new life, one that I wanted but that I hadn't allowed myself to hope for because I just didn't want to go through the disappointment.

"I was seeing everything completely different—at work, the possibilities. It was the most productive time ever for me workwise. I never made as much money. I was winning every deal. The guy I work with said, "Dude, what is up with you? I can't even keep up with you when we walk around the city.' I couldn't wait to wake up in the morning, to live life."

Keith and his wife, Adrienne, had avoided discussing the future; the subject was too frightening. They had backed into family life without sharing any larger ambitions for how they hoped to live. Adrienne had wanted children; Keith, like most men with CF, was sterile. Yet they had found a doctor who reasoned that since his CFTR malfunction was in the vas deferens, the tube that transmits sperm to the ejaculatory ducts, Keith might have what he calls "swimmers," and after eight attempts at extracting cells from his testicles and using in vitro fertilization, he'd become a biological father—twice. Still, the couple had never really looked ahead. "Life was like you were living from paycheck to paycheck, not in a monetary sense but 'How Keith was feeling,' " he explains. "Not reliable. It was not reliable to make plans.

"Adrienne and I completely recalibrated our conversations. What is it that we can accomplish in life now? Moving into a better school district? Getting a second car? We were actually making plans."

After the sixteen-week comparative study was completed, subjects who met prespecified criteria were eligible for an "open-label extension" or rollover; that is, everyone who wanted to continue on the drug was given free multimonth supplies, pending the outcome of the trial. Johnson recalls signing for boxes of VX-770 and greedily stuffing them into his backpack. His company was sold at a premium after a bidding war between Hewlett-Packard and Dell, and he and Adrienne started looking at split-level houses in Irvington, New York, near a school with an excellent reputation. He started showing up both weekend mornings for soccer, exercising so much that he lost weight—a curious anomaly, since his digestion was the best it had been in decades. "My exercise tolerance was so high, I was finding any excuse I could to do any type of cardio."

Johnson blew 62 percent in November. He took his doctors out to celebrate. He and Adrienne bought a house, making a deposit with the proceeds from his stock options, but had no money left for living room furniture—the "Gift of the Magi" room, he called it. His daughters,

Claire and Danielle, called it their indoor soccer field. He began, in a dream, to feel unchained from the rigid sense of responsibility that he could never miss a treatment. "You know what?" he says. "I could miss a TOBI [tobramycin, an inhaled antibiotic for treating pseudomonas infection]; I could miss an albuterol. I could go out at night and come home and not do my meds. And it was not being irresponsible. It was just this freedom. I . . . I just couldn't believe it. It was just so powerful."

"There's no way," Johnson told himself, "I ever *not* want to have access to this med."

❖

The second mock was held farther off-site, not in a distant city but at the Hyatt at Boston's Logan International Airport, overlooking an isolated runway where the previous night Air Force One remained parked, bathed in klieg lights, surrounded by sharpshooters, while a hundred-vehicle motorcade swept President Obama in and out of downtown for a school appearance and a fund-raiser. In addition to making it more convenient for the company's consultants, the move lowered the number of senior executives in the guarded conference room, though not the tension.

The speaker slate was reshuffled from a month earlier. During that first grilling, Vertex had lost a keynote speaker when ProEd's statistician attacked his credibility. The company was relying on a highly regarded expert in the treatment of hepatitis C and liver cancer to deliver an overall assessment at the end of its presentation, but the consultant, pelted with allegations that he'd compromised his professional integrity by speaking on the company's behalf, abruptly gathered his papers, stood, and said, "I think I have a plane to catch." Every company needs independent experts to support its claims, but now the team would rely solely on the testimony of Dr. Ira Jacobson, the principal investigator on the ADVANCE trial, who agreed to address the burden of the disease and the need for new medicines, but not telaprevir. Chief of the Division of Gastroenterology and Hepatology and a distinguished professor of medicine at New York–Presbyterian Hospital Columbia University Medical Center and Weill Cornell Medical Center, Jacobson had replaced John McHutchison in the ambiguous role of lead clinical investigator/paid ally when McHutchison, lured finally into industry, went to work for

Gilead in 2010. Lest there be any question of Jacobson's impartiality, he enjoyed a similar relationship with Merck.

Kauffman delivered the introduction. Emmens had noted that when speaking with Wall Street, Kauffman was almost preternaturally calm and unruffled—an informed senior figure who never got defensive or strayed from his data, conveying honesty and patience no matter how aggressive or trivial the challenge. Kauffman briefly outlined the contours of Vertex's presentation. Jacobson, he said, would offer a brief background of the disease and the current treatment landscape. Next, he—Kauffman—would return to provide an overview of telaprevir's development: how it was studied, first in the lab and then in nearly four thousand patients. Two other Vertex physicians, Dr. Shelley George and Dr. Priya Singhal, would then discuss the drug's efficacy and safety, respectively, before Kauffman provided the risk-benefit assessment himself at the end.

Sponsor presentations invite obvious skepticism, and Kauffman was eager to show not only that Vertex had produced a breakthrough medicine but also that it took very seriously its commitment to transparency—up to and beyond the requirements set by regulators. Drugmakers are expected to study drug-drug interactions, or DDIs, the myriad ways in which medicines combine to produce additional effects in the body. Patients with hepatitis C, especially those whose livers are already cirrhotic or who have cancer or have received transplants, take numerous other drugs; those coinfected with HIV take even more. Many develop ascites—their abdomens fill with fluid—requiring diuretics. The disease also correlates with an increased risk of adult-onset diabetes. As specific as a drug might be, it is no match ultimately for the devilish intricacies of whole-animal biochemistry, and many new drugs come to grief because they can't be administered with the other medicines that a patient is already being prescribed.

Kauffman told the panel that Vertex had conducted a comprehensive clinical pharmacology program characterizing the potential for DDIs with telaprevir. He showed a slide—one of sixteen hundred prepared for the mock—listing DDIs with more than fifty other prescription medicines. It had studied cross-reactions with opioids (narcotic painkillers) and immunosuppressants. It found that telaprevir interacted with the estrogen component of oral contraceptives, possibly reducing their effectiveness. Though the company's thoroughness was laudable, Weet and others worried about the risk of being *too* diligent. "It was positive that we had done a lot of DDIs," Weet recalls, "but you could also say it was negative that we had done a lot of them. We saw a lot of interactions, and it caused a lot of concern."

It was up to Singhal to present Vertex's case regarding telaprevir's greatest safety risk and clinical vulnerability: rash. Whether the discussion of rash at the AdComm remained scientific and grounded or, as Emmens put it, became "emotional," would reflect on Singhal's presentation, and she faced potentially the toughest questioning. A recent immigrant from India, poised and precise, she had arrived with her husband, a fellow doctor, to study at Harvard, and then joined Vertex as senior director of worldwide patient safety. Now she delivered a case history, with graphic slides, of the company's response to the rash issue, with special emphasis on the most severe skin reactions.

After the first cases were reported, Vertex had appointed an outside adjudication panel to investigate. In parallel, the company launched several investigations to try to understand what was causing the rash and how it could be managed. A special search category was established to ensure that all severe adverse events were thoroughly characterized. Singhal noted that the external panel in three cases suspected Stevens-Johnson syndrome (SJS), a life-threatening and horrifying skin condition in which cell death causes the top layer of skin to die and shed. Only one case was definitive, although it occurred eleven weeks after the last dose of telaprevir, while the patient was still on peg-riba as well as other drugs, and the panel considered it unrelated to telaprevir. The team, led by a Harvard dermatologist who would be available for questions at the AdComm, also investigated eleven suspected cases of another severe skin reaction given the acronym DRESS syndrome. Again, only one case was considered definite, its etiology unclear.

Science, Murcko likes to say, is an interactive process. The second mock played a lot better than the first, but there was endless room for improvement. The slide deck for the AdComm, already bulging, grew as the session proceeded. Each slide was prepared to answer any question that could possibly be asked, but as new people were brought in to hear the presentation, new questions arose, or there was a new spin on an old issue, and people chased off in search of more data to put together new slides—slides that Kauffman and the team would have to be able to summon instantaneously.

The problem was the wealth of information itself. Vertex had done so many experiments over the eighteen years of the program that the scale and scope of its inquiries went beyond the ability of its keynote speakers to stay on top of them all. If they surely had the answers the panelists would be looking for, locating those answers was something else. With a bullpen of up to two dozen other scientists seated in rows along the wall to fill in the gaps, even Kauffman's, George's, and Singhal's careful curating and narrative precision were becoming subsumed in the flood tide of data. It was during the simulated question-and-answer session after Kauffman's summary, according to Weet, that the problem reached a familiar stress point.

> There were still people brought up to the podium to give detailed answers to questions who Peter didn't trust to speak. He thought maybe they were gonna give too much detail. Maybe they fumbled over an answer, and he didn't like that. I was sitting next to him at the time, and he said, "I don't want that guy on the podium." And I said, "We're gonna get Bob to answer as many of these questions as possible." He said, "Okay. Let's make sure." I said, "I'm on it." Then another question comes up. *"I thought you said that guy wasn't going to be on the podium!"* I said, "Peter, there are some questions where we have to have the technical experts. Tox. Clinical pharmacology. Bob can't answer everything." He was getting very, very nervous. And, of course, people were nervous because Peter was in the audience.

❖

Approaching the annual meeting of the European Association for the Study of the Liver (EASL) in Berlin during the last week in March, Emmens was "jazzed." He had a breakthrough drug, abundantly tested and studied. He had the organization he wanted, and every head was down, focused on the launch. There were no premature high fives, and Emmens could feel the camaraderie surging with the size of the stakes. "When your company's on the line, and it's a drug that just about everyone in the company worked on, you become a team," he said. "A lot of the petty stuff goes away. You don't have to run around with a vision. You know what our vision is now? '*Holy shit, we'd better not screw this up!*' Right now the vision is to make the company sustainable, and it's really simple. In the next couple of years, we have to make more money than we spend. So that's my financial plan. We have to make enough money to sustain the level of research that we think we can sustain."

It was just this singular focus that the board had felt was lacking when it ousted Boger. Yet Emmens's main concern was that Vertex could become *too* concentrated on—dependent on—hepatitis C. "That's the only thing that worries me," he added. "This is really hard stuff now. You need another telaprevir now, or the sustainability goes away. HCV is a booster rocket, designed to catapult the company into the upper atmosphere, and then dump into the ocean."

Toggling back and forth between the reality of Vertex's situation now and its vision of itself a decade from now, Emmens transitioned smoothly. One moment he was the chief internal cheerleader for the drug, rallying the troops at a lunchtime presentation—webcast companywide—where he showed a stop-action movie of white-coated workers assembling a Boeing 757 from scratch at hyperspeed and then wheeling it onto a runway—the Vertex name and logo Photoshopped in purple on the fuselage. The next moment he was publicly extolling Boger and his vision, rededicating Fort Washington as the Joshua Boger Innovation Center and Fort Washington II as Joshua Boger II, or JB-II. A moment after that, he was waving the banner on Wall Street for Vertex's high-risk, high-reward, science- and patient-based, anticonsolidation, David and Goliath business plan and company narrative. He explained:

Every time that I say that this place is about research, it's not to placate the employees. The only future in this business is breakthrough products for sick patients who aren't getting adequate therapy.

That's the *only* thing that's going to work long term. So from a strategic standpoint, to me, one of the things I love is that that's a very consistent message.

I came on the board here because I believed that. When I was at Shire, I felt that search-and-develop was great, having no research was great, but I thought it was something that was not sustainable. You're going to have to go earlier and earlier to capture your products to a point where you're gonna pay the same prices as if you had your own research. The only thing you have is that you can be a wide net. So when I came here, my messages were: we need a wide net—lots of different projects, in different areas, in different sciences. But, also, we have to do both things well. We are a science company, but that does not mean that we will not go outside for more science and more projects.

I think one of the problems with pharmaceutical companies is that they try to narrow that net—we're going to be in two, three, four areas. When you mess with biology, you don't really know what you have, and you end up in areas that you didn't start in. So to say that is crazy. It's an assembly-line mentality. Nothing squashes creativity more than telling creative people what they're gonna create. We didn't go after epilepsy; it just happened. So here's a drug now in later-stage research that was tried for psoriasis. It did not work well then. But here we are.

Emmens knew and admired Merck's Frazier—"a really smart guy"— but doubted that, even with the right CEO and a return to its culture and principles, Merck would survive the giantism afflicting pharma. In hepatitis C, he anticipated that Merck's strategy would be to "make the two drugs seem the same." Since Vertex would benefit from the extent to which doctors and insurers differentiated between telaprevir and boceprevir, Merck had an interest in blurring distinctions. Frazier and the company also needed to justify the Schering takeout, so boceprevir was a priority asset. Emmens respected Merck's commercial organization but didn't fear it. If he were in Frazier's shoes, he knew he would be concerned about Vertex, but he'd be worried materially more about his anemic pipeline, the paralyzing scale of his operations, the bureaucratic politics permeating Merck's labs, the cultural and reputational overhang of Vioxx, and Wall Street. Emmens:

> My original premise for joining the [Vertex] board was that I believed the future of medicine was in medicine . . . *Duh* . . . and that drug companies are called drug companies because you need drugs . . . *Duh* . . . and the only reason you ever have a product is its perceived value. You can't jam it with a sales force. Perceived value has to be, basically, inherent in the product. We're in a market that is now squeezing—monetarily—down on an industry where me-too is not going to make it; incremental is not gonna make it.
>
> I mean, I'm amazed we don't use generics first in almost every disease. I have high blood pressure. All the drugs I take—four of them—are generic. That's fine. They work. So we shouldn't be paying six dollars a pill. What we should be paying for is companies that create a mechanism to continue to bring very, very good breakthrough drugs. If you can't do that, you shouldn't be in the business.
>
> The insane thing you're seeing in the big companies is a lack of sustainability, a leveling; and they're saying, "Okay, we're gonna do a stock buyback, we're gonna pay a dividend, and we're gonna cut our research." The market continues to want to see progression of earnings and a progression of dividends, and if you don't have those two things, the stock goes down. Well, that's a bad spiral. If you don't show me productivity in research that's gonna make the top line go up over time, if you're burning the furniture to heat the house, then pretty soon the house is gonna get cold and you're not gonna have any more furniture. These massive research departments have not been able to sustain it.

❖

The pre-EASL buzz on Wall Street wasn't about Vertex versus Merck but about Pharmasset's nuc-nuc combination, touted by the company as the most promising all-oral therapy across all genotypes of hepatitis C. Biotech fund managers, having tasted this very magic once before—with Vertex—pounced on the story.

Grabbing a chapter from Vertex's playbook, it seemed, Pharmasset submitted a research abstract to EASL describing interim results of a small Phase II study in which fifteen of sixteen, or 94 percent, of patients receiving its cocktail had undetectable levels of virus after fourteen days. On March 7 EASL released the abstract online. Though the data were embargoed to the press until the Berlin conference, the abstract, grossly preliminary, sufficed to trigger a land rush. Shares of Pharmasset stock, VRUS, rose 24 percent one day and 8 percent the next, to $66.92.

Pharmasset's gain was Vertex's loss, as analysts recalculated the opportunities in hepatitis C. A few downgraded VRTX shares, which fell as much as 5 percent, to $47.07, on the announcement. Vertex's news at EASL—positive results of a "quad" therapy, combining telaprevir and VX-222 with peg-riba—suddenly looked less than inspired: a limp second-day story off yesterday's headlines; a placeholder, not an advance, in the race to the promised land of an all-oral therapy. "A play, as Cumbo would put it, "but not *the* play."

Inside Vertex, Pharmasset's coup registered. Ann Kwong, promoted to lead Vertex's hepatitis C franchise, worried that Vertex still had no late-stage nuc in its arsenal. So did Smith. But the overall response was notably skeptical: not blasé, but been-there, done-that. "I almost feel bad for them," Michael Partridge said. "They have no idea what they're facing." The more that its drugs were studied, and the more the reality set in of how grueling and risky and time-consuming a series of buildouts Pharmasset now needed to navigate to get its drugs to patients, the less impressive its nucs would no doubt look. Emmens, especially, professed no worries about who would ultimately win the hepatitis C market. He was certain it wouldn't be one company, like Gilead in AIDS.

Vertex sent an eighty-person contingent to EASL, including the entire ET except Emmens and Ken, to show the flag. "This place eats money," Emmens noted. "It's too many people, but we have to hear everything, be everywhere. The company is in the balance." Emmens normally shows little strain, but he was suffering his third attack of diverticulitis in twelve months. Not too many years earlier, he'd finished marathons ahead of competitors half his age, but now, at sixty, carrying his tray in the cafeteria on the fourth floor of JB-II, he shuffled like a post-op patient. More and more, he was feeling his age. He planned to take some time off over the weekend.

On *NBC Nightly News* on March 30, the opening day of EASL, anchorman Brian Williams read a one-paragraph health item. "News tonight about hepatitis C, the liver infection that affects more than three million Americans," he said. "A new drug, expected to be available in the next few months, doubles the cure rate when given along with two other drugs that are already prescribed. In people with the most common strain of hepatitis C, the cure rate goes from about thirty-five percent to about seventy percent when this new drug, called Victrelis, is added to the treatment."

Victrelis was Merck's trade name for boceprevir. No drugmaker was mentioned in the report.

# CHAPTER 11

## APRIL 27–28, 2011

Kauffman would have preferred to go ahead of Merck, to set the tone for what bio-business writer Adam Feuerstein called "Hep-C-apalooza." The eighteen-member advisory panel, meeting over consecutive days in the Great Room of Building 31 at the FDA's White Oak campus, would recommend, by vote, whether to approve telaprevir and boceprevir. Since licensure of both products was all but assured, the larger struggle would come down to the relative merits of the package insert, or prescribing information (PI)—the label. More than a point of pride, a label dictates how and to whom a prescription drug can be sold, and Vertex aimed to have its field force launch its medicine with a superior PI. Kauffman believed that going first provided an edge.

Boger still argued that the FDA's white paper on EPO alone meant boceprevir should be rejected out of hand. But the team, without him, had moved on to the view that it faced a superheated two-treatment market. Vertex's hearing was scheduled for Thursday. Its forty-plus-person contingent flew to Washington on Monday, checking into the Doubletree by Hilton Bethesda, a forward base to rehearse, rehash, watch Merck's presentation via webcast, and wait. In Cambridge, Emmens, Smith, Partridge, Ken, and their teams readied for the aftermath.

Mueller was, as always, central to the overall effort, but the hearing had implications for every part of the company. Megan Pace brought a group to handle communications. Wysenski and the entire senior commercial team except Cumbo bird-dogged the on-site preparations, as did representatives of Tibotec. Amit Sachdev, a former FDA official and now commercial lead for Canada, commuted from home to help navigate issues of history, personality, and political tone. It was a rare week when he didn't have to be in Cambridge and Ottawa.

On the science and clinical side were two dozen researchers and doctors who had lived and breathed telaprevir—a few, like Kwong, for nearly fifteen years—and whose professional mien did little to conceal their fervor. Dr. Camilla Graham, vice president for global medical affairs, formerly had a busy hepatitis C practice and a position at Harvard Medical School. When she first heard about VX-950, she'd cold-called Vertex, saying: "You have to hire me." Since at least Sinclair Lewis's 1925 novel *Arrowsmith*, in which a doctor who goes to work for a drugmaker is denounced as "gone bad," physicians who work in industry have withstood doubts about their split loyalties between business and medicine. Graham, who still saw clinic patients at a Harvard teaching hospital, saw no contradiction, a positive deviant in both settings.

Weet, as liaison to the FDA and show-runner, so to speak, was attuned to the forces at work and the group's chemistry, and he worried about overpreparation. "I like to see it buttoned down more," he says. "I like to see, in athletic terms, a light practice. You don't have to run five-mile loops the day before the game." But during rehearsals Tuesday, the presentation was less than crisp, and when he talked with ProEd's people, they pressed for another full-dress session. "The day of the Merck presentation everybody wanted to kick back and watch the presentation," he recalled. "We were sitting over lunch, during a break, and ProEd's leader said, 'At five o'clock, I want to break out and have another rehearsal.' Darryl Patrick [head of early drug development] pounded his fist on the table and said, '*You're killing these people! Enough already!*' If you're talking about the dynamics in the room, it was just about ready to explode."

By any measure Merck was enjoying a remarkably friendly day before the committee, stoking Vertex's frustrations. Before the markets had opened, Merck announced it would buy back up to $5 billion in stock, following the trend of using cash to "return value to shareholders." Its keynote speakers represented what Wysenski called "the old Schering crowd," led by Janice Albrecht, formerly vice president of hepatology clinical research for Schering-Plough. Not the Merck of lore, the team nonetheless had a drug with a 63 percent to 67 percent SVR rate and a reserve of goodwill with regulators; what it couldn't prove with data it appeared to obscure and evade.

In its briefing document to the panel, the FDA staff had flagged several key concerns. The examiners criticized Merck's practice of adding EPO during its clinical trials, clouding the dropout rate; its attempt to redefine nulls—those patients who failed previous treatments—as people who failed only to respond to a four-week lead-in with the standard of care; its proposal to allow longer treatment for patients who still have detectable HCV RNA at eight weeks but who go undetectable at twenty-four weeks, a complicated new type of "response-guided therapy" with only scant supporting data that it would result in the patients' being cured; and the paucity in its studies of blacks, who don't do as well on peg-riba. As Albrecht and her speakers argued each point, the panelists pushed back, but seldom with anywhere near the degree of outrage the Vertex group thought was warranted.

Merck's failure to submit important data elicited the morning's sharpest rebuke. Lynda Marie Dee, a Baltimore lawyer advocating for patients, chided the panel and the FDA for tolerating Merck-Schering's failure to study how boceprevir affected sick patients already taking other meds, particularly antidepressants, which along with painkillers provided the only way many of them could tolerate peg-riba. "There are no drug interaction studies," she said. "I'm shocked we don't have studies on depression. I know there's a race with all these drugs, but I think it's irresponsible that these studies haven't been done before now."

Despite its qualms, the AdComm voted 18–0 to recommend approval. "To go to sixty percent to seventy percent [SVR] really seems like a dream come true," said Dr. Lawrence Friedman, chairman of the department of medicine of Newton-Wellesley Hospital, outside Boston. "I think this is a major advance, so I'm very enthusiastic about this drug." Friedman jabbed lightly at the complexity of Merck's regimen—the lead-in, the repeated testing of viral RNA, the shifting end points—saying, "You need to be somewhat of a Talmudic scholar to prescribe it." Patrick Clay, a pharmacist and director of clinical research at the

University of Missouri–Kansas City, vigorously defended Merck's lapses, declaring he wasn't worried about its shortage of data. "Given Merck's history, I have confidence in them," he said after announcing his vote.

At the Doubletree, the Vertex team stewed. When the AdComm finished its work at about five, a core group broke out for a light rehearsal, then went out for dinner. Weet recalled:

> I think that the overall sense was that Merck just got away with murder. The FDA briefing document to the Advisory Committee had all kinds of baggage attached to it. The anemia: they'd confounded their studies by treating patients with an unapproved drug for the anemia so that you couldn't tell what boceprevir's impact was on the anemia entirely, because they were treating these people with EPO. You can't be recommending the use of an off-label drug. We didn't know how the panelists were gonna handle that.
>
> The drug interactions piece: They had done nothing. So they got this finger wagging: "Shame on you, Merck, we expected more of you." The whole thing around response-guided therapy. The whole thing around nulls—they didn't have nulls. They created their own definition of nulls out of this pretreatment; if you didn't respond, they said, "Well, you would have been a null." Kauffman's yelling, "*That's unethical!*" because we had done a study with people who had actually failed forty-eight weeks of standard of care. Bob was just tearing his hair out, but they got away with it.
>
> So there was a certain view in the room that we were witnessing a crime. The hematology expert, who was there to really interrogate them on the anemia, didn't say a word, the entire time. They got off scot-free on the anemia. They got off scot-free on the nulls. They got off scot-free on RGT. They got off scot-free on their DDIs except for this finger-wagging. But, they're Merck. Some ways we were upset, but other ways we were comforted: "Hey, maybe this won't be so bad."

Emmens, after spending the day roaming into and out of the Charles Sanders conference room in JB-II, where a shifting crowd of about a hundred employees munched cold pizza and watched the webcast, was more troubled than he let on. An 18–0 vote meant Vertex could not allow itself to do worse. AdComm votes were nonbinding, but a unanimous recommendation ratcheted up the pressure on Vertex to match Merck's outcome, lest Merck claim to have the better drug.

❖

As a strong storm system rumbled Thursday morning across Montgomery County toward Silver Spring, the National Weather Service issued a tornado warning. The sky around the modernist, university-scaled White Oak campus was pewter. Inside the Great Room, Kauffman, dressed in a dark grey suit, blue Oxford shirt, and burgundy tie, rocked softly on his heels on the podium, awaiting the next question. As anticipated, rash was the hot topic, and Singhal's presentation had gone a long way toward reassuring the panelists that telaprevir-related rash was moderate in most cases and that doctors could manage it by following the company's protocols.

Panelist Patrick Clay sat at the back of the large three-sided conference table. Longish brown hair parted high, he fidgeted with his microphone. He'd grown up in Southern Louisiana Cajun country and worked as an electrician's helper on offshore oil rigs before getting serious about school and earning a doctorate in clinical pharmacy. He worked eighty-hour weeks and had an expansive résumé, including numerous publications based on his work treating low-income patients with AIDS. He thought the most challenging part of his job was predicting drug interactions and—his confidence in Merck aside—was a persistent advocate for proper DDI studies conducted with real-world drugs and antiretrovirals. Clay leaned in.

CLAY: I want to commend Vertex on the array of drug-drug interaction studies they've already done as well as the ones they've planned. And in your ongoing and planned future studies, I saw on ClinicalTrials.gov that you also are doing one in people who failed telaprevir and continue in that study. So I want to commend you that that's already up there and listed. But I want to focus in on some of your safety issues.

In the material provided by the Food and Drug Administration, they talk about your metabolite PZA, and they mention PZA is associated or a metabolite of niacin. But I'm more familiar with PZA as a metabolite of pyrazinamide, used in the treatment of tuberculosis.

You're describing a side effect profile that we expect to see from pyrazinamide. You've got increased uric acid, including incidences of gout. You have fever. You have anemia. You have thrombocytopenia. You have a rash. I'm not that familiar with the presentation of the rash in tuberculosis treatment.

But I guess this gets to the fact that you're giving single therapy to people, and I was curious if in your clinical trials, did you perform a PPD at baseline to see if someone was PPD-positive? And if so, had they already taken medicines for the treatment and prevention of tuberculosis?

It was not a question that Kauffman had anticipated, given its arcane nature. Other members of the Vertex team, scattered among the audience of about three hundred, Googled Clay on their smart phones, tapping to see where he might be coming from or what he might be aiming at with his line of inquiry.

KAUFFMAN: We did not provide that information. We did not necessarily test patients. It was really up to the clinicians who enrolled patients to decide if they were eligible for the trial. There was no specific inclusion or exclusion criteria related to prior tuberculosis status.

I'll point out, though, that the concentrations of pyrazinoic acid—although it is a metabolite of telaprevir—are much lower than the concentrations that are present when pyrazinamide is used as a therapeutic agent.

CLAY: I don't doubt that. But the side effect profile really looks like you're giving full-dose pyrazinamide. And I'm just curious how much information is provided to the investigators so that they could say, well, we need to test for TB in people before we give them this medicine. I guess maybe it gets to—I didn't have a great deal of information about how much of your drug ends up being pyrazinoic acid, and so maybe that would be helpful to me to know that. I don't really need to know how much of your drug ends up being pyrazinamide or pyrazinoic acid, because it looks like there's a lot of pyrazinoic acid on board here.

Russ Fleischer, seated farther down one wing of the table amid a lineup of senior FDA officials, delivered the agency's case for approval. Fleischer had greeted Weet warmly and publicly before the proceedings. He detailed Vertex's investigations into the most severe cases of rash, discussing at length rare adverse occurrences called SCAR events, and he remarked that a small number of patients on telaprevir—about 1 percent—had increased bilirubin, a yellowish bile pigment indicating a breakdown in a component of oxygen-laden hemoglobin, indicating possible anemia. After the chairwoman invited clarification questions, numerous hands went up. Again Clay importuned.

CLAY: I guess the first is just a comment. I have a hard time hearing presentations about dermatology, and your making an abbreviation of a multitude of symptoms and calling it SCAR, because in my mind, a scar is a certain dermatological condition. So that's neither here nor there. I have a question that relates to how well the study was blinded. These were blinded studies done?

FLEISCHER: 108 and C216 were blinded; 111 was open label.

CLAY: Okay. So my question to that is, you're conducting a blinded study in which you had a significant difference in rash from one group to the next. So I guess that's okay if nobody knew that there was the likelihood of a rash occurring.

I was just curious, when the sponsor submitted their material, did they discuss in there how it was managed at the investigator level?

FLEISCHER: The rash?

CLAY: No. The determination as to whether the rash was possibly or probably related to study medication.

FLEISCHER: I think that was in the rash management program. It was in the protocol describing how rash should be assessed and managed, that was given to every site, yes. But Bob can—

CLAY: No, no. I'm not asking about management. I'm asking about the determination in the reporting to the sponsor as to whether or not the investigator felt the rash in and of itself was related to the study drug.

Vertex team members were puzzled and confused by what Clay might be implying. Was he saying they should disqualify patients who developed telaprevir-mediated rash from participating in the trials because the rash presumably gave away who was on the drug, implicitly unblinding the trial? Or were investigators not supposed to talk about it? His words were inscrutable. But his tone was prosecutorial, raising flags. Kauffman, up on the balls of his feet, stepped in, joining with Fleischer to dispel Clay's concern, whatever it was.

KAUFFMAN: Yes. During the blinded phase, of course, no one knew what the treatment assignments were. The investigators used their judgment to decide whether it was related or not to the treatment that was occurring.

CLAY: But rash was included in the investigational drug brochure provided to the investigators when they were considering taking on this study?

KAUFFMAN: Yes. And by Phase III, there was general acknowledgment that rash was associated with telaprevir, and it was certainly in the investigators' brochure.

CLAY: Okay. That's fine . . . My next question actually relates to a different way—maybe it also relates to blinding as well. We talked about an increase in bilirubin within the first two weeks in a fair number of patients. When that sample is drawn and separated, as you need to separate out your samples in order to do viral loads, you're going to see a nice, pretty, yellow color. I'm wondering, did your placebo color your plasma yellow?

KAUFFMAN: I'm not aware whether it did or not.

CLAY: Because you're going to spin it down in a serum separator
  tube to send it off, and you're either going to have a yellow
  tube or not. If you did not blind—or unblind and mandate unblinding at the site level for your lab processing personnel—I'm
  questioning the validity of your blindness.

KAUFFMAN: I mean, I think in most cases the bilirubin levels were
  not markedly elevated. Therefore, it's highly unlikely the plasma
  would actually be icteric [yellowish, jaundiced].

FLEISCHER: Only about four percent were above grade three, so
  the majority of those were grade one and grade two. And so—

CLAY: Yes. Icteric is a clinical presentation, as I understand it. Now,
  my disclaimer, I'm just a pharmacist. But I have processed ataza-
  navir drug before, or atazanavir-containing plasma before, and
  the patient didn't have to be clinically icteric in order for me to
  see the yellowing of the serum.

By now it had become widely known among the Vertex group in
the audience, and in the packed conference room at JB-II, that Clay had
received research funding from Merck. That in itself was common for
panel members and Clay received support from many companies. Still,
Wysenski, Weet, Pace, and others were anxious that his attack, though
relatively inconsequential, could invite other panelists and the public
to question whether Vertex had compromised its findings by fudging
the impartiality of its trials. Emmens knew the relationship between
the FDA and an advisory panel was unpredictable. Safety now trumped
all. A few years ago, just one case of Stevens-Johnson syndrome sank a
drug, even though dozens of drugs on the market cause SJS. Recently
an advisory committee voted 9–0 for approval, and the FDA turned
it down. That had never happened before. There were many ways this
could go . . .

*It can blow at any seam.*

While the rest of the Vertex crowd muttered about Clay, Kauffman
rose to his full height. He rocked forward slightly, balancing like a crow
on a wire.

KAUFFMAN: Bilirubin levels were also elevated in the placebo
subjects, as you saw. It's not very likely that someone, without
knowing what the treatment assignment was, would really be
able to identify a treatment assignment based on the color of
the plasma because elevations occurred in the placebo subjects
as well as in the telaprevir subjects.

CLAY: Fair enough, forty percent versus twenty percent.

KAUFFMAN: But those are both high, so in any individual case, it
would be hard, I think, for someone to ascertain that. And I
think, also, just to go back to your question about rash, as you
saw, the rates of rash were quite high in the control arm as well as
in the telaprevir arm. True, they were higher. But overall across
the study, rash was occurring quite frequently.

Therefore, again, because of that, we think it's less likely that
a patient could be unblinded as to their treatment assignment
just by the occurrence of a mild or moderate rash because thirty
percent of the control patients had mild or moderate rash. So we
felt that, as best we could, that the blind would be maintained.

A cheer went up in the Charles Sanders conference room. Kauffman
had rebuffed Clay with science, patience, and logic, while others at the
company were reduced to rolling their eyes or cursing under their breath.
"And on the seventh day," someone blurted, "there was Bob."

❖

Mueller thought the rash discussion reflected the fact that there were no
big issues, and he was undismayed by Clay. "I don't jump out the window
because someone has an opinion," he said. During the public comment
period after lunch, the same patient advocates who Wednesday urged
the panel to approve boceprevir lined up to support telaprevir. Near the
end, a billing supervisor for a Pennsylvania health care provider, Kelly
Ann Mann-Hester, told the committee that she got hepatitis C from
blood transfusions after childbirth and that she was cured by telaprevir
in a clinical trial after having failed six separate yearlong treatments with
standard drugs. She was diagnosed in 1993.

"A well-meaning physician at the time told me I would not see my son graduate from high school," Mann-Hester said. "He had just started kindergarten. I walked out of that doctor's office that day and told my family that wasn't an option, obviously . . . I've done interferon, interferon with ribavirin, pegylated interferon with ribavirin at half strength, pegylated interferon with ribavirin at full strength, and now with the telaprevir. Telaprevir gave me the clear."

For seventeen years, Mann-Hester had expected to die young. She didn't plan for retirement. Now she needed to. Standing at a microphone in the center of the room, she went on: "I did get the telaprevir rash. I had it on my hands, my legs, and my feet. But it was of very little consequence to me, and would it have made a difference if you were telling me it would save my life? Absolutely not. I would do everything in my power to save my life.

"I was in a place where I was living to die. In my mind, I thought this was going to be the thing that would take me. I had already accomplished my goal, which was to see my son not only graduate from high school but from college. And so I had no hope left, that I thought I was going to live with this disease to die. Now I'm living *until* I die, which is a whole new concept for me, because now I have many windows and many avenues available to me that I did not have before. So I wholeheartedly am asking you to please approve this drug for the general public so that other people can tell my story at some point in time."

As the panel finished its questioning throughout the afternoon, the mood lifted, markedly. Lawrence Friedman, a gastroenterologist who had written more than one hundred textbook chapters and who teaches at both Harvard and Tufts medical schools, saw telaprevir in historic terms, and he went on record to lavish high praise on Vertex even before the voting, which was conducted by secret ballot, with panelists afforded time afterward to comment. As part of its presentation, the FDA provided a reinterpretation of Vertex's SVR rate, raising it from 75 percent to 79 percent. Friedman hailed the revised figure. "Considering where we started with non-A, non-B hepatitis, I think that it's a stunning achievement that we will be able to cure nearly eighty percent of naïve patients and probably the same number of relapsers, two-thirds of whom will only have to take treatment for twenty-four weeks," Friedman said. "We've almost completed a transformation of genotype one into genotypes two and three, which I think is just a remarkable success story.

"The other important aspect of this drug is that the protocol for using it is relatively simple, and it's familiar because what we're doing is basically using the same milestones we've used for peginterferon and ribavirin and grafting the new protocol onto that. And the drug that is causing us concern in terms of its side effects is only used for twelve weeks. And there's some wiggle room because if you have to stop it a little early, you might not sacrifice success.

"So I think there are so many positive aspects of this drug, and for those of us who have been in the field, this is a very exciting moment."

John Alam's determination to shorten treatment, sparked nearly a decade earlier by his dying father's illness, as much as any one factor drove Vertex's clinical strategy. Friedman's point-by-point endorsement many years after Alam himself had disappeared into the mist of what Ken Boger called "the ghosts of Vertex executives past" testified to his efforts and his urgency. A chorus of accolades built swiftly as the first seven panelists explained their "yes" recommendations. One by one, they thanked Vertex for the thoughtful elegance of its presentation and praised telaprevir—a "tremendous advance," as Dr. Doris Strader, a Vermont gastroenterologist, called it. Patrick Clay, sitting impassively, was the lone question mark.

Clay explained that he too voted yes, but he downplayed his endorsement, reminding the panel of what he saw as the larger imperative: a better cocktail, soon, while other lives depended on it. "The benefits far outweigh the risk in this, and it is yet another step," he said. "And that's all it is. There's still a long way to go. This is a marathon, not a sprint."

Nearly all the other committee members effused. Lynda Marie Dee, the patient advocate who'd chided Merck for its purposeful obscurity, raved about Vertex's careful attention to clarity, transparency, and clinical utility. As Kauffman, Pace, and the company's image-making team had anticipated, a drug is one thing, a drugmaker's actions another. Dee said: "You know, there's a discussion often among activists about whether drug companies should ever get A's; we do a report card in one of my groups . . . I won't say that this is an A, but it's very close to it. I'm very grateful for such clear data, for such concise rules about how to do this, such manageable toxicities. It was a great, really, really excellent application."

At about four o'clock in the afternoon, an enthusiastic 18–0 vote granted Vertex all it needed and more: a final, unimpeachable seal of approval; the guarantee of a strong label at launch; official parity with Merck; a sense that the last line of the blockade was down and that the ships, at last, after more than twenty years and a total investment of nearly $3.6 billion, could start steaming through. Goldman Sachs immediately sent a note to investors, raising its estimate of Vertex's chances of approval to 100 percent.

For Vertex, the AdComm was a coronation—its most visible benediction yet, in the arena that mattered most. As for the world beyond, Clay's point, though voiced begrudgingly, was real. Now that there would soon be available a type of drug—a protease inhibitor—that in combination with the standard medicines could double the cure rate and halve the treatment time for patients with hepatitis C, *what then?* Dr. Debra Birnkrant, FDA's director of antiviral products, posed the question, asking panelists to consider what impact the advent of a dramatically improved standard of care would have on the scores of molecules, singly and in myriad combinations, already in human testing. Having to face the rest of the industry, she seemed to want both guidance and cover. Birnkrant:

"Thinking about phrases we've heard based on the data we've seen over the last two days—that this a game changer, a paradigm shift, and a new era—can you comment on the impact of these drugs on current and future clinical trials with regard to standard of care or control arm? . . . The future is here. It's not really that far away, and we have to make some tough decisions. And that's why I wanted to get some input, so that when we go back to companies, we can state the obvious but then have the backing of this committee."

Two hours later, the team was back in Bethesda, crowded into the wine bar at the Doubletree. Mueller, hoisting a flute of champagne, delivered a long, grateful toast, recognizing one by one the people in the room for their contributions, including the group from Tibotec. Not normally this generous with praise, he was magnanimous, laughing heartily, plainly moved. "Usually I say just something is pretty good," he said. "This was stellar."

❖

One of Cumbo's first decisions after becoming interim vice president of sales was to reschedule Launch Week, moving the traditional sales meeting/pep rally up to the beginning of May even though telaprevir wouldn't be approved for several more weeks. The chief risk was that Vertex's field force would train without knowing what was in the final label, but Cumbo wanted his representatives to be ready to start call-ing on doctors—and outhustling Merck—on "Day One." Meant both to educate and energize, launch meetings combine the intensity of a maximum-stakes pre-exam crash course with the high-octane buzz of a casino. A recent Hollywood comedy, *Love and Other Drugs*, included its own version of a Pfizer launch meeting, featuring go-go dancers, the Macarena, and sales leaders who sleep with their trainees.

The commercial purposes of the daylong events—informational sessions, role-playing, KOL lectures, compliance training—were self-evident. Vertex needed to align its reps with the rest of the commercial team, which was rolling out, in waves, a broad educational campaign about hepatitis C and HCV testing; a twenty-four-hour patient hotline; an assistance program to help patients get and, if necessary, pay for its drug; and a city-by city marketing campaign for telaprevir under the trade name Incivek (In-see-veck). Yet, as ever, questions of identity, of values and vision, predominated. Now that Vertex had a drug and a sales force, it could ill afford to have its reps act like, as someone commented, "coin-operated machines."

Ken Boger planned to retire as general counsel in September—ten years after joining Vertex during the upheaval after 9/11 and the crash of the immunosuppressant VX-745. Standing on the low stage in the Amer-ica Ballroom at the Westin Copley Place hotel in Boston's Back Bay in a seasoned sports coat and tie, in front of the undulating digital backdrop that a few reps half-joked might induce seizures, he looked like a holo-gram of someone's tough but affectionate uncle, lovable but not to be messed with. He cushioned himself against the pounding rock music that the audiovisual team cranked up to introduce new speaker segments by concentrating on his terse, five-hundred-word speech. Smiling broadly, he began:

"I'm actually very happy to be here today . . . even at eight in the morning. I mean, at this podium, in this city, and at this company, which is on the verge of beginning to achieve a dream that was twenty years in the making. I say 'beginning' because our founder—and I understand you heard from him yesterday, and he's my little brother, and that's okay—he said, with characteristic hyperbole, that in starting this company that he wasn't just interested in making drugs, he wanted to change the world. I'm gonna get back to that, but it is, in fact, part of our goals at Vertex. It's fundamental to our corporate identity, and it's fundamental to our brand.

"But let me digress for a minute. Who do you suppose was the most admired company in America in 1987, according to a *Fortune* magazine poll. Any guesses? Right, Merck. I'm not going to pick on Merck— actually I am; it's gonna be a lot of fun. So who do you suppose was the most admired company in America in 1988? Merck. In '89? Merck. In 1990? Merck. In '91? In '92? In '93? All Merck. In 1993 *BusinessWeek* called Merck a 'national treasure.' "

For those like Ken who had been with Vertex from its earliest days, the specter/example of that Merck still burned—a sustaining inner flame. Boger's decision to leave Merck in 1989, at the height of its glory, was their shared backstory for everything since.

"So now let's fast-forward to 2011," Ken continued. "Neither Merck nor any other pharmaceutical company was in the *Fortune* magazine top *fifty* most admired companies, other than J&J, but people who make powder and Band-Aids don't count in this analysis. That was despite the same language you see from 1993 about integrity, transparency, and patient focus being used by Merck executives throughout the two decades leading up to the present. An integral part of Merck's corporate brand was trust, and they lost it, even though they spent a lot of time talking about trust and patient focus.

"The patient wasn't there when Merck launched a phony journal in 2002 to 2005 in a format designed to convince people it was a real peer-reviewed journal—just to pump their drugs. The patient wasn't there when Merck made a hit list of doctors to be discredited or intimidated for raising issues about Vioxx, and withheld that data about Vioxx-related adverse events. There are a lot of other examples, not just from Merck, but across the industry over the last two decades."

Post-Vioxx, Boger's desire to differentiate Vertex from Merck had taken on a deeper moral gravity, and Ken had been both the company's wise man and its enforcer. He felt the salespeople needed to understand what Vertex had at stake, not just reputationally, but in what it could become.

"So going back to the founder, who said he wanted to make drugs *and* change the world, that doesn't mean that personal goals don't play a role. One of our earliest advertising slogans was 'Ambition will cure AIDS before compassion will.' I really loved that. It wasn't necessarily the most effective ad, but I loved it. It said a lot to me.

"It doesn't mean that financial goals don't play a role. The first part of George Merck's famous slogan from the 1950s was to never forget that medicine and the drug business are for the patient. The second part of that slogan said that if you remember the first part, profits will follow. What the founder meant was that we focus on serious disease, and significant unmet medical need, with the principal objective of transforming patients' lives. It's a fundamental part of our corporate identity, it's a fundamental part of our corporate brand."

Vertex prides itself on having zero tolerance for legal and ethical lapses. Ken's twin brother, Jack, dean of the University of North Carolina Law School and a renowned anti–death penalty champion, tells a story that Ken says illustrates how companies should handle violators. An old political grandee at the school advised Jack on his first day on the job that he should grab the first person he saw and throw him down the stairs: after that, there would be no more trouble. Ken wanted the company's newest hires—in what had become a disreputable business of bribing doctors with cash and favors, practices both government and industry were now cracking down on—to understand just how their actions would reflect on Vertex.

"Now, how would we lose that?" he asked. "The same way that most big pharmaceutical companies have lost the trust of consumers, of patients, of regulators, and of society. By mistaking a commitment to the patient with a PR slogan. By losing sight of the fact that a brand, to have real value, describes reality. It's who your company is, it's who you are. And that's determined by what you do, not by what you say.

"How does that relate to compliance? Because whatever you think of the particular rules in our industry governing your behavior, they are in fact all intended to benefit patients—your grandmother, your grandfather, your children, your brother, your sister—and most of those rules do benefit patients. When we ignore them, or we twist them out of shape, we're acting in a manner that's inconsistent with an intention to benefit patients, and with every act we're then destroying a piece of that corporate identity.

"So I'd like to put out a few fundamental principles—they're pretty simple," Ken concluded. "The first one: talk is cheap. If acting for the patient on a day-to-day basis isn't painful at times, then you're probably not doing it. Second principle: five smart people in a room together can talk themselves into anything. Stay focused on the patient—your mother, your father, your grandparents, your children, or people like that—and don't become one of those five smart people, justifying anything. And last, to paraphrase my dear departed grandmother, 'Tell the truth. Don't stretch the truth. And don't get ahead of the truth.' Those things pay off long term actually in your personal life, they pay off long term in your business life, and in the new pharmaceutical future that we want to be preeminent in.

"Thank you."

❖

In the giant brain of the executive team, if Ken was the superego, Smith was the ego. Joining the same month, in the midst of the maelstrom, they helped maneuver Vertex beyond its fractious early days, Ken by being chief negotiator and conscience, Smith by strutting. After Smith first started, and some complex financial issue arose, Boger and Sato would say to him, "You'll figure it out, Ian. You're an athlete." Smith did figure it out. He had played Wall Street's appetite for magic like a virtuoso, and with the launch of telaprevir his job entered a new phase.

Smith revved up the field force using an onstage question-and-answer session with a popular regional manager from Philadelphia. Still athletic, he wore a blue shirt and no tie, his black trousers riding low on his slender hips. He was nursing a cold. Like Ken, he was concerned that the field force didn't fully understand the company's mission, and so he delivered a staccato, personalized financial tutorial on Vertex.

"Instead of having a revenue line, we used Wall Street to fund the company," Smith explained. "This is where it keeps coming back to you guys. Wall Street's done. *I'm* done. So you're going to be bringing in the cash flow for us now.

"The business strategy for us was money and medicines. It is as simple as that. You keep your company wide. You reduce your R&D risks. You have more projects going on and you cross your fingers and you hope.

"And when you open that envelope, you start to see stunning results. You start to see the Vertex swoosh. You start to see fifteen percent relative improvement in breathing in children. It's absolutely astonishing, and it might happen a couple more times this year. That's the uniqueness about Vertex. It keeps giving itself a chance. This company has something that a lot of companies don't have. We create hope in a lot of different diseases."

By becoming an operating company, Vertex was shedding its brash demeanor as a late-stage development company. The center of gravity for bringing in revenue to feed Mueller's R&D beast shifted from Smith to Wysenski, from stock sales to profits, and the new face of the company was sitting in front of him. Emmens was right: everything changes the day you go profitable. Smith advised them to concentrate on doing what needed to be done, but not to ignore Vertex's tribal sensibility and history. He urged them to join the fear and fun.

"As you go into the field, you're working for a company that really does pride itself on doing things different. Now there are certain regulatory confines; I understand that. So don't do things too differently. But think about how to get this done. Be creative. Enjoy the ride. Take a look around. This is a unique company.

"You've got to live with that hope in this business. You've got to take the risk. It's too easy to say no when you get there. We're not going to say no. The culture of this company is to keep swinging. And the way that we keep swinging is you guys drive cash flow. Because I'm *not* raising any more money.

"My challenge is in the next three years, go beat the Wall Street money. This really is possible. Think about that. Twenty years of putting money into this business, you can pay it back in three. It might just be."

Smith gulped some water, gazing out over his audience. He turned away, ambling toward a small table behind him, at the rear of the stage. "A favorite line of mine," he said, chortling, "is 'You call this a knoife? *This* is a knoife.' " On the twin jumbo screens flanking the stage flashed the familiar film image of the big-brimmed Australian reptile hunter Crocodile Dundee facing down a street punk. "Every story needs a hero and villain," he said. "We've got Incivek versus Victrelis." Someone murmured "victory-less," the new in-house rag on Merck's drug. Smith reached the table. "I'm gonna go to my prop here, and I have cleared this with Ken Boger, our general counsel. This is my knife."

From under a black velvet sheath Smith lifted a machete. It measured more than half the length of his arm. He waved it, grinning, and went on: "I'm just gonna hold this while I talk about Incivek. There's no threat here, but Incivek, we believe based on our data, beats Victrelis in every possible way. The cure rates are significantly higher, seventy-nine percent versus sixty-three percent. Response-guided therapy. We treat a broader patient population. And what I personally think is the biggest difference between the two drugs right now, and hopefully this will come through in the label, is simplicity.

"To drive this company, we need to do well in this market. Wall Street consensus this year is fifteen thousand patients. Fifteen thousand divided by about one hundred reps is one hundred fifty. You individually need to think about how you get one hundred fifty patient starts this year. Each rep. I hope you do that, I really do, because what that will do is give the company a great story to tell Wall Street, it will drive value, it will drive cash flow, and we will build the company.

"Now, I'm being a little pushy, so I'll put the knife down. But you can win this market. We do have a sharp weapon in our hands, and it needs to be used in the right way. This is a drug that should truly differentiate itself from Victrelis. I really believe we have the chance to build the company. The mark may be the stock price. We're building a big company. We're already the fortieth-largest company on the NASDAQ 100. I have hopes that in a couple of years we're on the S&P 100. We're chasing Gilead in terms of its market cap. We really have an opportunity."

By the next night the reps were primed. Vertex threw them a party at the top of the Prudential Tower, the ungainly Johnson-era skyscraper looming over the Back Bay. The venue has one of Boston's best views, in no small part because it's the only place in the city from which you can't see the "Pru." There were several craps and roulette tables where the reps noisily gambled for tokens while chugging the evening's featured drink, the Vertex Swoosh.

Emmens had recovered from his diverticulitis, and though he didn't drink, he kept up a lively, rotating conversation away from the center of the room, near the elevators. He liked to stay off on the side, away from the attention, with more opportunities to learn things. As a sales lead in his early days in the industry he'd once had to fish a coed group of trainees from a hot tub, but this group was battle tested, as he was, and except for their being inordinately exuberant at the gaming tables, he was relieved to see the reps acting like adults.

❖

Victrelis received FDA approval on May 14. Even with the stock buy-back, Merck shares had languished since Frazier took over as CEO at the start of the year. The drug was the company's main chance to show Wall Street that it could exceed expectations, which jumped in light of what industry observers agreed was a hugely generous PI—no black box safety warning, no onerous Risk Evaluation and Mitigation Strategy (REMS). Anemia was listed alongside other side effects like fatigue and nausea, not singled out. Merck got permission to include null responders. Vertex's label surely would be no worse, but it also couldn't be much better.

Merck priced its drug on a sliding scale, based on how long a patient, after a four-week lead-in period with peg-riba, needed to remain on

Victrelis. The company planned to grab as much market share as soon as possible and so it priced aggressively. At about $1,100 a week wholesale, the overall charge for treatment-naïve patients would be around $31,000, and for previously treated patients around $35,000. If someone needed a full forty-four-week regimen, the cost could be as high as $48,000.

Three days later, while Vertex awaited approval for Incivek, Merck announced a strategy that most reporters and analysts thought gave it a critical edge. It doubled its marketing muscle, partnering with Roche, its arch-competitor in hepatitis C, to copromote Victrelis. Selling competing versions of peg-riba, the two companies together controlled the market, visiting frequently every one of the five thousand doctors across the country who treat 90 percent of patients. Though Victrelis hadn't been tested with Roche's Pegasys, which commanded about 85 percent of peg-riba sales, Roche salesman now would call on their customers and promote "triple combination therapy" by mentioning Victrelis.

Cumbo was flooded with anxious calls from his reps. "I had to talk a few of them off the ledge," he said. Wysenski, initially shaken, went to work trying to learn more about the terms of the deal. She was encouraged to learn that there was no exclusivity; the Roche reps could also mention Incivek. Emmens acknowleged that Vertex had been thrown a curve but professed no worries, reasoning that the Roche reps, accustomed to bashing Schering's Pegintron, would feel little compulsion to boost a rival's product and that the battered Schering reps, who made up most of the Merck sales force in hepatitis C, would feel rightly that they'd been sold out. Bundling sales of Victrelis with Pegintron had been their sole marketing advantage over Pegasys. Now it was gone.

Not since Merck's Vioxx and Pfizer's Celebrex were launched head-to-head a decade ago had a pharmaceutical battle intensified so swiftly. The size of the opportunity in hepatitis C—some analysts were predicting a $10 billion market within a few years and a $20 billion market by the end of the decade—kept hiking the stakes. Cumbo was counting on his people to be more motivated and stronger psychologically than their counterparts and so it was of little consequence how many more reps Merck tried to throw at him. In a sales war, the goal was domination. Winning wasn't enough. "You break their spirit," he says. "That's what you've got to do with the reps. Representatives—the way I see it, and I was a rep for a long time—it's a very emotional-type thing. You have ups, you have downs, and when you're down it's sometimes really hard to pull out of it. Merck had got some really good, talented, passionate people. Those are the ones you've got to break."

❖

The following Monday at the start of business the FDA notified Vertex by fax that it had approved Incivek. The label had no serious issues. The blockade was down. Within an hour, a twenty-four-hour nurse hotline was activated, and the sales and patient support operations went live. Outside Chicago white-suited line workers started final packaging around the clock, determined to have the drug on loading docks within twenty-four hours, to reach the first pharmacies by Wednesday.

After a brief flurry of champagne toasts across the various departments, Emmens, Mueller, Smith, Wysenski, Pace, and Partridge converged on the square windowless room outside Smith's first-floor office in JB-II for an eleven o'clock conference call with analysts. The table was littered with scripts, whiteboard markers, Dunkin' Donuts cups, Diet Cokes, and BlackBerrys. Kauffman, traveling, was on his cell phone in an airport, available to answer questions.

"Two things to hit on," Partridge coached. "The DDIs in the med guides; throughout the label—with food . . . with food."

"Stay away from all comments on Victrelis," Smith advised. "We're very happy with our label. That's it. Any questions on consensus projections, I'd best take them."

The most urgent question, of course, was price. After Emmens and Mueller laid out the particulars of how Vertex had discovered and studied Incivek and elaborated on the breakthroughs the drug represented to patients and physicians, Wysenski announced its decision on how and what to charge for its first product.

"We believe that patients who need Incivek should be able to get it regardless of their ability to pay," she began. Vertex had assembled a commercial program not unlike "need-blind" admissions, which as practiced by wealthy colleges and universities amounts to: we're going to charge premium prices but we say to you that if you qualify and we accept you we'll make sure you can be here, even if it means we pay your way. Once you received a prescription for Incivek, Wysenski said, you could go to a website and find a menu of copay assistance and free-medicine programs. Dressed crisply in black slacks, a paisley blouse, a lime-green rain jacket, and a chain necklace, she elaborated:

"Our copay assistance program is designed to cover up to 20 percent of the total cost of Incivek for the full course of treatment for people who are covered under commercial providers. The copay program has *no* income restrictions. For people who are *uninsured* or *who are rendered uninsured*, and also have an adjusted gross household income of less than $100,000, we plan to provide Incivek for free.

"Now to Incivek pricing." Wysenski paused; Emmens and Smith performed silent drumrolls. "We've set the price of Incivek at $49,200 for the entire twelve-week course of therapy.

"I'd like to comment on the factors that we considered in our pricing decision. Broadly speaking our mission is to discover, develop, and launch medicines that cure or significantly advance the treatment of serious and life-threatening diseases. This is both expensive and risky, and we are committed to spending the money needed to enable us to tackle more tough diseases—this includes hiring and retaining the best scientists and medical team possible. More specifically, we took into account the fact that unlike HIV or hepatitis B, hepatitis C can be cured with a single course of treatment. We also considered the costs of other medicines and procedures used today to treat and care for people with hepatitis C and how well those treatments work."

By decoupling price from access to drug, Vertex presumed to diffuse any controversy while reassuring doctors that Incivek was a premium product. As the colleges and universities long ago discovered, with certain products, like education and medicine, consumers accept and trust that high costs confer higher quality and greater value, and so Vertex had no reason, other than access to strapped government dispensaries, to compete with Merck on pricing.

The analysts, following up, requested numerous clarifications. They had based their valuations for the most part on a lower price. Even if they had gotten the cost of Incivek wrong they wanted to be sure they could defend their previous projections about how VRTX could be expected to do in the months ahead, even if it meant rejiggering the prescription rates they used in forecasting income. The stock price started the morning pointing downward, but throughout the conference call it recovered, rising more than 1 percent. As the discussion wound down, Wysenski and Pace fist-bumped, then, as the participants started collecting their materials, Partridge took a call from an investor, Adam Koppel.

If there was one fund manager most closely, and emotionally, tied to the company, it was Koppel. As a managing director of the Boston-based Brookside Capital, a Bain Capital–owned hedge fund, he currently held about five million shares. Koppel had made his career on Vertex, first buying VRTX after meeting Boger in 2004, then moving in and out of the stock over the years. In 2006, when the share price dropped into the high teens, he bet big, making it Brookside's largest holding by far. Its peak position was about $280 million.

Koppel came to investing from medical research and consulting. After finishing an MD-PhD in neuroscience at the University of Pennsylvania in the 1990s, he took an additional year to earn an MBA before going to work for McKinsey, advising biotechs. What he looked for in companies was the "DNA" to replicate success, and he believed Vertex was the first biotech since Genentech to have it. Koppel got to know Boger, Smith, Alam, and Graves during the heady, triumphalist period of Incivek's clinical program, and he loved the company's swagger and way of thinking. Loyal to the old guard, he at first gave Wysenski a rough time, attacking her commercial strategy as thin. He thought the jury on Emmens was still out.

Koppel was "so shocked," he said, by the Merck-Roche alliance that he felt he needed to hear from Emmens directly. "Can you just help me understand what they're thinking?" he asked.

"I believe they're on their back foot now in the market," Emmens said. "So they're gonna attack rash. Roche gives them an additional voice. It doesn't change the drug. We have the better drug."

Koppel was eager to know if Vertex was in the market to buy a late-stage nuc, to counter the threat from Pharmasset, BMS, and Gilead, where McHutchison and the clinical team had several nucs and were running an array of clinical trials, testing them in various combinations. "We won't discuss business development," Emmens said. Smith, out of earshot, mentioned that Koppel's fund also owned 10 percent of a company, Idenix, that was promoting a nuc.

"Roche doesn't have any skin in the game," Emmens told Koppel. "Its reps have no incentive to say boceprevir is better. I don't think you can take a rep and turn his brain off."

Pace ushered Emmens to a small room in the communications suite directly after the call, to begin a round of press interviews. His time in the spotlight had arrived. Whether he truly thought the Merck-Roche tag team represented no real material threat to Vertex or was simply unfazed by a weaker product, Emmens was compelled, as Boger had always been, to challenge industry behemoths. "We be small," he liked to tell company employees, "but we be fierce." He told the interviewers what he'd told the analysts: Vertex had the better drug, more effective, simpler to use; the pricing decision was "appropriate" given the relative value to patients, physicians, and payers; the Merck-Roche alliance was bluster. "It's a relatively small number of prescribers," he told Reuters. "I don't think we need to shout too loudly." Looking ahead, he said, to the next wave of competition for supremacy with an all-oral treatment, Emmens noted that it was easy to lose one's head over preliminary results and that most experimental drugs fail.

"The new standard you have to compare yourself to is tough," he said. "It's us."

❖

"Welcome, everybody, in this evening hour," Mueller began, addressing an invited group of about a hundred—those who had worked most closely in bringing VX-950 out of the lab, through the clinical and regulatory gauntlets as telaprevir, and finally to patients, as Incivek. He stood near the checkout counter in the JB-II cafeteria, the largest space in the building, hours after the markets had closed. More champagne was passed around. Most of their coworkers were back at their desks or headed home. "I think it is a fantastic event, and I'm so excited I can't tell you. I cannot still believe that it happened, but it did."

"I can't believed that you hugged me today!" Emmens shouted from the sideline.

"I think that is the most astonishing accomplishment and achievement that we have as a company, and now we are moving really in the right direction," Mueller said. "I want to introduce you to a new type of tradition that we will have in the future. This monster is a bell: the transformational life bell." Mueller uncovered an old firehouse bell, bought on eBay, mounted on a rolling cart, and painted purple. He explained that throughout human history, a tolling bell has signified, among other things, hope, freedom, accomplishment and unity, and that from now on, Vertex would ring its bell whenever it received regulatory approval for a new product.

"When we really think about what we did with this wonderful Incivek approval today, we gave hope to millions of people who have now the freedom to achieve a better life and a cure. I think that this is really fantastic. The accomplishment I really want to give back to you."

No one in the room—not Kauffman, Murcko, Kwong, Kieffer, Hurter, Condon, Weet, or the many others who contributed key insights or volunteered extra weekends when a problem needed solving—no one could say definitively which of them had been indispensable, making the decisive difference, enabling Vertex against withering odds and the best efforts of its many competitors to keep going, push upfield, take its shot on goal, and score. Murcko calculated there were at least fifty major contributors without whose timely work the project might have stalled. In his toast, Mueller recognized them all collectively.

"I want to say just a couple of words to remind everybody. When we started out it was a very, very difficult active site to approach. Protease inhibitors are a tough thing to do and not many are out there. I think our scientists and our environment did a phenomenal job to tackle that type of approach. Then it looked like you never could make the drug. The first kilogram cost $2.5 million. To do any sort of clinical trials was almost impossible for a small company. It was a bold decision to move forward.

"And then we learned that nobody can swallow the drug because no one can formulate it. I must say we did a phenomenal job to come forward with the presentation as it is today: two tablets daily three times a day. This was not a given. With the clinical trials we had those long periods in front of us, forty-eight-week regimens, which is a huge, huge problem for a small company to do. It's costly, it's time consuming, it's difficult. I must say the clinical guys revved up phenomenally, to change the treatment paradigm. This is not easy to accomplish. There were tons and tons of interactions with the regulatory agencies needed to accomplish this. We built in the last year the most sophisticated manufacturing network that basically provided us a supply chain that is unheard of in industry. Cost of goods now is orders of magnitude less than it was initially. This is a huge, huge difference."

Vertex created Incivek, but Incivek also created Vertex. Mueller was its main beneficiary, inheriting a small world-class research organization with a powerful scientific culture; expanding it in conjunction with a bold, committed clinical group; bringing in inspired CMC and manufacturing; assembling a lean, $500 million-a-year outfit with a record of success and a promising pedigree. He was the scientific leader of a maturing drug company just beginning to make its mark on the world, and he knew the future depended not only on his ability to advance and incorporate new methods but upon the alliances he built. He finished by praising their united spirit. "You cure people," he said. "This is the high art of what we are in. This is a phenomenal ongoing journey. And it's not just Vertex alone; it's Vertex in the context of a broader community. We are all working together in trying to tackle the tough diseases that nobody else can tackle."

Mueller pulled the rope on the side of the bell. Its deep, mellow sound filled the cafeteria. He pulled the rope again, beaming, and again, until the sound of the bell was drowned out by the exultation in the room.

Notable by his absence was John Thomson. He sat at his desk downstairs, plowing through his email. Thomson had just returned from China. He wanted to get home to see his wife, Jackie. In the tolling upstairs, few were reminded that among the major intellectual turning points that resulted in Incivek, it was the initial belief that such a drug was possible and that Vertex, with its severely limited resources, could produce one, that ignited all the others. No one in the early days believed more fervently or worked as hard or proved as much or argued more determinedly against giving up. Thomson remained the truest of true believers in Boger's original vision, and Incivek was as much his achievement as it was the company's.

Thomson had a history of avoiding corporate celebrations, but now, at fifty-three, he had other frustrations just as urgent as when he was a self-described "protein jock" spending long nights in the cold-room on the company's first project. What he'd signed on to do at Vertex was to build a sustainable organization: a church, so to speak, built on Boger's vision. "I said, 'I hear you're starting a company," he would recall telling Boger. " 'Is it all theoretical, or do you need some scientists?' " Vertex had two breakthrough drugs and a standout pipeline, but when Thomson looked beyond the rest of the decade—to where the blockbusters of the 2020s would come from, and how a global Vertex would become the hub of a larger network of interests—he worried that the company was unprepared and too cautious. He especially worried that it was moving too slowly in China and the Pacific Rim.

❖

Vertex planned to celebrate the launch by bringing together the entire company for an extravaganza in the fall. Emmens wanted a timely, smaller party to reward the Cambridge site. The next Wednesday after work, eight hundred Vertex employees poured into the ballroom of the Hyatt Regency. Confetti cannons stood ready in the corners as the crowd drifted among the bars and food stations. Emmens spoke first, discussing the significance of Incivek to the company.

"We're not just launching a drug," he said. "We're really showing the world what Vertex is, and who we are, not only what we're doing now but what we're gonna do in the future. It's a big deal. We're here to become sustainable. That's when we have income that covers our expenses, which are now $3.2 million a day." He was quickly becoming admired in the company for his brisk yet thoughtful confidence and understated defiance. With reporters, he had compared Vertex's battle with Merck to David versus Goliath. Now there were two Goliaths. Emmens retold the biblical story, reassuring the crowd that size didn't matter—indeed, that they were more agile, more emotionally engaged, better positioned than the Philistines and their legions.

"David believed and was very good at what he did. He took one stone, and he hit Goliath in the head—well, he also cut his head off. So we're out there now slinging our stones. They're big and scary, but we're very good at what we do, and very focused."

Patient activist Kelly Ann Mann-Hester, who spoke so compellingly at the AdComm, nervously walked onstage and took the microphone from Wysenski, who had invited her. Mann-Hester had watched her husband die of hepatitis C in the 1990s and was a mother of four and a grandmother of five. She reprised her story of personal transformation, from living to die to learning how to live *until* she died. "I'd tried everything," she recounted. "When I began Incivek, I believed I was doing it for altruistic reasons. I honestly thought, 'I'm not gonna get cured. I've had seven treatments. It won't be me. Maybe they'll learn enough from me to help save somebody else.'"

Since being cured in 2010, Mann-Hester had made her mission "testing and treatment." It didn't matter if there were new cures for hepatitis C if you didn't know you were infected, like the estimated three million Americans and hundreds of millions around the world who remained in the dark about their diseased livers. She improvised a group experiment to demonstrate. "Raise your hand," she said, "if you have had a tattoo or a body piercing; if you've ever used IV drugs or if you've ever shared a needle with somebody or got stuck with a needle by accident, especially if you were a health care worker before 1990; if you had a transfusion or dialysis, or were born from someone who had a transfusion."

Several dozen people raised their hands, but hesitantly, looking around. Even here, announcing you might be infected with HCV posed risks.

"Now out of those hands, how many of you have been tested for hepatitis C?" she asked.

Only a few hands went up, a minor fraction. Mann-Hester seized on the troubling disparity. "That's my point, that's why I'm here. I'm doing this for that reason. Education starts at home. Test yourself. Treat yourself. Find out you're good to go so that this company can be good to go further. We have to remove the stigma from hepatitis C because that's one of the reasons that keeps people from getting tested, and there's no reason for it anymore. I'm living proof. I'm standing here as someone who was cured by Incivek, so I know. I want you to own your future, and I want you to help people own their future."

The last speaker was Boger. Looking over the milling crowd, recognizing only about half of them, he began, "Hi, everybody, I'm Joshua Boger, unemployed person." Most new hires didn't get the joke, but they laughed alongside the veterans. Boger was the founder who officially stepped down two years ago but was still a director. The painful circumstances of his departure had mostly healed with the realization that Emmens went out of his way to credit Boger wherever and whenever possible while also successfully sticking to first principles. None of the torchbearers had put out his or her torch. The culture seemed to be holding despite his prior apprehensions. In Emmens's speeches, internal and external, Vertex remained "Josh's dream."

As Boger had told the field force by satellite during Launch Week, the most common question he heard now from well-wishers was: Did you think this would ever happen? "This is exactly what I thought would happen," he said. "But I also thought it would take an enormous amount of effort and an enormous number of talented people. And there wasn't any reasonable expectation that those people would come together to do that.

"Bringing a major drug forward, bringing it out of discovery and into development, then out of development and into the market, bringing it to patients and changing their lives, is . . . *the hardest thing that humans do*," Boger said. "Putting a man on the moon? Not . . . so . . . hard. Engineering is an amazing discipline, but it's also a discipline that succumbs every single solitary time to time and effort. If you give me twenty years and a half trillion dollars, I guarantee I can put people on Mars and bring them back. I guarantee it. Any of you could guarantee it. But making a new drug that changes people's lives is almost beyond human understanding.

"So the second question I get often is, 'So, what's next?' 'You cannot imagine what's next' is what my answer is. You don't create an organization like this, fall over the finish line, even with a success as big as Incivek, and say, 'I'm done now, this is it.' "

Whatever the company's desire to downplay its earlier brashness, Boger felt that now, more than ever, was the time to act and sound bold, both to inspire and intimidate. Was designing and developing and bringing to patients a life-altering medicine—not a cancer "breakthrough" that extended life expectancy in trials by a few months, a drug that cost too much and did too little, but a novel compound that actually saved a life or restored a quality of living that had seemed beyond dreaming about—the hardest thing that humans do? Maybe not. But it was impossible for any of them to name anything harder. Who, at such a moment, would want to try? They burst into thunderous applause. A blizzard of confetti exploded from the corners. Boger, to borrow a famous trope from *Jerry Maguire*, a film about a high-powered sports agent who crafts the ultimate living room pitch to win over a woman who already has succumbed to his fervent mission and charm, had them at "Hi."

# CHAPTER 12

JUNE 6, 2011

At ten o'clock on Monday morning, Ken sailed into the small conference room and took his seat at the head of the table. At the opposite end, poring over his PowerPoint with Smith, was Dr. Christopher Wright, vice president for clinical development. With Kauffman on the road promoting Incivek, Wright was fast being groomed to become Vertex's top clinician. A neuroscientist by training, he was another defector from the world of Harvard medical research, a muscular MD-PhD in his early forties who sported a modest Afro, long sideburns, and fashionable designer glasses.

The protocol before the disclosure committee was a twenty-patient, Phase II, double-blinded, multidose study to evaluate the safety, tolerability, efficacy, and pharmacokinetics of the corrector VX-809 alone and in combination with VX-770 in CF patients with two copies of the delta-F508 defect. In other words: the first preliminary data from a tiny cohort on whether combination therapy could do for the great majority of people with CF what VX-770 alone was doing for the small G551D genotype.

Mueller and Wysenski huddled on one side of the table; on the other, Partridge, Olson, Megan Pace, and Dawn Kalmar, who had joined Vertex from Genentech as a product spokeswoman and who would compose the press release. Anticipation was appropriately high. Failure or ambiguous results in this study would represent a serious reversal, since Vertex had already showed that VX-809 by itself offered no improvement.

The hope now was that VX-809 would repair a significant portion of misfolded CFTR and get it to the surface, where VX-770 would then hold open the channels long enough to substantially increase the flow of salt and water across the membrane, lubricating the airways and other affected tissues.

"We're currently scheduled for a premarket release Thursday," Ken began, putting the team on notice that within the next three days Vertex would have to decide not only what the study had revealed, but, more to the point, what the company could safely conclude from it and accurately say about it.

Wright summarized the top-line data, reading from his laptop as the others followed along. Well tolerated, the cocktail produced no unexpected safety findings—always welcome news, since the safer the regimen, the more dosages can be increased. The primary end point of the study was to observe a drop in sweat chloride. When VX-809 was first given to patients as monotherapy, there was a significant drop relative to baseline; when VX-770 was added for seven days, Wright said, there was a further reduction consistent with Vertex's theory that you could double the electrophysiological effect. Clearly the molecules were synergistic. In 20 percent of subjects—two patients—the combined effect was profound, sweat chloride dropping steeply.

Although not on the same order of potency as the Vertex swoosh or the other company-altering "envelopes," Wright's presentation unambiguously proved the concept. The company didn't know if the combined effects of the drugs could relieve lung disease—the FEV1 data after the one week that the subjects were taking both drugs were scant and inconsistent—but the study provided ample promise for moving ahead, with the potential to increase dosages to gain more potency. The mood of the group was decidedly upbeat but sober: excitement marbled with caution and relief.

Smith zeroed in on the super-responders. He did a quick extrapolation, envisioning the shape of the curve. If combination therapy could dramatically boost CFTR function in even 20 percent of those sixty thousand patients worldwide with at least one copy of the delta-F508 mutation, about twelve thousand more patients might benefit from Vertex's cocktail. "If you can take this from three thousand to fifteen thousand, that would be huge," he suggested, brightly. Mueller, seated next to him, chafed. He reminded Smith that the study was far too small to validate any such estimate. "I would keep it as simple as we can," Mueller said. Vertex should say the concept worked in an exploratory cohort and that the company was optimizing a final regimen before going into Phase III.

Smith persisted. He proposed emphasizing the super-responders in the press release. Olson sharply disagreed. "This says nothing," he said.

"I agree," Ken said.

"I also agree," Wysenski said. "I don't want to overstate the data."

"The big whoopee is, even if we see with these patients what might be a billion-dollar curve, it still might be misleading, which could give a lot more patients a chance for a cure," Mueller said. "I would leave it the way it is right now without any further interpretation."

"The proof of concept is there," Ken concluded. "The combo works. No one dropped out due to adverse events."

Thus were resolved, in favor of not getting ahead of the truth, the tone and content of Kalmar's press release, which claimed modestly that the data from the study "open the door to the possibility of treating people with the most common form of cystic fibrosis," as Mueller was quoted saying. At the CF foundation, Bob Beall was similarly circumspect. "These data, while early, provide important new information about the potential to address the basis defect" among most patients, Beall commented.

VRTX skidded on the news. By midmorning, as the analysts weighed in and investors digested their mostly gloomy assessments, the share price plunged more than 13 percent. JP Morgan analyst Geoff Meacham, who as a PhD candidate had studied CF, expressed the majority view that the sweat chloride data were less robust than expected. "While the data are a step in the right direction, the synergy is very modest, and the data are by no means a 'game changer,' " Meacham wrote in a note to clients. He questioned whether the cocktail would "materially alter the course of disease [or] its management." Shares finished the day down 7 percent to $49.30, the largest one-day drop for the company in almost two years, erasing some of the 60 percent upsurge over the past six months.

Geoff Porges, as ever, led the bulls. He thought there was "more than enough activity to get excited." Like Smith, he was particularly encouraged by the two super-responders, believing that further experimentation could only identify other subpopulations who, because of residual CFTR at the cell surface, would benefit from a potentiator. Even the sickest patients had some CFTR activity. This was the new world of personalized medicine: you find those individuals, by drilling deeper into their genetic makeup, who are likely to benefit most from your drug. And so even beyond the hopes and aspirations of Vertex management, Porges was utterly convinced that Vertex had already cracked open the treatment of cystic fibrosis. The combination data had to be viewed as another striking advance.

"What I say to investors is, 'VX-770 is going to be like sunscreen for cystic fibrosis patients,'" he explained. "'If you've got really light skin, you get a huge amount of benefit from sunscreen. If you've got dark skin, you get less benefit from suncreen. But you should all wear sunscreen if you're going in the sun.' If 770 only improves your CFTR function by 3X instead of 10X, that's still better than no X. What I believe is that someday every CF patient is going to be on 770, but no one's really thinking that. Also, when you find a treatment for a rare disease, the number of people with that rare disease gets bigger and bigger."

Smith and Porges spoke by phone after the announcement. Ken especially was predisposed to worry about how much Smith, through his relentless courting of Wall Street, and its courting of him, could become influenced by its perspective, captivated by its assumptions, led by its needs. Porges's excitement about the combo outdid Smith's, but they were in close agreement that Vertex's CF franchise offered perhaps more long-term value to the company and its stakeholders than HCV. Vertex had patent exclusivity stretching to 2025, with no competitors in sight. With patients clustered for care around a small number of specialized pulmonary centers, two dozen salespeople could cover the world. Future costs were relatively negligible. The upside was enormous.

Faced with the question of what to conclude from the study, Smith was willing to go far, but still not yet as far as the Bernstein analyst. "He wants to know," Smith said, "why we're not telling everybody we've cured CF."

❖

Merck made much in public of being first to get approval, as if its reps would seize an edge by reaching doctors first. Its salespeople had pressed physicians before the AdComm to begin the four-week lead-in with peg-riba early so that their patients could get on Victrelis immediately at launch. But the company planned poorly. Perhaps due to the uncertainties over its label or a more general aversion to risk taking, it decided to wait and hold its launch meeting post-launch. Instead of being out in the field, racing from practice to practice, grabbing lattes and sandwiches at Starbucks to bring to doctors, then hosting after-hours informational dinners at hotels and restaurants with KOLs for local physicians, nurses, and patient counselors, Merck's field force was locked up in meetings. "I'll take that week," Cumbo said.

"That's roughly twenty-four-hundred calls that I'll make that they won't. Not only that, we loaded up on programs. This week we have forty-seven satellite webinars and a hundred promotional programs. That's just in five days. Plus I told my people to lock up all the speakers for the next week as well, so when they come out of their meeting all excited, you break their spirit."

Emmens dismissed the turbulence in the share price. The sell-off after the release of the combo data was a "clean drop"—not caused by any troubling data. He just wanted to ensure that the stock remained high enough so that Vertex couldn't be bought. As analysts furiously mined the first spotty weekly sales estimates compiled by the industry's most influential monitor of pharmaceutical commercial activity—IMS Health—Emmens cautioned the ET not to put any stock in the information for at least a couple of months, when IMS assessments tend to become more credible. "We have a company to run here," he told them.

A week after the CF disclosure meeting, Vertex bolstered its strategic position in hepatitis C by acquiring two early-stage nucs from Alios Biopharma Inc. of South San Francisco. Rather than buy the company to leverage an asset, as it did with ViroChem and VX-222, it paid $60 million up front for worldwide rights to two of the company's preclinical hepatitis C candidates. In announcing the deal, which could be worth up to $1.5 billion to Alios if both compounds were appoved, Mueller said Vertex would begin clinical studies with the molecules in an all-oral combination with Incivek and VX-222 by the end of the year.

Emmens left early that afternoon for a forty-eight-hour blitz of Washington and New York. In DC, Sachdev toured him around the Capitol, where he met with congressmen and senators to discuss their support for government reimbursements and for screening baby boomers for hepatitis C. On a visit to NIH in Bethesda, he met for two hours with Francis Collins, its director. Collins, codiscoverer of the CF gene and a prophet of genomic medicine, was understandably pleased and fascinated with Vertex's success with VX-770, but he was just as enthusiastic about the route the company had taken. Collins was frustrated that drug industry research productivity had declined in the past fifteen years and, as he told the *Times*, "it certainly doesn't show any signs of turning around." Earlier in the year he persuaded the White House to give NIH $1 billion to start in-house drug discovery. Whether or not the government could succeed where private industry had failed, the Obama administration believed the problem was too serious not to try.

Emmens flew to New York and checked in at the Four Seasons on Fifty-Seventh Street, between Madison and Park. Partridge and his team had chosen the location—designed by I. M Pei; peopled by wealthy shoppers; "a remarkable luxury experience, even by New York standards," according to its promotions—to host an investor breakfast. The next morning Emmens arrived downstairs feeling suitably refreshed. Joining him onstage, looking substantially more haggard, were Kauffman, Smith, and Fred Van Goor, the lead biologist on the CF project, who'd taken a red-eye from California. Though Incivek had scarcely been launched, Vertex wanted Wall Street to see the larger picture, and the invitation promised highlights of the company's presentations at the just-concluded European CF meeting in Hamburg, which few of the hundred or so fund managers and analysts had attended.

"Our discussion today is about cracking the code in a complex disease," Emmens began. "We think we've done this with cystic fibrosis, and that's why there is such excitement inside the company about our recent combination data. We have Phase III results with our lead compound for a subset of patients with this disease. These results are dramatically positive. We think we can apply what we know about the science underlying the disease to be able to help many more patients. That's a very big deal and that's an underappreciated opportunity.

"We don't have all the clinical data to prove this yet, but we're in a unique position to proceed. We've cracked the code and we're spinning the dials. We have the stethoscopes out and we're listening to the clicks. We believe it's a matter of time before we break this disease wide open and make a really big difference for a lot of people.

"From a business perspective CF is a very traceable and tractable market for a company that wants to stay, as we do, lean and very efficient commercially. CF treatment is concentrated among specialists. It requires a very small commercial footprint to reach patients with CF and the people who care for them. In fact we don't call sales forces like that salespeople. We call them enablers. The scientific core of our company is making a difference in the treatment of serious diseases like CF and hepatitis C, and these turn out to be disease opportunities where reimbursement is more favorable, where you don't need to build and maintain a large sales force, and you can build a global operation in a relatively straightforward manner. That's how we think about building our business."

He turned over the microphone to Dr. Susanna McColley, head of the pediatric cystic fibrosis center at Children's Memorial Hospital in Chicago. Dressed in a tapered white knit suit with black buttons, a gold chain necklace, and stiletto heels, McColley introduced her slide show with a personal anecdote. She had decided as a post-doc at Johns Hopkins in the early 1980s that she wanted to do clinical research in CF. Her advisor at the time told her there was no such field. Since the gene was discovered, she'd watched her patients suffer and die amid fading hopes for a cure. After detailing the VX-770 Phase III findings, McColley said: "This is truly the most exciting clinical trial data I have seen. I'm not trying to be hyperbolic here. I don't have a financial relationship with this company. But after twenty-two years waiting for something, with all these promises made back in 1989, it's very exciting."

McColley said little about the combination data, which remained preliminary. But she discussed findings from the just-concluded DIS-

COVER study of VX-770 in patients with two copies of the delta-F508 folding defect—Keith Johnson's trial. "There was not a clinical benefit of VX-770 as monotherapy in this group," she reported, explaining:

"I want to stress again you wouldn't expect to see that, because of the way this mutation works. There were some patients, though, who did have a pretty big drop in sweat chloride. This is likely due to the fact that there is an occasional delta-F508 CFTR protein that winds up on the cell membrane. That's speculative. But this supports going ahead with the studies combining 809 and 770."

Smith invited questions. In answer to a wandering query from Morgan's Geoff Meacham about the combo data that seemed more likely to impress his clients than to elicit new information, Kauffman reprised McColley's message that Vertex was steaming ahead. "We want to maximize the correction so we can get as much CFTR on the surface as possible and then add 770 on top," Kauffman said. "If we can do that we're gonna see a more dramatic effect than we've seen here."

Meacham followed up. "Did you harvest any bronchial epithelial cells from patients?"

"You can't really do that on living people," Kauffman said.

McColley had watched her patients' life expectancy triple in thirty years, but she had "just lost an eleven-year-old," she explained, indicating that there remained much more to be done. An investor asked if she had any G551D patients for whom she wouldn't prescribe Vertex's drug. "The pressure on clinicians is going to be with the zero-to-six-year-olds," she said. "One hundred percent of US states, a lot of the EU, Washington DC—all have newborn screening now. So we're identifying this mutation by four weeks of age. The pressure is going to be: 'Can't we crush the pill and give it to our baby and won't that prevent her from developing CF lung disease?' The answer to that question is, 'Quite possibly, yes.' But first you need safety, pharmacokinetics, etc. I can't really in good conscience give a drug to a young group of patients for whom it hasn't yet been studied."

"Just to comment," Kauffman inserted, "we're designing a study now beginning in infancy to cover the age range three months up to six years. It's in process."

"It'll never be fast enough for my families. I just want you to appreciate that," McColley said.

"I understand."

Smith left five minutes at the end for discussing the launch of Incivek, about which the company had scarcely more granularity than the analysts. He invited questions, but before anyone could raise a hand, Emmens jokingly piped up, "How's the launch going?" There was uncertain laughter. They knew he couldn't really comment: Reg FD restrictions, for one thing; lack of real data, for another. "Sorry," he said, "I couldn't help it."

"How *is* the launch going?" Smith asked.

"This my sixteenth drug launch in my career," Emmens answered. "That means I'm old.

"There's two ways to look at it. One is, you're gonna look at prescriptions, and you're gonna look at them too soon, and you're gonna project too much. What I tell you is I listen to the sales force noise and the feedback that we get. They go out at one hundred ten percent. How long that stays is a very good indicator. If that's not there, there's something wrong.

"There's nothing wrong. We're getting great feedback. That's all we can say at this early stage. It's going great."

❖

Keith Johnson got a phone call that Monday from the codirector of the cystic fibrosis center in the department of pediatrics and pulmonary medicine at Beth Israel hospital, Dr. Maria Berdella. He was at the W hotel in Union Square on business. "She said, 'They've terminated the study,' " he recalls.

"She said, 'Look, I know you've done really well on this med, but they say it's not meeting their end points. They're just terminating the study.' I said, 'What do you mean, terminating the study? What do you mean, not hitting their end points? Why take it away from *me*?' I thought I was on it forever. It would be FDA approved, and somehow we would find a way to get it. I thought somebody somewhere would look at this and say, 'Okay, this guy gets the drug.' Somebody would be my advocate. There would be a mechanism in place, because why wouldn't you want this for a patient?"

Ken had anticipated the disappointment facing Johnson and others who had rolled over in the study, and he was "really ripped about the way we're talking about it" inside the company. Smith wanted to disclose that almost a third of the patients rolled over, suggesting a robust response to the drug. But Ken worried that the criteria used in determining who got to stay on the drug after the initial evaluation period were badly designed and thus the data would ignite false hopes. Subjects in the first part of the trial qualified for the open label extension based on either a 10 percent actual rise in FEV1 or a 10 percent drop in sweat chloride—*at any single point in the study*. "You could have been twenty points below at the end of the study, but if you had a spike up ten points during the study you got rolled over," Ken observed. Sweat chloride can't be influenced by other factors. But FEV1 "is not exactly a precise measurement," as he explained. "You have these spikes on a regular basis.

"As proof of that, we had a significant number of placebo group people who qualified for the rollover—because of an FEV1 spike. We have a whole bunch of people in that group and I had to fight to make sure that we were not going to disclose any of that, because I thought the disclosure was misleading. It turns out that we have now disclosed that we've terminated that rollover group—because it's not having any effect. In other words, exactly what I was worried about becoming the reality. So I want to get that information out. I pushed to have that released because I want doctors to know that at least that little bit of data that we have *says it doesn't work.*"

Ken feared that Vertex faced a dangerous dilemma if there was any ambiguity about whether 770 alone could benefit patients with two copies of the DelF508 defect. Despite persuasive preclinical evidence that the drug could be effective in people with other gating mutations, the FDA, moving cautiously, had indicated it was likely to approve it only for the 4 percent with G551D. With such a narrow label, the looming question of what physicians and payers would do regarding the other 96 percent once VX-770 was approved weighed heavily on the company.

"Because what's going to happen here," he says, "and I've already been told this by a bunch of doctors in the CF area, they're gonna give it to people with delta-F508 mutations. They say they're gonna do that, but what if the insurers don't pay for it?—which in my humble opinion they shouldn't. I don't believe that my tax dollars should go to funding Medicare and Medicaid drug payments for CF patients who have homozygous delta-F508 structures because it doesn't work. But they're gonna do it anyway. And they're going to come to us and want us to do an expanded access. My answer, though it's not controlling on the company—but if it were—is, 'No.' There's no data that suggests that this works so we're not gonna give it to you. I think it would be irresponsible to do that. I might as well give you apricot pits."

Johnson was sure there was no way he had been taking a placebo. He knew his body. Since 2004, his chief goal each year was to go twelve months without needing IV antibiotics. Every year he'd failed, going six months, maybe seven, before a pulmonary exacerbation. The last time he was hospitalized was eighteen months ago. VX-770 had turned his life upside down and now, abruptly, he couldn't get it—"a putting Keith on an island type of thing," as he described his jagged emotions: loss, desertion, separation, exile, solitude, hopelessness, rage. He was told he had to return the drug, and Adrienne joined him at Beth Israel. "When we're done with the visit," he recalls, "I just leave. It's just unraveling, and every hour that slips by it's getting further and further away from me and I'm coming to the realization that we're gonna have to live the way it was before. I'm not saying I wish I never had this experience. I'm glad I did. But every hour it's further and further away in the rearview mirror."

Vertex had devised a policy to enable selected patients, based on what used to be called "compassionate use," to receive VX-770 before approval. In May, the company put in place an expanded access program that was reviewing about one hundred patients in the United States who had FEV1s below 40 percent but weren't "in groups where we think the risks are too great," Ken said. It wouldn't, for instance, provide the drug to a patient on a respirator. "It's unlikely they would benefit. Bad things happen to those people on a regular basis and you wouldn't be able to prove that it wasn't because of your drug." Ken:

> In CF you have some pretty compelling stories about people who
> are in rapid decline, and there's a drug that can help them, and it's
> not approved, so what do you do? Our position has been in the past

and continues to be that if we focus on a single patient, it's almost always compelling. What you have to do is focus on the greatest benefit for the greatest number of people. That's not what patient advocates want to hear, but that's probably the most important thing a company can do. Our view is, we need enough data about a compound so that we believe we have a pretty good handle on its side effect profile and its toxicities.

The company's interest has got to be to the larger group that might be harmed if the drug is delayed, or not approved. We need affirmative data that the drug has a certain profile that we have a handle on so we can assess whether to give it to a particular patient. The second criteria is that we need to be certain when we open up expanded access we're not allowing expanded access for people who otherwise might be in a clinical trial we are either running or planning to run, because that will gut the trial process. If we allow people the certainty of getting the drug in an expanded access program versus the uncertainty—because there's always a placebo—of getting the drug in a clinical trial, they'll always opt for the expanded access, and the clinical trial program will die. The FDA understands that, and they take the same position.

Selling advanced data storage technologies gave Johnson a keen appreciation of both the scientific process and the perils of information-sharing. He was avid for privacy, spurning social media because of their invasive mining practices. He read carefully the endless releases he had to sign as a patient, and respected Vertex's position, particularly the clauses in its release form designed to keep patients from broadcasting their experiences and feelings. Though he was too healthy to qualify for expanded access, and staggered by his re-reversal of fortune, he wouldn't have disagreed with Ken's logic or his arguments. "I just think about how irresponsible and reckless it would have been for me if I would have gone onto a forum or something like that and said, 'I'm on 770 with double delta-F508, and it works. Everybody, you're cured,' " he says. "You have to be careful about the expectations you set in a situation that is seemingly hopeless to begin with. So I have loads and loads and loads of respect for that because I actually think it could interfere with the study and the science."

At Beth Israel, after surrendering his meds, Johnson blew a 52 percent—near his original baseline. Indeed, ever since his peak FEV1s of 60 percent a year ago and 62 percent in November, his numbers had been declining. He dismissed the newest drop-off as part of the usual variability. "That was a bad day at the office, really, which will happen sometimes." After his initial feelings subsided, he resolved grimly to fight through his abrupt new impasse, although he had no idea how. What he did know—the same thing that troubled Ken—was that FEV1 spiked and dipped, and there was much else to consider in deciding if a drug for cystic fibrosis was effective. "Who's to determine what my baseline baseline is?" he asks.

"What does five percent mean to me? It means everything. It means getting up the stairs from the subway with no problem. It means being able to play with my kids, most of the day. Even five percent means life. It's a different category of life."

❖

## JULY 28, 2011

"When I see the dailies," Partridge said merrily, "I think, 'Okay, who's guarding Wilt? You may want to put your hand up in his face a little more." For the last couple of days, the Incivek-Victrelis sales figures dribbling out of IMS Health were stunningly lopsided: 92–8. Partridge was riffing on the all-time National Basketball Association single-game scoring record of 100 points set in 1962 by Wilt Chamberlain in Hershey, Pennsylvania, against a hapless and vertically challenged New York Knicks squad. If this was an indication of the ultimate outcome of Vertex versus Merck in hepatitis C, the fight could be over before it started.

Partridge posted himself at the whiteboard for the Q2 earnings call. After two decades of reporting on pipeline progress and clinical data, Vertex at last had product earnings to talk about. The Wall Street consensus for Incivek was around $30 million in sales. Porges, at the low end, projected $20 million. A "whisper number" of $40 million made the rounds, but it included inventory, and most company-watchers dismissed it as hype.

Smith entered tapping on his BlackBerry, grinning. He had kept Vertex flush with other people's money for a decade by raising their hopes; now he wielded results. Before reporting the figures and restating the company's guidance on expenses, he told the analysts: "The opportunity that exists for Incivek is very large, and with this new opportunity we are giving careful consideration to managing reinvestment and R&D for future growth, yet achieving earnings and cash flow both for shareholder value." Smith then announced that total revenues for the quarter had increased to $114 million, from $32 million a year ago, based on net product revenues from Incivek of $75 million—crushing the consensus figure. About half the total represented inventory build, "channel-filling," not prescriptions. Vertex's cash cushion had gone below $500 million but it finished the quarter with more than $600 million in the bank. Based on the sharp uptake of the drug, Smith forecast conservatively, the company would be "significantly earnings positive in 2012."

In the twenty years that Wall Street had been trying to determine what Vertex might be worth based on models and projections, here was the first shred of real data, representing less than a month and a half of sales of its flagship drug. The absurdity was lost on no one involved in the call, but that didn't stop an urgent spike in speculation. Citigroup's Yaron Werber asked Smith to help him extrapolate.

"So we're talking about maybe six weeks, and you sold seventy-five million dollars," Werber started. He did a few back-of-the-envelope calculations. "You said half was retail demand, and then you said that retail stocking on the order of around thirty-seven, thirty-eight million is already out of the channel. So that's within three weeks. And then there would have been inventory on top of that. It almost seems like I can back into a doubling of the rate of sales within the first three or four weeks of July over the last quarter. Am I thinking about this correctly?"

"Directionally, Yaron, I think you're thinking about it correctly," Smith said.

Ken, listening in on the call, worried that Smith's comment was "on the verge of a projection, a forecast," in which case Vertex might have to issue a press release confirming it, in order to conform with Reg FD. He concluded no harm was done, but it was part of his job to keep watch over Smith's close, ambiguous relationship with the Street. Ken:

> I think Ian has actually pulled back a bit. The guy who's really out on the edge right now is Michael, which is okay because that's his job. He's the guy that wants to answer every question that Wall Street has, and if we don't have the answer we can speculate with them about it. I think Ian's big task is coming up. I think he's gonna be *very* hard pressed, emotionally, not to speculate on earnings and sales.
>
> Ian absolutely talks the right way about not being driven by Wall Street on a quarterly earnings basis. What I haven't seen yet is what's actually the only thing that's important, which is how you act, not how you talk. He presented something to the ET and Matt jumped all over it. I had gone crazy over one slide. The slide was four charts that showed what Wall Street's projected earnings per share were for 2012 and 2013. And the headline of the chart was: How Can We Meet Wall Street Projections?
>
> That is exactly what I'm talking about when I say walk the walk. Talk all you want about not being driven by Wall Street, but as soon as you start *thinking* about how you're going to meet Wall Street projections, you're off the reservation. I was really gonna go after it, and then, before I could say anything, we get to that page and Matt says, "Let me say something before we get any further. I don't ever want to see a slide like this again."

Emmens, like a team owner watching his franchise build a solid first-quarter lead, relished Vertex's position, but he was attuned to any disruption in the game plan. He, too, like the analysts, had to project ahead. The next step for the company was to think about connectivity, growth, and scalability worldwide. All three required costly investment. Yet no CEO can ignore for long Wall Street's betting culture and relentless insistence on mounting returns. Now that Vertex was on a steep path to profitability, Emmens wanted to satisfy investors and analysts, but the

Street always had been—and would continue to be, especially through the launch—secondary, the touts setting the odds, the fans in the sky-boxes caring more about the point-spread than the actual outcome of the game. The solution was to win big, crush the spread, leaving the team on the field to stick to its plan, running audibles when opportunities arose, immune to the pressures of speculative second-guessing and the fortunes at stake on the sidelines. Emmens:

> What do you do with the money? Investors are biased toward "give it back to me." We want to reinvest the money in the business. Here you get to this disconnect of timing. For the investment community five years is a millennium. And in the drug development business that's probably a third of the time you need to develop a drug. You're always going to have that tension. So we have to build the trust with these folks that their money staying with us is a good thing, and that we can make it work. That idea has been discounted by the lack of productivity of the big guys.
>
> At Shire, we went out and bought an orphan drug company. The market went berserk. They hated it, because that's not what you do. Exactly right, but strategically we needed to do other things. We needed to get into research, but we also needed to learn about that, so we bought a self-standing business. Strategically it was a way for us to have a research platform. The market didn't get it. We lost fifteen percent. I was a piece of shit for, I don't know, three months. Now it's a third of the business and the top contributor to growth in a company that's growing twenty-five percent on a billion-dollar quarterly base.
>
> You have to outthink the market. You have to think strategically and opportunistically at the same time, which are often at odds.

It was early, of course. Any number of things could—and would—go wrong. A patient could die, then another, inviting the FDA to put a hold on the drug. The supply chain could falter. Government payers could run out of money. A day earlier, the Seattle biotech Dendreon, another company whose stock had soared along with expectations as it launched a promising new drug, shocked investors by announcing that sales of its prostate cancer medicine, Provenge, were much slower than expected because doctors were worried about getting reimbursed. As Wysenski had predicted, any chink in the smooth connection between the user, the customer, and the payer—even a doubt or hesitation—could be catastrophic. Provenge cost $93,000 for a one-month course of treatment. Doctors unwilling or unable to shell out the money up front were concerned that insurers might not reimburse them on time, if at all. Shares of Dendreon plunged 62 percent—proving again Wall Street's fear of any uncertainty arising from a drug launch. Noting the tendency for the industry to rove in financial lockstep, and for investors to become easily spooked, biotech columnist Adam Feuerstein of the website TheStreet predicted: "The ill-effects of the Dendreon debacle will unfortunately spread beyond the company's stock price to infect the entire biotech sector."

Then there was Merck, which during its earnings call the same hour as Vertex's announced it would cut about thirteen thousand jobs by the end of 2015, adding to the twenty thousand it had shed since acquiring Schering. "The pharma industry is starting to look a lot like the auto industry," a commentator noted. Though Merck reported over $12 billion in sales in the quarter, analysts drilled CEO Ken Frazier about Victrelis, which accounted for just $21 million. Looking for a positive catalyst in hepatitis C, they were disappointed to find Merck at the trailing end of a 75–25 percent split in the market. "There are headwinds for this industry, and the Merck outlook was indicative of the situation," Damien Conover, analyst at Morningstar, concluded. "This just confirmed what people thought, with the environment forcing Merck to do cost cutting to adapt."

Despite the ebullience at Vertex, which would spill into the next week even as the stock peaked at $52 before sliding to $43 along with the rest of the sector among broader economic fears, the spreading summer doldrums weighed on the industry, which more and more seemed stagnant, terminally self-afflicted. As Boger observed in the early days, no matter how different it tried to be, Vertex would be held to account by Wall Street not only for what it could control but also for a myriad of intangibles beyond its reach. "I'm not worried about us," he said. "I'm worried about others falling on us."

❖

Washington was a separate matter. The federal debt-ceiling "crisis," manufactured by the Republicans who took control of the House in the 2010 elections, drove the country to the brink of default until, late on the night of July 31, Obama and congressional leaders of both parties broke the stalemate. The ceiling, which had become a bargaining chip in the debate over government spending, was extended in exchange for $900 billion in immediate across-the-board budget cuts. A "super-committee" was assigned the task of coming up with a second round of deficit reduction and a "trigger" was adopted that signaled that if Congress failed to enact the cuts the result would be sweeping reductions in military spending, education, transit, and Medicare payments to health care providers. By the end of the week, credit rating agencies began downgrading US government securities for the first time in history. Stocks plummeted.

Vertex was aloft, yet VRTX got slammed, plunging 17 percent. Partridge, alarmed by the disconnect, sought to reassure employees who, just as they expected their 401(k)s to pop, watched VRTX get swept down. Conflating the broader anxious investment climate with what would soon be dubbed the "Dendreon effect," he encouraged them, in an article for the company newsletter, to be clear-eyed about what was happening. He wrote:

> Amid the broader market decline, stocks of health care companies have taken a disproportionate hit recently, in large part because investors believe that the debt-ceiling agreement will result in near-term, potentially significant cuts in health care spending. The health care sector had outperformed most other sectors since the beginning of the year, making it relatively easy for investors to sell these stocks, book a profit and walk away.
>
> To put this in perspective, Human Genome Sciences (just launched Benlysta, a lupus drug) lost nearly 25 percent last week, while InterMune (launching Esbriet for idiopathic pulmonary

fibrosis) lost 29 percent. Important to note, Vertex is still up about 23 percent on the year, while the NASDAQ biotech index is down by 3 percent. In the NASDAQ 100, Vertex is the tenth best performing stock this year (beating such household names as Electronic Arts, Starbucks, Apple and Gilead). So, despite the downturn, we're hanging in there.

We have reason to be optimistic about the months ahead as well. The Incivek launch has so far exceeded Wall Street's expectations, and recent IMS data suggests to many smart investors that this out-performance will continue. We are only a few months away from regulatory submissions of a second drug, VX-770, for cystic fibrosis. Our value is clear and measurable, and we are well positioned to see investors come back to the stock once the dust settles from the panic selling.

Smith viewed the dip as a short-lived phenomenon, an imperfect storm. He calculated there were three forces affecting VRTX: the stomach-churning uncertainty in the capital markets, generating a flight from equities; broad calls from chief investment officers to pull money out of any company launching a high-priced drug because launches are high risk and many fail; and, disassociated from the other two, Vertex's internals. Those, he believed were solid, starting with the launch. "Count the scrips," he said. "The scrips are coming in, day after day. We continue to advance toward the filing of the cystic fibrosis drug, which we have great confidence for. We have greater clarity with the FDA about how to advance VX-770 monotherapy into other mutations, thereby broadening the benefit. We're coming up on data with JAK 3. And financially we're just getting a capital structure that is giving strength to do things in the future. So inside the walls of Vertex the company is just getting stronger and stronger."

What worried him most, short term, was a severe adverse event with Incivek. Until the company could treat tens of thousands of patients, any such reported event, even a one-off, would be statistically meaningful and reputationally a major setback. Early on, it could stop a drug in its tracks, so however promising Incivek's ramp now looked, everyone on the ET worried about one. "I don't want to blow up a patient," Smith said. "We've built a big tanker here that doesn't turn in tight circles. Our operating costs are close to $1 billion annually. And we need this launch to keep going. We need a good strong top-line for 2012. We're living by this drug."

Where Smith and Emmens were in strong agreement was on the company's long-term outlook and the need to balance strategy and opportunity to outthink Wall Street now and over the next several years. The challenge came down, as Boger had posited to the Harvard Business School team, to managing portfolio risk. Looking a decade out, Emmens couldn't envision Vertex's expanded product line, in what new diseases, but he thought he had an idea of what it might look like in say, 2016. "It's probably what you're seeing," he said. "If we have the best anti-inflammatory for inflammatory diseases, including RA, it's a six-billion-to-ten-billion-dollar product. Flu could be a billion or two per year. Epilepsy could be five hundred million bucks, but the market has no clue about that, because when there's no model the market doesn't know what to do with you."

Of all the scenarios clouding Emmens's forecast—besides, of course, an unanticipated late-stage failure—the most ominous was the one where Vertex, beaten to market by an all-oral treatment for hepatitis C, suffered a sharp drop-off in sales of Incivek before its JAK-3 inhibitor or its flu molecule, now just starting to be tested in humans, reached approval. Each week the rapidly increasing prescriptions grew into a sales curve with a trajectory that would generate for several years at least enough cash to grow the company *and* reward investors. But what if an all-oral regimen came sooner, or the drugs furthest along in the pipeline failed, or were delayed? Three and four years into the future could open a yawning gap between what the company was bringing in and what it was spending.

Smith jauntily started most visits to Emmens's office by offering his preferred solution, also the preferred solution of every analyst: "So where's our late-stage nuc?" The Alios drugs trailed by at least a couple of years compounds being tested, singly and in combinations with partners, by Pharmasset, BMS, Gilead, and several other companies. If nucs would eventually overtake protease inhibitors, as they had done with AIDS, Smith thought Vertex needed to be far more aggressive. Emmens saw the future differently. His urgency was less about satisfying those who considered that scenario as a major threat than about positioning Vertex to be in the right place in 2020 and beyond.

"We're working on a midterm play," he said. "One of the things we could do, with a lot of money, is buy one of these companies and change the game. There's two reasons you do things, and you're often doing them both at the same time. You either do them offensively or defensively. We've done some defensive moves. We haven't done the offensive move yet. Ian wants to cover the bases, and he knows Wall Street will respond positively to that. But I think the market will be chopped up, with all these tie-ins. I'm not so sure I want to be in this market in 2017, to any great extent. I don't want to depend on it, because gradually you get chopped away. Even if we're the market leader, as these guys come in, you ain't going up, you're going down. And Wall Street kills you.

"I constantly fight with Ian on that. He wants today to make the market happy. I'm worried about ten or fifteen years out."

❖

Porges didn't rely solely on IMS data and Vertex's sales figures to generate his projections. A month after the launch he and his Bernstein colleagues had conducted two focus groups with ten high-prescribing physicians from across the New York area to gauge their preferences, which strongly favored Incivek over Victrelis. Now in mid-August, he wrote a note advising investors that the drug was doing even better than predicted. He said sales for the year—with only 222 days postlaunch— could top $1 billion.

Porges outlined three scenarios. If prescription growth froze, sales would total $725 million, slightly exceeding the Wall Street consensus of $700 million. If the rate of increase tailed off slightly, the figure would be about $900 million. If sales kept growing at their present rate, he said, the drug could hit $1.2 billion, which would make it easily the most robust drug launch in history, exceeding Celebrex, Vioxx, and Lipitor. Investors would be able to see which outcome was most likely by looking at Vertex's third-quarter results. There had been no resistance on price, and Incivek was outselling Victrelis 4 to 1. Cumbo's force was crushing the opposition. A few days later Canadian regulators approved Incivek.

"It doesn't get any better than this," Emmens commented again and again. If he was worried about the risk of complacency, he seemed more concerned about making people inside and outside the company appreciate how exceptional a position Vertex was in. More than anything, he believed the launch validated a formula for success that prized heterodox thinking, improvisation, boldness, a tolerance for the messy uncertainties of leading-edge science, and the nimble strategic vision that Boger and the other early adopters had had the moxie to dream up and see through.

> When you think of this industry, of all the successes there are, it's mostly opportunism, either scientifically or from a commercial sense. Very little of it can be planned. Every time we've tried to plan—say, "I'm gonna be the best at lipid-lowering agents, or I'm gonna be the best at hypertension, or something else"—it's never worked. Not once. Lilly was *the* company that sold antibiotics in the fifties, sixties, and seventies. Then they invented the antidepressant Prozac. So they had anti-infectives and Prozac, and they said, "We're gonna build those divisions. We're gonna be the best in the world. And we're gonna shut everybody else out." Guess what happened? Nothing. And they're sitting there today with their patents expiring, and thirty years of research, and no productivity. I could tell you the same story about a dozen other companies.
>
> I mean, what are the chances of our first drug being able to pay for our current size? What are the chances that you'd get through all the trials unscathed? If I told you ten years ago that we're gonna go up against Merck the very first day on our first drug, you'd say, "Well, you're gonna fail." If I told you you'd be the fastest drug ramp in history, you'd say I was insane. So when I look at this in a retrospective way, it's a pretty incredible story.
>
> The analysts always say, "What's your model? My model, I need to plug in numbers." Our model here is, "Let's just find people who

are really sick, see if we can mimic the disease in the laboratory, and see if we can make an effect on that disease." People ask what's our strategy. You know what I say to them? "Find sick people and make drugs for them." You get too fancy on the strategy you're gonna screw yourself up. It takes away from the opportunism. The strategy is simple: Are you in the health care business to treat patients with serious diseases, and not be afraid of really tough targets? That's good enough for me.

Just after Labor Day, Vertex announced the results from a Phase IIa study demonstrating that its JAK3 inhibitor, VX-509, substantially improved the signs and symptoms of rheumatoid arthritis, solidifying the company's late but competitive position in the race for an oral Enbrel. Almost half the subjects on the drug showed a better than 50 percent improvement. Unlike Pfizer's less specific experimental JAK inhibitor, which appeared slightly less potent but had completed late-stage testing and was on track for FDA submission by the end of the year, VX-509 seemed to have fewer side effects. With Enbrel, the original injectable for crippling joint disease and other autoimmunities, fading over time, the market leader was now Abbott's Humira, which with sales approaching $8 billion verged on replacing Pfizer's Lipitor as the world's top-selling drug as Lipitor's patent protection expired. Vertex said it would move the molecule into a larger, six-month Phase IIb study.

Eleven analysts published opinions on the protocol. Across the board, they concluded that VX-509 was at least as effective as any of the more than a dozen JAK inhibitors in clinical testing, as well as Enbrel and Humira. The Street ascribed little or no value to the compound yet, but most analysts saw considerable upside. Where they differed was on safety. Porges called VX-509 "a promising asset and significant driver of incremental value," adding, "we regard the lack of hematological side effects, and the potential for once-daily dosing, as being the main sources of potential differentiation." ISI Group's Mark Schoenebaum echoed: "the lack of neutropenia for VX-509 could be an eventual differentiator . . . we see only upside potential for the stock if the detailed data are convincing." Even Morgan's Meacham, perhaps Vertex's most influential doubter, was impressed: "We view the VX-509 data as encouraging . . . however, at this point, not enough is known about the clinical profile of VX-509 to identify a clear point of differentiation relative to other oral RA compounds."

A notable dissenter was David Friedman of Morgan Stanley. Friedman had joined the firm in 2006, left to manage a hedge fund in 2009, then returned a year later to launch coverage of small and midcap biotech stocks. When he'd initiated coverage of Vertex in April, the stock price was $56; skeptical, Friedman set a target price of $22. Now he seized on a small increase in a liver enzyme called alanine transaminase in patients taking VX-509, which also rose in the placebo arm. "From a safety perspective," Friedman advised investors, "ALT elevations seem to be the most concerning AE noted. Those elevations paired with the lipid elevations make the safety profile start to more closely resemble [Pfizer's] tofactinib."

The next Tuesday, Emmens spoke at Morgan Stanley's Global Healthcare Conference in Manhattan, to a room packed with investors. Friedman introduced him. In his remarks, Emmens made a point of noting that the Incivek launch had been successful, then added that Vertex planned to file for regulatory approval of VX-770 in the next couple of months—a repeat performance unprecedented within the industry. He turned to Friedman. "I think we've done a little better than you thought we would," he said.

"I absolutely agree."

Emmens kept going. He had as little tolerance for analysts as Boger had. His contract as CEO expired in May and he relished the knowledge that he would soon no longer have to bother with the standard industry script at such events, which, as Xconomy's Luke Timmerman observed, was designed not to offend at any cost and could be reduced to "politically correct platitudes about novel technology, helping patients, having a 'good working relationship' with the FDA, and constructive partnership talks." Emmens leapt at the opportunity to admonish a critic in front of an audience accustomed to excessive deference.

"Do you understand this company? I don't think so," he said. "By what I read, I don't think you understand our company. I'm going to do the best I can to prove you wrong again and again, because it's been fun. But when does it stop?"

Emmens appeared jovial, but his underlying frustrations with the financial oddsmakers were keen. He thought Friedman was reckless and made his feelings known more pointedly in a private meeting afterward with Morgan Stanley's CEO. His public remarks were so uncommon that they were reported by the Dow Jones Newswires, then picked up extensively by industry bloggers, a case of man-bites-dog but also a refreshing acknowledgment of the chasm between what went on inside companies and Wall Street's predisposition either to overrate or discount it, depending on your thesis. Friedman responded by saying he was "trying to do the work that we see."

Having made his point, Emmens was ready to move on: "Okay," he replied. "You do your job, we'll do ours. Keep it up."

# CHAPTER 13

***

## SEPTEMBER 23, 2011

With nearly twelve acres of contiguous indoor exhibition space, an eight-story media tower, and a video wall twice as wide as the one Vertex displayed at Launch Week, the Boston Convention and Exhibition Center in a former industrial corridor in South Boston normally hosted trade shows and professional meetings. The largest building in New England, it was one of two facilities in the city equipped to host 1,750 employees—about 90 percent of Vertex's global workforce—for a daylong "milestone meeting" followed by a formal gala headlined, it was promised, by two top-name musical acts. After registering and mounting the escalators to the glass-enclosed top-floor foyer overlooking the innovation district where the company's new home office was going up and, beyond that, the resuscitated harbor, early arrivals, half of them jet-lagged, milled around the coffee stations, speculating on the entertainment. The dominant rumor was Aerosmith, whose lead singer, rock legend Steven Tyler, was cured of hepatitis C in 2006 after eleven months on peg-riba. "It about killed me," Tyler said afterward.

The company had invited every employee and a guest and had paid for them to travel to Boston for the weekend, Emmens feeling strongly that it would be the last time before Vertex got too big that such an all-company outing would be feasible, much less affordable. As it was, the tab was $4 million. Emmens considered it well spent. Never again would Vertex be in a position to celebrate its past and present while laying out its future. The previous week *Science* ranked the company number one in the magazine's list of top employers in the biopharmaceutical industry, beating out the previous year's winner, Genentech. That Friday, Ken retired, signifying the start of a new era without any Bogers involved in the company's operations. It was a good time to acknowlege progress, take stock, paint a vision, back-pat, then feast, drink, and kick out the jams.

If Emmens had not been a hero to his troops earlier, his takedown of Morgan Stanley's Friedman in front of Friedman's clients and bosses instantly had become lore. He evoked the episode in his opening remarks. Roaming the stage like a talk-show host, riffing nonlinearly, he told them he had two recurrent dreams about the future. To illustrate his "bad" dream, he showed a slide of a mock magazine cover from GoogleUniversal's *BusinessWeek* dated September 23, 2036. It was a blow-up of the eighteen-story Fan Pier headquarters (on which construction had halted minutes earlier after a load of rebar fell on three construction workers, briefly pinning two of them) draped with an "office space for lease" sign and bearing the caption: "Vertex Pharmaceuticals; What went wrong?" "I hear a rumor that Morgan Stanley is taking over the building," Emmens said, pantomiming a hunched, churlish Friedman, "and there he is: '*I told you so.*' "

In his "good" dream, he said, he and Boger strolled together along a beach as old men, doddering, their trousers rolled, dousing for coins with metal detectors. Emmens acted out the scene. "What's that?" he asked Boger, his voice creaky and drifting. "A pen," Boger said, bending over slowly to pick it up. Emmens: "What's it say?" Boger: "Merck." "Merck?" Emmens said, as if struggling through the haze of time to recall something. "Isn't that the company we put out of business with Incivek?" Laughter erupted across the audience. But Emmens, once going, was gone, segueing into his next impression: Mueller as front man for an Elvis tribute band. "He ends every song, 'Zank you very much!' " More laughter. Then it was back to Wall Street. "Our plan in ten years is to buy Morgan Stanley." He smiled conspiratorially. "We'll have a layoff the first day: '*Here's your synergy!*' "

Every CEO, to motivate and inspire, needs foils. But Emmens had a larger, more serious purpose; he wanted the whole company thinking about where it was headed, not just in the months and years ahead but decades from now. With the ET arrayed on café stools behind him, he invited each to speak briefly and to provide a visual metaphor—a snapshot of Vertex in transition, linking past to future. HR director Lisa Kelly-Crosswell offered a close-up image of hands at a loom, weaving a tapestry, the human warp and woof of a growing worldwide health network. Wysenski put up a grainy black-and-white slide from the 1880s of the uncompleted iron lattice foundation of the Eiffel Tower, suggestive of the pinnacle to come. Sachdev put up two pictures—a racing bike and a racing motorcycle, a stripped-down monster—noting that Vertex would only speed up as it evolved. Smith, as he often did, showed sheep in a pen, his preferred analogy for corralling investors. Gazing out over the sea of tables, awed, he marveled that the company had made it this far.

Mueller, the team's ranking futurist, recalled the company's twentieth-century origins as a trailblazer of "rational" drug design, then spun a superseding vision for the next fifty years, one that was not only smarter and more target-based than the old hunt-and-peck drug discovery methods but more integrated across intellectual realms, disciplines, technologies, and continents. "Rationalism is good," he said. "Integrism is better." The new scientific reality, he said, *must* transcend symptomatic treatment in favor of a new paradigm: "repair, replace, restore, regenerate." Many of his slides were of fluorescing stem cells, artificial organs, virtual limbs, and bionic prosthetics. "Health is value," he said. "It's not just about pills anymore."

At night, the crowd descended to the cavernous exhibition hall, a cement-floored ballroom the size of two football fields lit up warmly from the rafters by ten-foot colored globes and giant luminous squid-like balloons. Nearly four thousand predominately tall women in minimal dresses and festively attired men, a spectrum of nationalities, mingled around constellations of low white couches near the arena-sized stage, cruised the bars and groaning food tables, or else loitered in the back amid a small carnival of Wii setups, photo booths, and hyperenergetic foosball games. Boger, wearing a tuxedo with a glittery gold bowtie, circulated with his wife, Amy, sculpted into a sequined cocktail dress, rock-steady on spike heels. When he took the stage briefly for a few off-the-cuff remarks—"One of the things we learned was, don't put your baby in the arms of a big pharmaceutical company"—the milling throng overwhelmed him. "It's the first time," a veteran scientist remarked, "I ever saw Josh look small."

For those few renegades remaining from the early days, the scene vindicated their sacrifice and commitment, but also represented a mismatch in time. They had tried to do everything differently, better, more imaginatively; had tried to redeem the drug industry's fallen, besmirched mantle; and had undeniably succeeded—for now. Yet here was a sea of strangers carrying forth their banner beneath a continuous slide show assembled from the day's training sessions, dynamic action shots of Emmens and Mueller, the company's new visionaries. Did the company still get the fullness of the challenge?

"I couldn't believe Ian said he could never imagine this day," Murcko said, struggling to be heard above the cloying reggae beat of San Diego rocker Jason Mraz, the opening act. "Of course we imagined it. Why did we leave Merck?" Murcko was seldom surly or resentful or bitter. He prided himself on his objectivity. But the day's cheerleading and celebration seemed to him to lack a crucial bit of Vertex's core memory: the sheer gleeful vengeance with which the founders resolved to upend pharma, a legacy of militant heterodoxy and willful self-exile.

Boger and the torchbearers had produced the successes that the company was toasting now, but Murcko thought in twenty- to thirty-year time frames, and for the past five years, he hadn't seen the same magic coming out of the labs and into development. It troubled him more than he let on. Discovery remained the hardest part of making a drug, and Murcko thought that Vertex, like the rest of the industry, was coasting on old research triumphs while failing to produce and sustain newer ones. The company hoopla seemed forced; the great risk and commitment of doing leading-edge science—and therefore the compelling personal need and greater triumph—were somehow underappreciated. "You can't go to parties and tell people you work for a pharmaceutical company," he muttered. "You work eighty hours a week your whole damn career, and people look at you like you're evil."

After Mraz finished his set, Murcko and his wife, Kathy, a musician and teacher, got up to go. The throng was just hitting its stride, getting louder, beating together in the way of all giant parties. Singer-songwriter Lionel Richie, who came to fame as a vocalist and saxophone player with the Commodores in the mid-1970s—before many in the crowd were born—took the stage, launching into a string of old hits. Murcko recalled feeling similarly when he was at Merck, and the company had the talk-show host Dick Cavett entertain a launch party. Richie, backed up by an eight-piece band, swung into the Commodores' 1977 single "Easy," an R&B ballad with country and western roots expressing a man's feelings about ending a troubled relationship:

> That's why I'm easy
> I'm easy like Sunday morning

The Murckos took the long escalator to the glass-walled foyer, watching the sea of action from above, munching chocolate chip cookies from one of the dessert tables set out for the revelers as they left to hail a cab or find their cars. "It's sad," he said.

"Corporate," Kathy whispered.

❖

Smith expected October to be "a really interesting period" for VRTX. Propelled by IMS data showing a boffo launch, the share price had climbed back over $50. The "short short" thesis—that Vertex would price the drug too high or screw up reimbursements or otherwise disappoint and frighten investors, as Dendreon had—was succumbing to the mounting pickup rate in prescriptions. Most on Smith's mind was the run-up to the Liver Meeting in early November, in San Franscico, shortly after the Q3 earnings call, a prime period of opportunity for those analysts who considered Vertex a one-drug business and were betting against its strategy for an all-oral regimen against hepatitis C.

"The 'long/short' now starts to play," Smith said, "because they say, 'Oh, Pharmasset has some wonderful data, and they're gonna have their drug on the market by 2015.' Frankly, they do have some good data. They do. They are our prime competition. And investors believe that as well. So we're gonna go through October leading into our earnings call over the long-term/short-term debate, because Pharmasset may be coming to take our market away from us come 2014 and 2015. And yet people are gonna say, 'What's the revenue number that's coming up in three weeks?' "

Despite its breakthroughs in CF and other diseases, Vertex remained what Smith called "an area under the curve company." Analysts valued it based on an imagined rainbow-shaped graph of revenue projections over time, those revenues coming almost entirely from sales of Incivek. Now that income was ramping sharply, the concern on Wall Street had shifted at once to the out-years, fueling Smith's argument with Emmens, the ET, and the board. "What we need to do is, instead of coming down at the back end, we need that line to continue," he said. "I think we have enough credibility with Incivek and 222 that if you add a late-stage nuc and commit to a large Phase II study, people will believe that works. And all of a sudden, you have an offense against all the other all-oral regimens that are being created. At the moment we don't have that."

Boger, quasi-exiled like a deposed king who retains the right to wear his robe at state functions but is banned from the war room, complained that Vertex was failing to exploit a rich opportunity, concentrating more on playing the Street's game than seizing its imagination. "Analysis will follow the excitement," he said. "Someone will generate the analysis to support the excitement." As ever, he believed you created value in biopharmaceuticals by painting a vision of the future, by enlarging possibilities and minting fresh hope, not by analyzing numbers to disprove someone else's discounted cash flow model.

"This is a source of frustration for me," Boger said. "I love Ian to death. But Ian does channel back the Street's viewpoint into the company. So when I suggest things from my impotent director's chair, like, 'Ian, isn't this fall a great time to have an R&D day, where we actually talk about all kinds of programs that they don't care about right now, not because we think they're gonna add earnings per share to their estimates but because we have to start building that story?' he says. 'No one will come. They're not interested.'

"You know what, that's not a good enough answer. They weren't interested in our original story either. Of course they're not gonna value it, because we never talk about it."

Thomson, too, bridled at the trend within the organization to pull back and rely on traditional avenues to build the business. Though Mueller supported him, he worried that Emmens and the ET were hampering the company's prospects in East Asia. Thomson's forays had reached the point where he was ready to recruit Chinese partners to develop either a narrow-spectrum antibiotic or the company's flu drug. The twentieth-century model for selling drugs in Asia was to partner with a Japanese firm for the entire region. But Vertex had come to grief over the licensing agreement with Mitsubishi for Incivek, and Thomson pressed senior management not to repeat the error.

"Japan has rights to telaprevir in Asia," he recalled, "which tells us that molecule may never make a dime in the third of the world's population that most desperately needs it. We're stymied to a large degree with our current biggest asset. Now, how much we learned from that for the future is something different. My viewpoint is go and partner regionally—in China, in Malaysia, in India, wherever—individually. It's more work. And it requires becoming familiar with the unfamiliar, which *spooks* the business development fabric of our organization." He went on:

> The Chinese pharma sector ranges from quaint, old-fashioned, pre-historic, scary, natural-product-oriented, provincial—all of those things—with elements of rapid growth. They're fast learners. And the rate at which they're evolving is dramatic. You go to Chinese companies and see this scruffy old set of facilities where there are questionable standards and equipment and capabilities, while over there is the new ten-story ultramodern research facility they're building. So they're absolutely in a state of transition; the good ones. The government is stimulating their getting into the modern fast lane of innovation. They don't know how to do it. Very few of them have taken to market a novel drug. They have very limited experience in developing new molecules, let alone a first-in-class or a first-in-humans, which is what we're leading in there with.

While Thomson champed to light a fire in Asia, Wysenski and others moved to plant a small but significant flag in Europe. European regulators signaled they would review VX-770 for *all* CFTR gating mutations, increasing the potenial market, and more than half the sales force for cystic fibrosis was slated for the EU. Weet, Wright, Kauffman, and their teams, meanwhile, put the finishing touches on new drug applications both at the FDA and the European Medicines Agency. With fast-track review all but assured, Vertex expected to begin generating earnings on its second drug by the second quarter of 2012, launching another curve, this one most likely nondegradable. True to type, most analysts paid scant notice to anything but the unspooling IMS data for Incivek.

❖

Coming off Milestone Day, Mueller plotted the next stage for generating future medicines. The weight of integrating the work of far-flung scientific operations, combined with a development flow chart resembling a New York subway map, required reorganizing the sites, the divisions, and the chain of command. On October 4, Mueller took three hundred researchers off-site to an upstairs ballroom at the Hynes Auditorium for Cambridge Science Day, an annual by-invitation meeting of the company's scientific leaders. He spoke for an hour, outlining development plans, rhapsodizing about blue ocean projects, setting goals, and announcing that by the end of the year he believed Vertex would become the first Western company to develop a new drug in China. At lunch he sat with the site heads, Murcko, and a few others, further outlining his ambitions. "Stellar failure," he told them, "is better than mediocre success."

The more Vertex lunged ahead internally, the more Wall Street tried to yank it into place. The next Monday, IMS released sales estimates showing that total Incivek prescriptions for the week ending September 30 dipped almost 3 percent from the previous week, suggesting that sales were flattening and stoking uncertainties about whether demand for the drug would continue to rise. Coincidentally, Pharmasset said it would expand a midstage trial of one of its nucs, PSI-7977, adding two new arms: one monotherapy for twelve weeks, the other in combination with ribavirin.

Vertex knew the IMS figures were wrong. Managed-markets lead Jeff Henderson had the shipping documents showing how much of the drug had left the loading dock in Illinois, and they demonstrated continued week-to-week improvement up to the present. Partridge, pressed by Smith to investigate, had long experience with IMS's methodology. Tracking sales of more than one million prescription products, the company samples daily from different channels, then conducts some statistical algorithms to generate an estimate of the amount of drug being sold. "If you add up the dailies, you don't get the weekly figure, not even close to it," Partridge says. Vertex contacted the company. It discovered that one of the major distributors, Caremark-CVS, had stopped reporting its prescription data for Incivek sometime in late September, zeroing out 12 percent of prescriptions.

It was a clear, warm, end-of-a-long-holiday-weekend morning across the region, Day 19 of the peaceful occupation of Zuccotti Park near Wall Street. With satellite demonstrations beginning to sprout in other cities, the depredations of the financial system, real and perceived, rumbled across the economy. New York mayor Michael Bloomberg told reporters at the start of the Columbus Day parade that he didn't anticipate any effort by the city to remove the occupiers: "The bottom line is people want to express themselves, and as long as they obey the laws, we'll allow them to," he said. "Bloomberg said we can stay indefinitely! Big win!" @OccupyWallSt tweeted triumphantly—prematurely as it turned out.

VRTX started trading down sharply and Vertex pressed IMS to correct the problem. Partridge's phone was jammed with anxious investors and analysts wanting to know whether the company's sales and revenue projections could be relied on, but with the third-quarter earnings season bearing down, Reg FD barred him from commenting. Meantime, a panel of doctors at a Cowen investor conference in New York told the audience that they thought all-oral therapy would be available sooner rather than later, affirming, it appeared, the growing industry chatter. By late morning, VRTX tumbled. Normally, about 1 percent of the company's two hundred million shares changed hands daily, with long-term investors disregarding the general noise, but now a steep slide took over. Three hundred thousand shares traded in a few minutes, driving the share price below $38, altogether erasing more than $2 billion in value in the time it took to decide on a strategy to stem the damage.

"IMS owned up to the last two weeks in September being a little bit off, but they weren't acting with any urgency," Partridge recalled. "So we said, 'Look, this is creating a big disruption in our business. You guys are creating us all kinds of problems, so what can you do to clear it up for everybody? It has to be in a broad disclosure format because we can't do selective disclosures. And, by the way, you're creating this mess, so we'd appreciate your speaking on behalf of your own product, so we don't have to issue a press release to say what we *think* is happening.' "

Here—with the question looming over where the next pool of patients for Incivek would come from just as Pharmasset upped the ante, not with data but confidence enough in PSI-7977 to try to leap ahead with two small, but provocative, Phase II studies—was an opening for the "long/short" thesis you could drive a freight train through. Based on three shreds of disconnected and questionable intelligence, it became immediately popular to think that doctors had already started holding back—warehousing—patients in anticipation of the next wave of treatments, that Incivek was fizzling, and that Pharmasset was the future.

Emmens, despite his irritation, conceded the allure for investors, the elegant either/or/but-certainly-one-of-them simplicity of the long/short play. "You make it us against Pharmasset, and you take one or the other. What a great trading opportunity that is. You think they're gonna come and take our business in two years, so you say, 'I can bet on that, and I can hedge a little with Vertex,' since if Pharmasset blows up we'll go up like crazy. If you're in hep C, that's what the bet is right now."

VRTX plunged 9 percent by the end of trading, with five times the average volume. Pharmasset soared to an all-time high. After the stock markets closed, Bank of America–Merrill Lynch biotech analyst Rachel McMinn, usually bullish on Vertex, wrote a call note downgrading her target price. McMinn still recommended the stock as "a buy" but in cutting her target to $65 from $72 she took $3 billion out of her valuation, sparking a second round of selling when trading resumed on Tuesday. Throughout the morning Partridge took more urgent calls from longtime fund managers reporting that they couldn't stop the portfolio executives around them from selling VRTX. Reg FD sealed his lips. His futility was self-evident.

"We couldn't comment because our quarter was closed and we couldn't preannounce earnings," he recalled. "We told people, 'We can only say what we said before—which we really believed in.' We couldn't tell them what was going on with IMS. We know IMS was looking into why their data might be inconsistent, but there was nothing more we could tell them."

Smith hoped investors believed in the long-term value of the company. "You're effectively playing the trust-me card," he observed, "but you don't know where the trust-me card kicks in." He pressed Emmens to step in to protect the stock and Emmens phoned the CEOs of IMS and Caremark, urging them to act quickly to correct the numbers. The directors were arriving the next day for a regular all-day meeting, and though he already had urged them to filter out Wall Street's heedless volatility, which in many ways was just as damaging as the dark-arts trades and destructive lending practices that aroused Occupy Wall Street and most of its supporters, Emmens worried that the board, too, needed to be reassured, if not about the actual figures, by an awareness that he and Smith were on top of the situation. He summed up the rising disconnect in his cover letter to the other directors:

What a great time in the history of Vertex. Execution by our team has been virtually flawless. With a blockbuster launch and another breakthrough medicine about to be submitted for approval, we are working on our future, with ten clinical projects encompassing eight different compounds, seven of which already have proof of concept. Beyond a great launch, everything else has been on time and nothing has failed in the clinic yet. This situation is unprecedented in our industry. As they say in the understated German way, "Not bad."

The only frustrating part is that Wall Street is focused elsewhere. The markets are bearish, highly volatile, and as a result myopic. I think they will ultimately catch up to our breadth and true value.

It wasn't until midday Wednesday, while the directors met in the fourth-floor boardroom in JB-II, that IMS notified the company that it would issue a bulletin regarding sales figures for the last two weeks in September. The after-hours announcement stated that the figures for the figures had been wrong (scrip estimates actually rose 5 percent for the week ending the twenty-third and 4 percent for the week ending the thirtieth) and that IMS was still missing sales data which might result in updated estimates. VRTX stabilized at a floor of about $40. Emmens, with an agenda that included several hours of discussion about finding a new CEO before his contract expired and a scheduled report from Vertex's advisor on the company's takeover risk, was not mollified by the correction. "This is crap," he muttered, scanning his BlackBerry during a break. "This is not our business. This in not what we're in business to do."

Even with a depressed share price, Vertex was fortunate that the lust for mergers and acquisitions in biotech—what its advisor called "trades"—had cooled as Big Pharma tried to absorb those it had made during the frenzy of the previous decade. The company was safeguarded by its relatively high price and by the faith of the large funds—Fidelity, Wellington, Brookside, and a few others—that held 90 percent of its stock and whose managers, however bemused to be losing money on a portfolio leader, hadn't panicked. Even at $40 a share, Vertex's valuation was above $8 billion, meaning that, with a premium of 50 percent or more based on its rapidly improving balance sheet, strong cash position, and promising pipeline, it would cost another company $12 billion to $15 billion to take it out—a sizable enough trade to force an aquiring CEO to think long and hard.

Whereas three weeks earlier with Friedman he had been jovial, now Emmens was grim. *Where did it stop?* Wall Street's negativity toward the possibility of success in biotech had clouded its judgment so severely that a single misreported data point had ignited a stampede. It didn't matter what you did, he thought, only how much fear the naysayers could whip up on any given day. Twenty-two years of ferocious innovation and backbreaking science and you still faced being downgraded $3 billion overnight because of an isolated, incorrect, one-week sales *estimate* combined with the sudden perception that an untested rival would someday soon eat your lunch. Vertex remained lucky for the moment, but Emmens worried deeply about the overall trend, and he wasn't sanguine about the company's ability to withstand a future raid on its independence if the situation didn't reverse.

"It's really hard now," he muttered. "Their biggest concerns, every one of them, we've showed them that's wrong. You *can* make a pipeline; proof of concepts *are* working; we've *got* a blockbuster; it's *not* a soft launch; we're *getting* the patients; we're *getting* the price. So it's very frustrating to be hitting all those marks, and they're off somewhere else. I don't care about the share price, but this is an incredible bargain for somebody to look at it in terms of enterprise value. I think there's a disconnect now between the value the market gives you versus what it would be to another pharma company, in terms of profitable assets.

"This company," he said, toting the math, "you would be buying immediate income, a billion in cash by the time you closed, and pipeline. If we're a nine-billion-dollar company, at fifteen billion they can give you a fifty percent net premium, and the next day still make it accretive. *Duh.* What you have to do is convince your shareholders that the company is undervalued, that the market is screwed up." He recalled a scene from *Jaws.* "Maybe you can. Or, 'Maybe the shark go away.' You beat on the water, or maybe he comes up and bites you."

The next morning, following IMS's product bulletin on Incivek, VRTX rebounded nearly 7 percent to $43 a share—20 percent less than it lost on Monday. Emmens, Smith, and Partridge hoped the end result would be to reaffirm Vertex's credibility while disparaging the validity of outside estimates, but the damage was done. Porges, who'd also challenged the IMS estimates based on his own focus groups and expansive modeling of how the medical community was adopting the drug in stages, told Reuters the episode should teach investors to wait for reliable data. Seizing the opportunity to defend the long thesis on Vertex, he sniped competitively at BOI-ML's McMinn, his erstwhile ally in the tug-of-war with Morgan's Meacham, Citi's Werber, Friedman, and the rest of the shorts.

"This is the nature of any drug launch," Porges told the newswire, "and no pharmaceutical or biotechnology company marketing analyst with any measure of experience would make judgments about the performance of a launch, and a product's long-term potential, based on one, two or even three weeks of market audit data . . . We trust that the investment community will be discouraged from doing the same after this correction."

❖

"The future," Vertex's takeover defense banker liked to tell clients, "gets repriced every day." New information roils both the math and psychology of every investment decision. CEOs of big companies prowling for acquisitions ask themselves whether they're willing to bet their jobs on a trade, because if they're wrong, their company's stock will go down, and they'll be fired.

By the next Friday, big drugmakers scouting Vertex saw the future as anything but a race between two small companies competing, zero-sum, for primacy in hepatitis C. Abbott announced, based on very small midstage trials, that it could have a twelve-week all-oral four-drug combination therapy on the market by 2015. Using data culled from a pair of studies with just forty-four patients, the company projected SVR rates as high as 90 percent, with annual sales of about $2 billion. Two weeks before the Liver Meeting, Abbott stock—ABT—bobbed nicely for a couple of days while Vertex, Pharmasset, and Gilead all retreated as Wall Street absorbed the increased likelihood that several companies would slice up the opportunity in hepatitis C, as Emmens predicted.

Murcko took Friday off to move his mother from a hospital to a rehabilitation center in Connecticut. His regular monthly meeting with Mueller was scheduled for Monday, and he emailed Mueller before he left to let him know he'd be out for the day. "Never heard a thing back from him," he later recalled. "Nothing, which I thought was weird. Just didn't make any sense at all. I thought 'Jeez, he must really be preoccupied with something.' I didn't know what."

If disruptive innovation means a state of permanent revolution inside those companies that promote it, the challenge ahead was how to scale Vertex's creative DNA and expand its frontiers while trying to build a productive, sustainable R&D pipeline. "The issue, as always here, is, 'Are we fearless enough and can we execute well?' " Murcko said. For years Mueller had had up to twenty direct reports. He wanted to streamline and simplify his leadership structure, bring some people up, shift some around, move functions, eliminate redundancies. In an effort to create powerful and aligned R&D organizations, he named two heads of research—Cambridge site head Mark Namchuk for North America and longtime UK site director Julian Golec for the EU—to spearhead the discovery of future medicines. He promoted Chris Wright to run global medicines development and affairs, putting him in charge of all clinical duties including regulatory affairs. In the process, he decided to disband Murcko's Disruptive Technology group and reintegrate it piecemeal into the functional lines.

Monday afternoon Murcko walked into Mueller's office for his meeting, carrying a stack of research papers and brimming with a long list of ideas. For years he'd been pushing the company to think harder about neurodegenerative diseases, and he had been pleased on Science Day that the keynote speaker was a pioneer in treating ALS—Lou Gehrig's disease—an opportunity now for follow-up.

Mueller preempted him. "Your position has been eliminated," he said. Murcko groped for an adequate response. He had no trouble seeing the logic of the move. Most midsized companies outgrew the need for chief techologists and his independence and influence had been shrinking since Mueller terminated his new project group in 2008. But emotionally Mueller, whom he'd struggled alongside for nearly a decade, seemed absent, conveying no trace of friendship or familiarity, much less sympathy or remorse. He was reading from a script.

"That's apparently fairly common," Murcko recalled. "He didn't want to talk about anything else. The conversation lasted four and a half minutes. I asked him a couple of questions. One was, 'Is there anybody else in my area who was being let go?' I was worried about my secretary, Vicki. He said, 'I don't want to talk about that.' I asked whether there was anything . . . actually it was more of a statement. I said, 'To my knowledge there was no paper trail of any kind of any dissatisfaction with anything that I've done. Is that accurate?' He said, 'Yes.'

"I think on the script there was something to the effect that, 'This in no way is a reflection on your performance,' and he offered to write me a reference or be helpful to me in that sense if I needed it. Then he handed me over to Lisa Kelly-Crosswell. She was one office over. She walked me through the procedure and suggested I go home. I said, 'Well, I want to clean out my office and say good-bye to people.' She said, 'Sure, we'll set that up.' I found out when I got back to my office that they had turned off my email."

Murcko drove home, switching swiftly into "detached, objective, scientist mode. It was all very, 'Oh, this is interesting.' " No one at Vertex except possibly Thomson had worked harder to realize Boger's vision. Murcko had a major role in finding every one of Vertex's breakthrough molecules now on the market or in the clinic: from codirecting the "project from central casting" in HIV; to modeling the core structure of the ICE inhibitor pralnacasan, which generated the company's experimental medicine for epilepsy; to being a coinventor of Incivek; to nurturing the CF work at Aurora and bringing Olson to Cambridge to shepherd it through; to championing the kinase program that made the JAK-3 inhibitor and driving the phenotypic screening approach that produced the flu drug. As word got out, as another torchbearer put it, that he'd been "kicked to the curb," emotions in the labs, offices, and cubicles flashed over.

Knives came out for Mueller, who himself could be cutting, not in person but behind others' backs. Everyone had seen him berate fellow scientists in public, in the hallways, when they weren't around to defend themselves. The coldness of his actions seemed to some of the old guard to betray a desire to assassinate those who dared still challenge him. Mueller had advanced Vertex's science, but he had inherited it and wasn't responsible for finding the drugs that had brought it to this point, diminishing his primacy in the eyes of those who knew the fuller story. "Peter is like an alpha wolf," a veteran researcher muttered. "He needs displays of submission. Mark wouldn't do the public submissive rollover behavior that Peter required." For those who valued Vertex's original spirit, the organization's apparent thanklessness opened a gaping distrust, not just with Mueller but the company itself, what it stood for. "There is no more Vertex," someone said, raising the specter of defections. "Vertex is dead. They should call it something else."

Mueller moved on decisively. "Honorable Colleagues," he began the reorganization email distributed companywide the next morning. In outlining major personnel shuffles throughout research, global development, and medical affairs, he stressed the need for sustainability and congratulated Namchuk, Wright, Trish Hurter, Tara Kieffer, and a dozen others who had been promoted into new leadership roles. At the end he wrote: "In conjunction with these changes, Mark Murko [*sic*], Jack Weet, Paul Caron . . . [three others were named] have left Vertex. Please join me in thanking them for their contributions and wishing them well in the future." The previous week, Weet, exhausted, had finished driving the new drug appliction for VX-770 through the FDA portal, eleven months after delivering the NDA for Incivek. Caron was the modeler who in the earliest days had predicted that the active site of HCV protease would be monstrously hard to inhibit—"a bowling ball."

Mueller and his team rolled out the changes in a series of presentations. Namchuk, a popular scientific manager who'd been with Vertex since before the Novartis collaboration, delivered a pep talk in JB-II. "We've got to do better than we are," he told a packed audience. "Our job is to invent first-in-class medicines faster and better than everyone else. We may be close to the top now, but I think we have to be even better in the next few years." Without mentioning the reorganization's casualties, he referred in passing to "legacy pieces left in that confused the new structure."

Namchuk spoke excitedly and at length about several research projects, notably a Cambridge-based effort to correct the underlying cause of multiple sclerosis, an autoimmunity in which the fatty myelin sheaths around the nerves of the brain and spinal cord are damaged, leading to scarring and a host of painful, debilitating symptoms. The hallmark of the disease is the loss of myelin, which, like insulation on a wire, helps the cells conduct electricity. Lose the sheath, and the nerve eventually dies. Vertex scientists had labored for two and a half years to develop a novel assay: basically an MS brain in a test tube. "We took all the component parts of the brain, took them apart, put them in a dish, gave them MS, saw if we could reverse it, and then put all the parts together and see if we did everything right," Namchuk said. Their goal was to find small molecules to stimulate cells already in the brain to fix the damaged myelin.

"They're sort of like the Red Sox pitching staff," he explained. "They're sitting in the clubhouse, eating chicken and drinking beer. And so what we have to figure out more successfully than [Sox manager] Terry Francona, is how do we get them to jump in, to do some good before it's too late and we're out of the play-offs. What we're trying to do is stimulate this growth and cause these cells to wrap around that nerve."

Namchuk showed slides of cell cultures saturated with rejuvenating myelin, spidery dyed lines showing that its lead compounds had activated the cells, which had "gotten up off the bench" and were trying to reinsulate denuded neurons. Like Fred Van Goor's movies of reactivated cilia in the bronchial cells of patients with CF, the images were visual confirmation of Vertex's approach and a driving totem for the chemists now optimizing the molecules, although any other parallels between the two projects at this point were thin. CF was a well-thought-out, clear, well-executed program, with a single, coherent team of people working fanatically, a well-understood target, and total support from the foundation. At the moment, despite Namchuk's tantalizing pictures and flair for analogy, Vertex possessed none of this in MS.

Murcko returned quietly on Friday to say his good-byes and clean out his office. Like other predevelopment candidates, the MS compounds were a promising but distant hope. One of them might come to market a decade from now, if ever. Mueller's challenge was to see beyond the curve, to the possibilities for the next generation of medicines and the one beyond that. Yet by his decision to jettison Murcko, he ensured that Vertex would have to meet it without the company's most seasoned, deepest-thinking drug hunter; without the person most fluent in emerging technologies and, more to the point, when they were ripe for exploitation and how to marshal them. Murcko was melancholy, writing in an email: "Honestly, I wonder how long it will take before at least some people start to whisper about how ineffective I was and how this was a kindness . . . it seems never to be the case that someone's reputation grows after they leave." Boxing up his books and papers, Murcko didn't complain to the steady stream of well-wishers, many in tears. He resolved to be a good soldier, vowing to himself, for the company's sake and in right-stuff spirit, "not to be an asshole." Later he said:

The not-being-an-asshole rule is really important to me, because I can imagine, being in a situation like this, you could allow yourself to say things that might be perfectly justified, and defensible in a court of law, but which afterward, nonetheless, you would feel like a rat for having said them. So I'm trying not to go down that road.

But I do have to wonder what it means when a situation like this is handled in this sort of way. Certainly I know people at other organizations who have been in similar situations who ended up landing better, where there was more of a conversation of where to go from here, what to do next, some sort of transitional period. None of which happened in this case, which strikes me as odd. It also strikes me as inefficient, because certainly there were things I could continue to do for Vertex in some sort of transitional role. There may be good arguments against that. Whether there's a larger lesson or not for Vertex I can't say for sure, but I've had a number of people say to me that they thought what happened in this particular case, not just to me but to the other folks who were let go, was brutal. And if that's an accurate reflection of what happened to us, I have to ask what that says about the organization. Hopefully nothing. It's just an aberration and not an aspect of some larger change in the culture.

❖

Two days before the Q3 earnings call, Emmens charted his succession. The directors wanted him to stay on as CEO, but he'd convinced them that what Vertex needed next was both an "operations guy" and a top scientist—someone who in the coming few years, before the next great commercial breakthrough after CF, could focus the company as it matured.

The board had culled through thirty candidates but so far had reached no consensus. Emmens resolved to announce a change by the end of the year. In particular, he believed Vertex needed a leader who could come in and adapt, rather than someone who would think of it as his—or, far less likely, her—company. Also, he thought, the situation required a strong counterweight to Mueller, not a taskmaster but someone subtle enough, with the right balance of emotional intelligence and scientific sophistication, to handle his complicated temperament. Emmens:

Peter's funny, because what you see is not what I see. He's very, very demanding. He's a control guy. That's what made him good in bringing the right products forward and making the time frame to market. But he models it. He works like a dog, big hours. The answer is never, "No, I can't do that." It's always, "Yes." And with *me*, Peter's also a very good employee, because when I say, "Peter, I'd like to do this," it's, "Yes sir." So I'm very careful about what I say to Peter, because if I say we're not gonna do cancer anymore, he might put up a little fight, but he'd cut the whole place out. Just like that. He's not who you think he is.

In a way, Peter is very easy to manage if you let him do what jazzes him, which is in the best interest of the company right now. I want somebody who wants to watch what he does, understand it, put it in the bigger context, and help him do it. If you do that, you'll keep Peter, you'll keep him happy, you'll keep him engaged. Peter likes to learn. He's always learning. If he ever gets bored here he'll leave. He's an unusual character.

❖

After the markets closed late on Thursday afternoon, October 27, Emmens, Mueller, Smith, Wysenski, Kauffman, Pace, and Partridge took their usual seats for the Q3 earnings call. Their going-in strategy aimed to blunt the momentum of the shorts with blowout earnings figures and a convincing commercial vision to show that Incivek sales weren't flattening five months after launch. "I think our tone should be giddy, not cocky," Emmens instructed. "Very, very confident."

Representing the company's operations in the first full quarter since the launch, the numbers conveyed multiple meanings. It was the first time in Vertex's twenty-two-year history that it had generated a quarterly profit from its own product sales: a potent rebuke to the IMS confusion, which had resulted in the data firm releasing three separate product bulletins correcting for misleading scrip estimates; a gauntlet thrown down to the analysts; a powerful vindication for Emmens, Wysenski, and their strategy; and a crystal ball into the future, specifically the next fifteen months, since what the Street would look for would be a weekly run-rate from which to extrapolate earnings through 2012.

"I'll tell you one thing," Mueller said, handicapping the call. "Tomorrow something will happen, and they'll jump out the window again."

"There's an optimist for you," Kauffman said.

"One of the guys did a study that R&D heads that are German, once they get their first drug approved, they're not productive anymore," Emmens told Mueller.

"But they're from Frankfurt."

As Partridge contacted the operator to start the call, Smith cut in. "These guys are never wrong," he said. "Don't rise to the bait." After a decade of finessing a mounting burn rate by stoking ever-higher hopes for the business—and two weeks of grinding through the volatility over IMS—Smith gleefully disclosed the third-quarter results. Revenues for the quarter were $659 million, a twenty-five-fold increase from a year earlier. Net product sales of Incivek were $420 million, beating expectations handily. Vertex declared a profit of more than $1 per share, with income totaling $221 million. Smith, who had pushed hard for the release, revealed internal data showing that Incivek sales had reached $40 million to $45 million per week in September and October, a run rate implying fourth-quarter sales of $520 million to $585 million, far above the current consensus of $411 million.

The September–October numbers were spectacular and unimpeachable, far higher than anyone had predicted, but indeed had leveled off, more or less as IMS had crudely, even erroneously, detected. Wysenski explained that the uptake among the highest prescribers had been so dramatic—seventeen thousand patients in all—that many had reached their capacity for putting new patients on the drug until the first wave came off. She said the company was focused on expanding the number of doctors using Incivek to reach the next wave of patients.

During the Q&A, Bank of America's Rachel McMinn seized on the apparent softening. "I guess I wanted to ask a little bit more about that one slide, the $40 million to $45 million in weekly sales," she began. "How are we supposed to look at that? Is that a gross sales or net sales number? And then, I guess, as we think about that for Q4, I mean, are we talking about $560 million in revenues if we just kind of use that as a thirteen-week number? Do we think about $2.3 billion in 2012 sales?"

Smith, ever alert to the need to inspire confidence while falling short of giving the analysts guidance about future performance, jumped in to clarify: "A lot of questions in there. First of all, let me give you the basis of the data in the chart, and then you can use the data as you wish. It is actually net revenues that we record . . . Now, as far as, should you then take the number that's between forty million and forty-five million in September and October and translate that through to both November and December? I'll leave that for you to do a forecast. It's not intended as a forecast. The reason we put this slide up was to get people—to help people get an understanding of the high volume of patients that are actually being treated, the high volume of scrips that are being written by prescribers.

"Our feeling," Smith said, "has been that this has not always been tracked the way that we're tracking it internally. And we believe that we have the most accurate internal metrics of how this launch is going, and we believe that because we monitor that on a daily basis. And so we wanted to put—give people an idea of where this launch currently stands. I think the best way to do that is on the revenues."

Wysenski, defending Vertex's commercial strategy, stepped in. "In response to the chart, Rachel, I think you're noticing that the rate of growth has changed. But I think that's more a reflection of the exceptionally quick ramp early in the launch. There are further opportunities for penetration into this market. There are many more patients to treat, and we are going to continue to do everything we're doing with a highly effective sales team to target those prescribers in that patient base."

Here, as Emmens had always worried, was the problem with trying to value Vertex by the area under the Incivek curve. In a more perfect world, you wouldn't bet a company's future on a disease where most of the patients didn't know they were sick, had no symptoms, and didn't need to be treated right away. Whether, how, and when the other three million Americans infected with the virus would show up for treatment remained anybody's guess. Up to one hundred thousand patients had been warehoused awaiting the approval of Incivek and Victrelis, creating pent-up demand, but now, with the promise of all-oral therapy, maybe for as little as twelve weeks, suddenly more likely as soon as 2015, doctors evidently were starting to warehouse again—or so it could be made to appear by those long on Pharmasset and short on Vertex.

VRTX climbed by 3 percent to $43.99 in Friday premarket trading while the analysts pounded out notes to clients. ISI Group's Mark Schoenebaum, whom Smith dubbed "The Pollster" because his contacts within biotech were omnivorous and unrivaled and he positioned himself as an oracular figure, raved that the Incivek introduction appeared to be the fastest drug launch of all time. He anticipated that cumulative sales could exceed $1 billion—the standard for a blockbuster—as early as the current quarter. Merck also beat analysts' earnings forecasts but announced anemic sales of Victrelis of just $31 million. Despite having Roche at its back and deeply discounting the drug with the VA and other government payers, the company had pushed its market share to 25 percent.

"In the closely watched bout between two new hepatitis C drugs," veteran biotech reporter Andrew Pollack wrote prominently in the *Times*, "it's a knockout in the first round." Yet by one o'clock, despite everything—despite leaping the threshold to profitabity and unheard-of sales numbers; despite crushing Merck—VRTX was trading at $41.15, a loss of $1.43 from the previous day, when none of these things were known to investors. The future gets repriced every day. Enough investors were worried that sales were flattening that the long-short thesis picked up steam even as Vertex produced quarterly results beyond anyone's imagining.

# CHAPTER 14

NOVEMBER 2, 2011

Brookside Capital's Adam Koppel, who'd made a career bet investing in Vertex when the stock tumbled into the teens in 2007–2008 amid abrupt fears of Stevens-Johnson syndrome and the specter of Schering-Plough, joined McMinn and about a dozen other analysts and fund managers around a conference table crammed with laptops in the Aurora board room in San Diego. Not the R&D day proposed by Boger, the visit was arranged by Partridge's group to showcase the company's work in CF. McMinn that morning again pulled back her price forecast, leaving Porges as the lone influential bull on the stock. By midafternoon, they would all board an executive coach to Anaheim, for the annual North American Cystic Fibrosis Association meeting at the Civic Center—for Vertex, as well as for most of them, a four-day nonstop whirl extending right into the Liver Meeting in San Francisco.

"Were you totally disappointed in mono 809?" Koppel asked site head Paul Negulescu, who led the session.

"I was totally excited," Negulescu said. "We're seeing what we expected to see."

Koppel asked several more questions aimed at nailing down the size of the market for VX-770, which Vertex recently announced it would sell under the trade name Kalydeco, and he was pleased to learn that in laboratory assays the scientists had established that the drug ought to be effective not only in people with other gating mutations, but in those with residual CFTR resulting in moderate ion-trafficking problems—up

343

to another 10 percent of patients. "Why not run a study for all comers?" he suggested. "Give the drug for a month and measure sweat chloride." Olson responded that payers and regulators would insist on more rigorous study designs.

Touring the labs afterward, Koppel and most of the others were dazzled—by the robotically controlled HBE cell bank, a decade in the making, that allowed the company to continually test new compounds in new combinations; by the patch-clamp assays that allowed the analysts, after fiddling with their pipettes, to inject compounds into rows of plastic wells and observe within seconds the spike in chloride gating activity on their computer screens; by Van Goor's movies. The tour only confirmed Koppel's impression that Vertex had a "backbone"—a machinery for replicating success.

He also was perplexed, riled by the shorts, reconsidering his faith in management, and rediscovering the limits both of his fiduciary obligations and his patience. In April, before Incivek was approved but after the AdComm, VRTX jumped within days from $48 to $55. Expectations for the second quarter on the market—Q3—were for $156 million, or negative 30 cents per share. Koppel had jumped in, buying another 1.9 million shares. Now the stock was trading at $36–$37, down every day since last week's earnings call. He blamed BOA's McMinn, Morgan's Meacham, and Citibank's Werber for driving the sell-off. Koppel knew them all and claimed to like and respect them. What irked him was that they had done the same thing once before, and in the long run their theses had proved wrong. Sitting the next morning on a bench amid the milling throng at the CF meeting, Koppel was restless, pent up.

"They're the three that created the uncertainly in '07, when it was over rash and '*Oh my God, there's competition!*'" he recounted. "With rash, there have been cases, but it was way overblown. And the competition in 2007 is not even the competition that matters today.

"Why is the stock so weak right now?" Koppel asked. "It's weak because of weekly scrips, which is ridiculous. The problem they're running into is, investors invest in the future, they don't invest in the current. The ET needs to solve 2015, and they haven't done it. Because there is no 2015 and beyond, the market is hyperfocused on week to week.

They brought that hyperfocus on themselves. If they can just show that they're even gonna be in the game in 2015, 2016, that they can sustain a billion and a half, two billion in revenues, and be one of the potential all-oral players, I don't see how the stock doesn't—not in a day, not in a month—rebase within six months in the seventies."

Porges, too, was at the Anaheim meeting, buttonholing Partridge on an upstairs escalator festooned with Vertex banners, reporting his sense of mounting "investor fatigue" with VRTX. Partridge wore a brave face but he was grim. "It kills me," he said, "that stockholders who've stayed with us are having to go through this." Even if the standard two-part thesis on Vertex was proving correct—that it would be first or best to market with its drugs; that the scientists, portfolio strategy, and culture gave it a platform and a formula for repeated successes—Wall Street withheld any credit for the second part. That left the company exposed and vulnerable. In the end, fund managers are stewards of capital, and can't be expected to retain their positions indefinitely, no matter how much, like Koppel, they consider themselves friends of a business or believe in its cause.

Judging by the gloom surrounding Partridge, Porges, Koppel, Smith, and the beleaguered investor team, it was easy to miss that most of the three thousand attendees shared a growing born-again fervor, and that the chief cause for their extraordinary excitement was Kalydeco (VX-770). CFF's Bob Beall was ebullient. He bounded onto the stage at a packed plenary session to James Brown's "I Got You (I Feel Good)." He recounted with a sense of awe the story of a woman who knew her daughter was on 770 because her cat no longer licked her when she sat on the sofa. He called Vertex's NDA filing "a day that was only equal to the announcement of the discovery of the gene." In the face of nearly universal resistance, Beall and his team had moved heaven and earth to get to this day. That the CF Foundation also looked forward to hundreds of millions in royalties from Kalydeco affirmed his stewardship, adding perhaps to his enthusiasm, though it was expected that the organization would eventually sell them off in order to preserve its philanthropic neutrality.

Smith roamed the meeting rooms and poster sessions with one eye on his BlackBerry, smiling inscrutably. The excitement in the hall and in the national media over Vertex's clinical findings, coincidentally published that week in the *New England Journal of Medicine* and carried widely on network newscasts, confirmed what he already knew from talking with the scientists and the KOLs—that the possible initial market for Kalydeco was more than twice what Wall Street was calculating. If you included the other gating mutations and patients who had enough working CFTR not to need pancreatic enzymes but who could also benefit from improved ion-conductance, that was perhaps another four thousand patients. Based on study data, Kalydeco might generate $1.75 billion annually, as far as the eye could see.

Koppel saw the same great prospects. What's more, he said, unlike with hepatitis C, "you don't have another fifty companies crawling up your ass." Other drugmakers were looking at CFTR now as a hot target and other experimental approaches were advancing in the clinic. But Vertex owned its market as only companies with medicines for rare diseases can, succoring those concerned with how it would solve its 2015 problem even as they continued to worry about the area under the Incivek rainbow. "The take-home from San Diego," Koppel said, "is 770 is a drug for twelve percent of patients."

And yet within the binary, seesawing long-short scenario that had overtaken Wall Street, Vertex and Pharmasset—which owned two promising nucs and little else—were now valued almost the same, their market capitalizations near $7 billion. Up the coast, Occupy Wall Street protesters in Oakland declared victory after peacefully shutting down nighttime operations at one of the nation's busiest shipping ports, escalating their movement and adding to traders' anxieties. Obama, plagued by low approval ratings and a paralyzed Congress, issued an executive order pressing the drug industry to resolve a growing number of shortages of vital lifesaving medicines, the latest scandal to rock the industry. In the churning uncertainty, worsening not just in pharmaceuticals but across the planet, Tom Wolfe's space-age mantra seemed more and more apt. It can blow at any seam.

❖

Vertex's commercial booth in the first-floor exhibition hall at Moscone West, two blocks south of Market Street, served as a showroom/encampment for the Incivek sales and marketing teams, as well as the HCV clinical group and the ET; only Emmens remained in Cambridge. Doctors and researchers, bombarded by competing video loops along the central alley, ambled in and out, navigating around tasteful low-slung white sofas and theatrically lit iPad stations. Like most of its competitors, Vertex drew passersby with lattes and flavored coffees served by attractive baristas. Market research manager Karolyn Cheng branded the ambiance "loungey"—stylish but informal.

Sunday, early afternoon, Smith and Mueller stood talking at the edge of the action. Mueller was animated. The Pharmasset buzz had grown deafening in anticipation of a three o'clock presentation of new data from a small midstage trial with PS-7977, its lead nuc. Mueller believed Vertex had an effective, competitive all-oral regimen for hepatitis C with Incivek, VX-222, and ribavirin, but the source of his confidence, especially with Smith, remained purposefully obscured, a matter of body language. He tried to convey that Vertex had what it needed, especially after the addition of the Alios nucs, to remain competitive in the disease without disclosing specifically any new data or compromising Smith in his handling of Wall Street. Mueller dismissed "this nuc hype bullshit."

Smith countered that if Mueller was right, it was "all the more reason" to buy a late-stage nuc to ensure that the perception of Vertex's 2015 drop-off was exterminated for good. As he saw it, the company faced a reckoning: buy a nuc for $1 billion and solve the problem, or save the money at the cost of continuing uncertainty.

Cumbo, nearby, palmed his cell phone, scanning the other booths: Merck, Gilead, Roche. The consistent message from the field force was that Incivek, its historic launch aside, had like all other medicines problems that hadn't shown up in the trials. Doctors reported less rash but more anemia and painful anal-rectal itching, forcing the company into action, working overtime to provide advisories and assurances to doctors and to the FDA that the side effects were being rigorously monitored and remained manageable. Cumbo had heard from his team that Merck's reps were spreading rumors of a spike in adverse events, including deaths, with Incivek, and though many doctors were reporting that the Vertex salespeople were the most customer focused they had ever seen, the war over the drug's reputation among the lower-prescribing physicians on whom it was relying for the next wave of patients had not yet been won. Sales were flattening.

More than four hundred people, including Mueller and Smith, packed the overflow room to hear Dr. Edward Gane of the New Zealand Liver Transplant Unit in Auckland report the results of Pharmasset's ELECTRON study on giant video monitors bookending an empty stage. A total of forty patients infected with HCV received a dual regimen of PSI-7977 plus ribavirin for twelve weeks. They also were randomly assigned to take Roche's Pegasys for either four, eight, or twelve weeks, or not at all. According to Gane, all participants achieved viral response by the end of the study, no viral breakthroughs were noted, and the only serious adverse events were in the Pegasys arms, none in the dual arm. Gane put up a slide that caused many observers to gape: a comparative analysis of the various arms, 100 percent SVR rates in all of them. What Pharmasset was claiming conjured the ultimate hope in hepatitis C: not a quad or a three-drug combo, but a pan-genotypic two-drug cocktail, a one-size-fits-all superdrug. The blockbuster takeway: Pharmasset's compound plus riba could cure hepatitis C in twelve weeks *without* interferon.

On closer inspection, the combination was no panacea. As hivandhepatitis.com, which evaluates experimental antivirals, reported, nearly alone among the media: "Because the researchers were uncertain whether the two oral drugs alone would work, they chose a population of relatively easy-to-treat patients who could be most easily 'rescued' if the experimental regimen failed: treatment-naïve people with HCV genotypes 2 (about one-third) or 3 (about two-thirds), and no cirrhosis; about 40 percent also had the favorable IL28B CC genotype. A majority were men, most were white, and the average age was about 48 years."

In other words: no nulls, no treatment failures, no harder-to-treat genotype 1 patients, few higher-resistance black people, no seriously ill cirrhotics, no elderly—the very populations Vertex went to pains in its pivotal trial to show it could cure. And, so far, no relapses.

After Gane finished his report, the moderator invited responses. A respectful silence descended for twenty seconds over the main hall and the overflow room. "You've stunned the audience," the moderator told Gane. The lead question from the floor spoke to a shared hesitation. Prior nucs designed to inhibit the HCV polymerase had gotten knocked out in large studies due to toxicity. Wouldn't PSI-7977 suffer the same fate? Gane could make no guarantees, but he optimistically reported that during the twelve weeks of the study he and his colleagues had seen no safety signals.

Enthusiastic applause ripped through both rooms. Smith, seated a dozen rows from the back, clapped appreciatively while Mueller, standing at the rear, kept his hands in his pockets. Most stunning was the instantly telescoping timetable. Pharmasset would doubtless move full-bore into Phase III. It was chasing the megamarket for direct-acting antivirals against hepatitis C much sooner, and with a much more potentially credible candidate, than Vertex and Merck thought possible even six months ago, on the eve of the first real breakthroughs against the disease in a decade. The pace of change had suddenly lurched by an order of magnitude: a breathtaking speedup and a boon for patients no matter which drugmaker prevailed.

❖

Giddy Pharmasset senior executives cruised the sea of fund managers and analysts, ten-deep at the bar, during the company's investor relations event that night in a contemporary art gallery a dozen blocks from the Moscone. They showed a short movie, *The Nucinator*, starring CEO Schaefer Price as a Schwartzeneggeresque avenger. Price played the leader of of the Nucleotide Resistance Movement, a heavily armed, emotionless, and efficient killing machine who leads the way to an interferon-free world for patients with hepatitis C by exterminating all types of HCV. The film had amusing, low-budget dialogue to go along with the cartoonlike *WowPowBam!* fight scenes: "*Competition crusher!*" and "*Now that's what I call viral suppression!*" The crowd cheered lustily.

Price owned 3 percent of the company. Later, in his remarks, he predicted an interferon-free world by 2014. Under the heading "Who cannot take interferon?" Price put up a slide of an iceberg, the vast submerged area representing a huge phantom patient population that, lest anyone still need convincing, he believed more than justified Pharmasset's overweight valuation. He talked passionately about "everyone treated with the same regimen . . . our drug being an all to everyone," and unveiled, Steve Jobs–like, the product: a lavender oval-shaped pill embossed with the company's logo and a *1*. "One pill, once a day," he said. "And it's pretty, too."

By the next morning, VRTX plummeted. Partridge saw there was nothing Vertex could do to stem the damage. He was worried, but not deeply so, even as it scraped 17 percent lower over the next twenty-four hours, to $30. Partridge was already getting calls from contacts attracted by the daily downward repricing of the stock, and they were the kind of desirable, long-term "value" investors who buy into companies selling at less than their intrinsic worth and whom he and Smith had been chasing for years. He was grim about disappointing shareholders and employees, but as he enjoyed noting, danger was also opportunity.

Smith, pacing the lobby, spoke by phone with Emmens, then grabbed a bag of potato chips and roamed the aisles of scientific posters at the far end of the exhibition space from the Vertex booth. He liked to get off alone and think in a crisis. He and Emmens agreed on the issue: "Do we chase hepatitis C?" Two schools of thought competed within the company: those who wanted to race ahead and those who thought the game was over. Smith worried about disgruntled investors and the hundreds of people inside Vertex trying to build the hepatitis C franchise, but the good news was: "CF is awesome. It's going to carry us through."

No one at Vertex was immune to the reality that, for patients, the twenty-year chase in hepatitis C was at last producing new cures with astounding speed and chances of success, especially relative to other diseases, and in the larger sphere this could only be considered a triumph for all involved. But all had their roles to play, and after the Liver Meeting they faced the certainty that the revenue arc for the next couple of years would be significantly lower and shorter than projected, ratcheting up the pressure on all of them. Patients were still lining up for treatment and so Vertex redoubled its program to sell Incivek hard now while targeting those groups—cirrhotics, nulls, people coinfected with HIV—who couldn't wait for the next wave of treatments. "Making hay while the sun shines," Sachdev said.

Vertex threw a formal IR event that night in a posh suite at the Four Seasons. A more staid affair than Pharmasset's, it was an opportunity to present new quad data that showed higher-than-anticipated SVR rates in some of the hardest-to-treat patients. The improvement went all but unregarded by the analysts, who had spent the day conferring with KOLs and had come to the conclusion that Pharmasset's story would surely result in wholesale warehousing of patients, especially if, as it appeared, the five-year campaign for testing baby boomers soon succeeded and diagnosis rates spiked. Patients were the coin of the realm. The analysts were taking them away from Vertex and giving them to Pharmasset. Koppel and Porges, investor and analyst, sat in the back checking the score of the Eagles-Bears game on Koppel's iPad.

"I hope this is the nadir," Wysenski murmured. Stepping off the elevator, she had bumped into Pharmasset's chief commercial officer buttonholing one of her salespeople. The final night of the conference, they were all beat, bored by the eternal road show *Sturm and Drang* and the peculiar savagery of the Liver Meeting, where it seemed you were either way up or way down. It could be worse: they could be Merck. Everyone was eager to get to the airport. Each night Sachdev had sat through several dinners lasting until one thirty. He tried to reschedule on an earlier flight so he could crash in his own bed, sleep for an hour, and have the next day to work and see his kids.

Kauffman, happy with the quad results and not disappointed to be flying out the next morning to join Smith and Partridge for a few days at a Credit Suisse investor conference in Phoenix, was the least downbeat. After the CF meeting, he too agreed that cystic fibrosis would bring more value to the company than hepatitis C. VRTX might be in crisis but Vertex was exceeding his ambitions. As for Pharmasset, he preferred to wait to see the data.

❖

Koppel sat spread-legged late the next morning on a couch in a solitary corner of the Four Seasons lobby, hunched over a coffee table, typing on his iPad, bristling with mixed emotions. Vertex had preoccupied him for more than a week. It was rare, and probably unhealthy, to become so personally invested in the gyrations of one bet, but the long/short thesis had swept everyone investing in either company deeply inside the fortunes of both. He was disgusted with Emmens and Wysenski. He didn't believe they had a midterm play and thought what they should do was quickly buy one of the two remaining companies with a late-stage nuc, Idenix, 13 percent of which Brookside owned. "Not only did they not put fear in the minds of the shorts, they supported the shorts. I've never seen a management team support the shorts in the way that they did," he said.

Porges arrived, crossing the room from the elevators with a seasoned look: funereal, but gleeful to be getting out of town. "There's blood in the water," he mocked ghoulishly, doubling down on his Australian accent for effect. "The piranhas are circling. But there's life in the corpse. Stock at thirty—fifty is the number we came up with." He sat down, chortling, "These guys fucking suck."

"This is what I did yesterday," Koppel said. He proffered the iPad. It showed a chart combining and comparing thirty possible scenarios based on three columns of data—probabilities of a host of likely contingencies. "I said there are essentially three different major questions with the company right now. What's the degree of the warehousing? Do they get a nuc or not, which essentially is a proxy for can they maintain their 2015 hep C franchise? And what are the scenarios for CF?"

"Time out," Porges interrupted. "I'd be interested in this, but the only part of this that really matters is this." He pointed to the last column.

"I agree with that," Koppel said, "but I need to know near-term what the cash flow generation is."

"Just take the October number and flatline it."

If putting a price on Vertex or any company by trying to model the future was bound to be futile, as the last few months had proven, neither of them could resist the exercise. Emmens's model of finding really sick people and making drugs for them might satisfy scientists but the Street demands valuations based on something more tangible, even if it's a speculative algorithm that blends and ranks various likelihoods, a garbage-in, garbage-out system.

"It could be worse," Koppel said. He explained how his team had graded the probabilities for what would happen to Incivek's revenues over the next three years in the event of ten different outcomes. "We have complete discounted cash flow (DCF) models behind each one of these. If it's severe, we're thinking it goes 2 billion, 1.7, 1.3, and then off the map. If it's moderate—and these are back of the envelope—let's say 2.2, 2.3, and 1.5. And if it's light, it's our original numbers before this weekend—"

"Yeah, yeah, that's fine . . .

"Do they get a nuc or not?" Koppel continued. "We charge them a billion and a half dollars in capital, and we say they can sustain from 2015 to 2020 at between 1.5 and 2 billion. If no, then it's just the cliff."

"When do you cliff it?" Porges asked.

"2016."

"I cliffed it at the end of 2014."

Koppel laid out a range of options for VX-770, from clinical and regulatory failure to approval in record time and blockbuster sales. Porges studied them.

"Okay, what do you think are the probabilities that it doesn't work in other gating mutations?" Porges asked.

Koppel brightened. "I think it's going to. We say it's a ninety-five percent chance it reaches the market and a sixty percent chance they get more than just G551D." He explained that after blending the eight most likely scenarios and crunching the numbers, his researchers came up with a price target of $48 a share—approximating Porges's $50. They were on the same page, if not yet reading from the same verse.

Porges put down the iPad. "This is so McKinseyesque. It's why I never worked in the organization. I could fire a rifle straight through this and blow it up," he said, pausing. "You haven't modeled what if Pharmasset fucks up."

"What are you gonna tell me, that you can model better than I can model?"

Porges rose to the challenge. "The big variable is realized net present on CF. Very important on the DCF."

"They're gonna have to spend a billion and a half and do something," Koppel said, returning to the lack of a late-stage nuc.

"Buy back the stock! If the company is such a bargain, buy it back!"

Koppel had consulted by phone with associates back in New York the night before and again that morning. He wanted Porges's bead on what he considered the likeliest simulation: "Moderate warehousing. They try to maintain the free cash flow—I give them a billion and a half dollars. And I give them the base CF plus the conductance."

Porges scratched his forehead, smiling skeptically. He returned to his earlier point. " So, how are we gonna incorporate in the model Pharmasset's primary development strategy not working?"

"Well, then it's just upside; but then it's not worth modeling. I'm not gonna make an investment thesis that the competition is gonna fuck up. Right now I'm just trying to find a base on the freakin' stock. I'm not frightened right now, but the bears have proven correct on all three of their theses. At some point as an investor, you're blind if you can't admit you've been wrong. I'm not wrong on the launch. I'm not wrong on how they compare with boceprevir. I'm not wrong on CF. But I'm certainly wrong on perception and I'm wrong on management's competence and poor decision making on how to sustain their hep C franchise."

Here was the rub: Emmens and Vertex *saw this coming*. Why they let it happen when Koppel, Porges, and others were urging them to take Pharmasset out before it got too expensive was beyond reason. They couldn't understand it.

"It's amazing," Porges said. "The downfall of just about every pharmaceutical company that ultimately fails is almost always scientific hubris. The guys at the top in the scientific department get positive reinforcement from scientific success, and then they overstate their own ability to make scientific judgments based on new information. That's what happened at Merck. That's what happened, certainly, at Amgen. You can almost argue that it started to happen at the end at Genentech. It's happened at Gilead. And it's probably what's happening here.

"Four years ago, I told them they should have bought bloody Pharmasset, when it was a billion dollars—"

"So did I. We all told them. But they think they're smarter."

"That's what I said: scientific hubris."

What neither of them knew was that Emmens, in fact, had pursued Schaefer Price, taken him to dinner in 2009, proposed that the companies collaborate, telling Price that the combination of VX-222 and Pharmasset's compound would be "a global segment killer." Price had been eager and receptive—until the next Monday, when Pharmasset's scientists inexplicably pulled out and Price didn't notify him directly. "You don't do that in CEO-land, by the way," Emmens says. Unlike a cash-rich behemoth, Vertex couldn't afford a takeout, friendly or otherwise, even if it wanted one, which it didn't. Emmens knew others blamed him in hindsight and it nettled him when the Street didn't assume he'd tried to do the obvious.

Koppel mulled Porges's postmortem. "I agree when you have success you think no one is better than you and you can do everything yourself internally. But I still think there's something special in that organization."

"On the ability to come up with innovation, I completely agree with you."

"But they're not being paid for it. No one cares."

"Their downfall," Porges said, "is that they failed to get the nuc, period, and that they failed to get Pharmasset. In the same way when I wrote the note that Gilead should have bought Vertex seven years ago, Gilead turned up its nose and said, 'Oh no, we don't really like telaprevir.' They'd be in a totally different position. They wouldn't be trading at seven times forward earnings."

"Right. Because they'd at least have the telaprevir free cash flow, and they'd have one or two other assets."

Lunchtime approached. A couple of retirees read the *Chronicle* in plush chairs by the gas fireplace, and groups of businesswomen and a few families clustered near the desk. Both Porges and Koppel had appointments, calls to return. Porges rose abruptly to leave.

"So in the same way Gilead thought it knew better, and that the drug wouldn't come to market, Peter Mueller thought he knew better and thought this drug wouldn't make it."

"Here's where we are," Koppel said. "We all want Vertex to succeed better than it's succeeding now. They need to get a nuc. Or they need to say they're going to milk telaprevir and get out, and don't spend the money on a quad. I do want to turn the card on the Phase II triple all-oral. Because it would be worth developing that if there is a real issue with genotype 1 on 7977. But the data better truly be pristine. The data better be really good efficacy with clean side effects."

Porges was philosophical. "Pharmasset is so far out on a limb—the fact that they've been pounding this 100, 100, 100 so hard. If that's all this is about, those expectations are off the charts. Pharmasset is saying we can treat everybody the same, regardless of viral variability or patient variability. The more variability you find in the absence of interferon, the more vulnerable Pharmasset's thesis is. Their thesis depends on curing patients independently of host response—completely independently."

He turned on his heel to go. "A shit show," he said.

❖

Despite the pain of the moment, the acid test of Vertex culture was its firm faith in science. After the Liver Meeting, Cumbo foresaw a period of "free play"—speed dating among the major players in hepatitis C, where they frantically sought to hook up either through collaborations or M&A. He trusted Mueller, Kauffman, and the R&D organization to remain all-in on advancing an interferon-free therapy, and though he had to reassure his people blindly, not knowing what Vertex's next move might be, long experience in an industry where failure and disappointment are norms had taught him to prize an even keel. Partridge, his own faith buttressed by newly inspired prospective shareholders, sympathized:

"It was bruising to the people who came into the company and worked so hard to make the launch a success—the people who are actually in the field, talking about Incivek, educating nurses, or being the sales reps for docs, or being the community liaison or what have you. All those people did a phenomenal job. Well, that ought to be rewarded. But the stock doesn't reflect that sort of success, and the headlines are negative. That's deflating, when you work that hard to produce an outcome, and it doesn't carry over."

Pangs of disappointment and recrimination flashed through the company. Like the field force, Ann Kwong, promoted to lead the hepatitis C franchise, was whipsawed by the uncertainty. For years, using HIV as her example, she'd warned that Vertex could wind up like Abbott and

Merck, pioneering developers of protease inhibitors shut out when Gilead rapidly overtook the market with a nuc-centered once-a-day cocktail. Avoiding her usual high profile at the Liver Meeting, she pressed Mueller and Wysenski hard, behind the scenes, to move more aggressively.

On the surface little had changed but the sudden explosion in competition, not a winner-take-all free-for-all but an industrywide hunt for multiple combinations of antivirals to match with the most appropriate patients. Merck, Roche, BMS, Novartis, Johnson & Johnson—all were actively realigning. Most desperate was Gilead. John McHutchison's team was running trials of six drugs in various combinations, but so far had discovered no category-changer to stave off the destruction when its major HIV patents all expired in 2015.

Less than three weeks after the Liver Meeting, Gilead Chairman and CEO John Martin announced that the company was acquiring Pharmasset for $137 per share—about $11 billion. Eyebrows were raised not just by the size of the trade but the whopping 85 percent premium. Two days later VRTX closed at $26.60, giving it a market capitalization of less than $5.5 billion—a nearly 60 percent drop from the heady days six months earlier when, on the cusp of the most robust drug launch in history, everyone on Wall Street was projecting that Vertex would sell $2 billion of Incivek in 2012. Now that was history. If any pharma CEO wanted to take Vertex out, the time had become unexpectedly ripe, but no serious moves were made, or even rumored within the company. With its hepatitis C revenues up in the air, the rest of Vertex's value, invested heavily in R&D, defied computation. It simply wasn't the kind of risk-mitigating trade industry behemoths were seeking now.

Emmens hoped the company's next CEO would be Dr. Jeffrey Leiden, who joined the board in 2009 and who also sat with him on the board of Shire. Leiden, fifty-six, was a rare hybrid, a distinguished cardiologist and professor of medicine at Harvard who joined Abbott in 2000, presided over its labs through the rollout of Humira, and went on to become president and chief operating officer of the company's pharmaceuticals group. Leiden left Abbott in 2006 to become managing director of a life sciences venture capital firm. Emmens respected him and considered him the only candidate to meet Vertex's specifications:

high-science background, major commercial and financial experience, strong judgment and leadership skills.

He chose Boger and two other members to direct the search, telling Boger, "You've got to be convinced. If you're not, I'm worried." Boger had lobbied to have a commitment to the Vertex values enshrined in the job description, and the board agreed. He thought scientists as a cultural type were generally undervalued for the breadth of their thinking, once telling a reporter, "It's much more likely that your top scientist can whistle the melodies from three sonatas than it is that your humanities-trained person can pass the science section of the MCAT." Leiden, an elected member of both the American Academy of Arts and Sciences and the Institute of Medicine, won Boger over handily at his interview.

On December 9 Vertex announced that it was initiating safety and tolerability studies with both Alios nucs. Effectively, Vertex decided to back off from its quad strategy for nulls and invest more heavily in nuc-based regimens favoring genotype 1. Mueller commented that the company hoped to advance rapidly into Phase II development during the second half of 2012, and that those studies would evaluate the compounds in conjunction with Incivek and VX-222 and possibly ribavirin. A play, if not *the* play in hepatitis C, the announcement was well received on Wall Street, where shares of VRTX began to recover, climbing back above $30.

Vertex's wild year wrapped up swiftly with the usual preholiday preparations, the wind-down to a forced two-week vacation at Christmastime before the lead-up to the Morgan Conference in January, when the cycle, enlarged by the expected launch of Kalydeco, would start all over. It was the season of prizes and surveys. On the website TheStreet, readers voted Emmens best biotech CEO of 2011, with 33 percent of the vote. But columnist Adam Feuerstein, taking matters into his own hands, retrospectively fixed the balloting, awarding a tie to Price, who came in last with 11 percent. "That's not right," Feuerstein wrote of the survey results. "Therefore, I'm using my prerogative as the overseer of this award to bestow upon Price special recognition for the extraordinary job he did this year. It's easy to make the case for Price being the true Best Biotech CEO of 2011. I don't mean to diminish anything accomplished by Emmens and the other nominees, but Price deserves praise for moving the field of hepatitis C therapy forward by a great big leap and also for helping his shareholders profit."

Undeniably, Price and Pharamasset had won the long/short war on Wall Street and made Pharmasset's shareholders and employees deliriously rich. Price himself made a reported $250 million. Whether the company's "great big leap" for patients was more than a figment of what the writer Robert Teitelman once called Wall Street's "promiscuous imagination," time would tell. What seemed clear was that Pharmasset's magic story, like Vertex's during the period right after it reported the twelve-of-twelve SVR rate with VX-950, had bedazzled Feuerstein among many others who wrote about the drug industry.

Partridge spent his first weekend at home in months making an ice rink in the yard with his children. He pushed himself to keep it all in perspective. "They were communicating about their early-stage data aggressively, like we used to do. Should we be surprised?" he asked. "So take the position that there's nothing you can do about it. There really isn't. You know, we've done great. We developed our drug. We communicated really well up to the approval and launch of the drug. We were able to leverage those communications to generate a high market cap, avoid a takeover, and raise lots of money. Great. Launched the drug; *that* goes really well. Have another drug coming up behind, great; that validates the whole pipeline strategy. But you know what, hep C is a difficult market. You're curing them, one time. You don't see them again. People don't have to be treated right away."

On December 15 Vertex issued two announcements. The FDA granted the company's request for priority review for Kalydeco (VX-770), with a target review date of mid-April; the board also appointed Leiden president and CEO, effective, at his urging, February 1. With another launch to prepare for, Leiden wanted to start right away. Emmens, the company said, would remain as chairman.

# CHAPTER 15

<br>

## JANUARY 10, 2012

The annual Morgan conference "was supposed to be a victory lap for Vertex," as the *Globe*'s Robert Weisman put it. More than twenty-five thousand patients had started on Incivek, nearly twice Wall Street's consensus. Instead, the buzz and headlines were all about Bristol-Myers Squibb, which announced it would buy Inhibitex for $2.5 billion, leaving only Idenix with a late-stage nuc still in play. Vertex tumbled "from king of the hill in the treatment of hepatitis C to yesterday's news in about six wild months," Xconomy's Luke Timmerman wrote.

After being introduced by Emmens, Leiden took the podium at a peak moment in the company's history, when it was flush with income, stockpiling cash, with a second launch imminent and a cascade of patient data on the way to clarify which diseases and clinical studies to focus on next. In hepatitis C, Vertex had quietly positioned itself to defend a major share of the future market. Though Leiden was eager to discuss CF and the pipeline, the likely keystones of his tenure, he also vigorously defended his flank. "There isn't one magic pill that will solve the problem," he said. "It's clear that the HCV space will evolve into different combination treatments for different patients. It's not clear what the best combination will be. What you want is to have the component parts in your company so you can put them together."

The continuity in leadership reassured employees and investors alike, though Emmens's accelerated departure revived a well-worn concern: What *was* it with Vertex's executive revolving door? How could Vertex sustain itself through a global buildout without more stability? The handoff from Boger to Emmens had smoothed the transition to profitability, but at what cost to the culture, and the people, remained unclear. Leiden plainly was very smart, but could he lead an organization where, as Emmens put it, "the smartest guy in the room is the guys in the room"? Hard-driving, focused perhaps to a fault, Leiden bristled with energy and optimism. He recognized that his job was to steer as much income as possible from the company's current drugs to nourish its research and development organization while giving shareholders a reason to stick with the stock. He could well envision, if he stayed five years, a half dozen more launches. It didn't get any better.

An AdComm for Kalydeco (Ka-lie-de-co) was scheduled for late February, as the fast-tracked preapproval machinery hurtled toward a mid-April deadline for FDA action. Three weeks after the Morgan conference—less than a day before Leiden officially succeeded Emmens and two days before the Q4 earnings call—the company learned by fax that the agency had approved the drug. "The FDA," Porges told Reuters, "is trying to demonstrate a willingness to move quickly with medicines that make a big difference, and this is a big difference maker."

The agency issued a statement to the effect that after two decades of false starts and disappointments a new paradigm had dawned. "Kalydeco is an excellent example of the promise of personalized medicine," Commissioner Dr. Margaret Hamburg said, praising the CF Foundation's aggressive venture philanthropy as much as the drug itself. "The unique and mutually beneficial partnership that led to the approval of Kalydeco serves as a great model for what companies and patient groups can achieve if they collaborate on drug development." Beall ringingly agreed. "We get their attention," he said of Vertex. "We're at the table with them."

At a crowded ceremony in JB-II, Mueller lustily rang the purple transformation life bell. The company priced Kalydeco at $294,000 a year, ranking it eighth among super-orphan medicines. Again Vertex ensured that no patients would have to go without the drug because they couldn't afford it. Wysenski spent more time on the analysts' call discussing its need-blind patient assistance guarantee than any other element of the clinical package. KOLs lauded the magnitude of the achievement. Dr. Bonnie Ramsey, a leading CF researcher at the University of Washington and Seattle Children's Hospital, compared the approval to the first moon landing. Vertex started shipping Kalydeco to pharmacies within forty-eight hours, with a prominent warning on the label: "Not effective in patients who are homozygous for the delta-F508 mutation in the CFTR gene." Olson, exhilarated, acknowleged also feeling "haunted" by the remaining patients with other mutations waiting for a corrector.

Smith led the earnings call with blowout revenue numbers showing Vertex in the black for the first time, with a $30 million annual profit for 2011, compared with a $755 million net loss the previous year. Sales of Incivek totaled $951 million. Now that the company was profitable, what it had to do to satisfy the Street was prove it wasn't at risk of lapsing back into the red. Whatever it hoped to accomplish until it pushed its next product over the finish line had to be achieved within the limits imposed by its earnings. To help the analysts value their investments, Smith said the company expected Incivek sales for the coming year to range between $1.5 billion and $1.7 billion.

McMinn pressed Leiden on what Vertex planned to do with the money. "The exiting CEO had previously articulated that R&D investment would be first into the pipeline; second, small M&A and deals; and third, share repurchases," she said. " Jeff, I was wondering if you could— if I could take your temperature on your view of that capital allocation strategy."

"I think Matt and I are in complete agreement about that," he said.

"Just a quick follow-up. Anything specific on HCV, or do you feel comfortable with your HCV portfolio as is?"

"What I like about our pipeline is we have many different swings at the ball," Leiden said; "with the two Alios nucs, which can be combined with each other; ribavirin obviously; our non-nuc, VX-222; and Incivek. I think when we look at our pipeline, we have the component parts to create a number of winning regimens."

If Koppel or anyone else imagined that a change at the top would impel Vertex to jump into the sweepstakes for a late-stage nuc, Leiden meant to shut the discussion down cold. Vertex was in for the long game. There was much more experimentation to do. Two weeks later its strategy seemed prescient when Gilead announced that six of ten null responders enrolled in a dual-cocktail study relapsed within four weeks of taking a combination of what was now called GS-7977—Pharmasset's lead nuc—and ribavirin. Gilead shares plunged 15 percent, while VRTX climbed to $37.60. "This will be a major blow to those individuals who had come to believe in the 'all-conquering' nuc thesis about hepatitis C," Porges told Reuters, speaking as much to his rivals and the companies he felt had scorned his counsel as to the investing public. "It'll be a blow to confidence in Gilead, their management of the investor expectations, and the value of 7977 and the whole Pharmasset transaction."

Vertex, presumed to be licking its wounds and muddling through a hasty leadership change, exploited the slightly receding headwinds. Back in the chase publicly, it answered Gilead's stumble with its own announcement a few days later of preliminary data from its all-oral trial in hepatitis C. The study found that thirty-eight of forty-six patients— 83 percent—had undetectable amounts of HCV after twelve weeks of taking a regimen of Incivek, 222 and ribavirin. With the nuc-based thesis still prevalent, a nuc-less triple cocktail might not wow the competition, but the results conveyed that Vertex intended to hold on to its market— "a small feather in Vertex's cap," Timmerman wrote.

❖

Keith Johnson followed Kalydeco's progress with a deep personal urgency bordering on obsession, strategizing incessantly about how to get access to the drug. In January, he learned about a clinical trial combining it with the corrector VX-661, found a sympathetic nurse, and in early February got a call to show up the next morning to be screened. He blew an FEV1 near 50 to qualify. On March 8 he arrived at the hospital to receive his meds. He recalls his expectations: "I don't care about 661 at this point. If it works better, great. I just want to get access to 770. I don't care about anything else. There's a real good chance I'm gonna get 770. All I want is the chance."

Johnson keeps a picture on his cell phone to illustrate what happened next. "So they hand me this: one pill. I said, 'Wait a minute. Where's the Kalydeco?' 'No, in part A, one in five chance of 661 only.' 'What do you mean, 661 only, where's the bloody 770?' Now I understand these drug addicts. For a split second, I have this mentality that I'm gonna break into this fucking place and steal all the 770 I can."

Part A of the study evaluated 661 monotherapy, twenty-eight days on, twenty-eight days off; Part B tested different combinations of dosages of both drugs. Johnson had a choice. "They said: 'You can back out of Part A and get into Part B if you want.' It was too much uncertainty. I thought, let me see what this is like. In the back of my mind I'm thinking, if I don't get placebo I think this is gonna validate my hypothesis where there's a subgroup of the double-deltas where there's a certain amount of the protein at the cell surface. So let's see what happens.

"The only thing I felt was tired. After about eight hours—unbelievable fatigue. The first day I was off it I didn't feel the fatigue. I didn't feel like I did on the other one. Maybe it was placebo. I don't know."

Throughout the spring, Johnson weighed his disappointment against Vertex's need to test its medicines rigorously in patients. Even if he could find a doctor to write him a prescription, what insurance company would pay for the drug? His only avenue was to build a case that he was an outlier and thus deserving of further medical observation and testing.

❖

In April Vertex received interim data from a midstage study combining VX-809 and Kalydeco in patients with two copies of the delta-F508 mutation. The results were better than expected. Though the findings were preliminary, subject to revision upon further analysis, the forty-six-patient trial demonstrated robust enough effects—significantly improved breathing, substantial weight gain, sharp sweat chloride reductions—to move quickly into pivotal trials. Worried that word would leak out as the data were shared with CROs and clinicians, the company decided to disclose the findings, releasing lung-function data to show how well the cocktail worked.

Partridge calculated that if the analysts remained true to their models, they would have to take the probability of approval for VX-809 within two to three years up to about 50 percent. The scale of the opportunity multiplied and became likelier all at once. "You're going from a couple of thousand patients to thirty-five thousand," he said. Running the numbers, doing a quick discounted cash flow, Partridge figured it would translate to a $12 to $20 jump in the price of the stock. The magnitude would reflect Vertex's credibility: "a barometer of how much they trust us."

VRTX went vertical, soaring on Monday, May 7, from $37.41 to $58.12—a 55 percent increase. The company gained back $4.2 billion in market value, the largest one-day rise in its history, effectively putting it back where it was during the optimistic weeks before the launch of Incivek a year earlier. On Tuesday, the price went to $64.16. ISI Group's Schoenebaum said Vertex could have a $4 billion-plus annual franchise in cystic fibrosis. Adam Feuerstein predicted cystic fibrosis revenues of $6 billion to $7 billion per year, equaling Gilead's revenue from all its antivirals, including its dominant HIV drugs.

The following Monday, Morgan's Meacham, who for a year had fanned fears about Vertex, boosted his rating from "neutral" to "overweight" and raised his price target per share from $45 to $82. He conceded, in a note to investors, being "a bit late" in making the revisions.

❖

"Pretty amazing stuff, that 809 data," Boger emailed to a correspondent. "Also amazingly straightforward; it just works."

He was rolling now. Vertex was thriving without him there to drive it ahead, yet its growing record of success and reputation for leadership and integrity enhanced his luster and credibility, capital to be invested in the arenas of medicine, business, government, education, political action, public policy, philanthropy, and the arts, all of which Boger took on at a gallop. On a rainy night in May, he and Amy stood like hosts outside the ballroom of the Boston Sheraton, amid the throng at the state American Civil Liberties Union's annual Bill of Rights dinner. The honoree was singer-actor-activist Harry Belafonte. Journalist Amy Goodman of the news program *Democracy Now!* gave the keynote. The Bogers were offering a $100,000 challenge grant, and he was on the program.

With the national election looming, he recently had spent a long Saturday at the state Democratic convention in Lowell, sizing up the early contenders against Republican Senator Scott Brown. "The second most important election in America for 2012," he predicted. He'd just agreed to head a $1 billion capital drive at Harvard Medical School and, after swearing not to go back into business, even as a consultant, he had become executive chairman of a start-up seeking a cure for the leading cause of childhood blindness. As ever he failed to see boundaries. With the fate of Obama's health care overhaul law and the country's direction at stake, he spoke, in his remarks to the crowd of 900 civil libertarians, to what he saw as the essence of the campaign.

"There are two competing historical theories about how to establish the delicate emotional state that drives economic growth," Boger began. "The first is what I will call 'European Feudalism' and is popular among many of those in power: ensure that everything breaks my way (. . . because I'm owed that), continue to advantage my advantages (. . . because that's the natural order), and then I'll consider investment, especially if I am guaranteed to win (. . . and am completely covered in the unthinkable case that I might lose, because, you know, I'm too important to fail).

"The economic and social paradigm alternative to this Feudalism I'll call, for short, 'The Bill of Rights.' As radical a social contract as has ever been written, the Bill of Rights outlines a remarkable thesis, simply put, that treating everyone with justice and respect is good for everybody in the end.

"The Bill of Rights isn't a cost we pay," he concluded. "It is something we believe in just because it is right, but it's also an economic development plan: together we are stronger and more productive than any of us can be separately or selectively. To be together most productively, we require the ground rules to make it clear that opportunity is open for all and that the power of the majority can never be used to cut some of us 'out of the herd.' The love we feel for the Bill of Rights is its protection of us all. All one hundred percent of us."

Boger had joined the ACLU in graduate school after the organization defended the rights of Nazis to march through a predominately Jewish suburb of Chicago, Skokie, because he admired the intellectual honesty of the position. Equating the Bill of Rights with a business plan and loving it because it defends "us all"—even the worst of us—was in his view not just a no-brainer but wise, progressive, patriotic politics. Indeed, he saw it as a prescription for growth. As he posited in a subsequent email decrying the country's willingness to squander its vision and hope by indulging short-term reward incentives:

"Boger Tax Plan: 99% tax on all capital gains under a year. 90% under two years. You want to be an investor? Invest. You wanna be a trader? Go to a casino."

❖

Michael Partridge and his wife took their children to a lodge in New Hampshire for Memorial Day weekend. Before starting a daylong hike, he checked his voice mail. Ty Howton, Ken's replacement, had phoned from Vertex: "If you're getting this message, you need to call me."

Partridge assumed the worst. The direst possibilities—a patient blow-up or program-killing tox result or government investigation or FDA stop order—were the most likely but by no means the only scenarios that could erupt to test a company. Howton had learned late the previous afternoon that the proportional analysis of the FEV1 data from the CF combination trial was wrong. FEV1 is measured in percentages. The lung improvement measures that Vertex had received from the outside vendor doing the analysis and disclosed to Wall Street and the public weren't absolute but relative—a frequent error made even by many doctors. The responses to the drugs were still impressive and consistent with the other markers, but the company needed to revise the data.

Cell service on the mountain was spotty, but Partridge was able to participate enough in an hours-long conference call that afternoon to return at night to the lodge to bang out a call script. Senior management met all day Sunday at Vertex. "Jeff was the voice of reason: 'This is what happened; this is what we have to do; we have to correct it before the market opens Tuesday,'" Partridge recalls. The reinterpreted figures showed that after eight weeks Kalydeco and 809 improved lung function at least 5 percent in 35 percent of patients and at least 10 percent in 19 percent of patients, compared with 46 percent and 30 percent, respectively, as announced earlier in the month.

Speaking on the update call were Leiden and Wright. Neither was a Vertex veteran known by the analysts. Wright explained that the revised result remained "a significant and clinically meaningful difference given such a short study with such a small amount of patients." Leiden stressed that Vertex was advancing aggressively into Phase III. Questioned first, as always, by Porges, he laid out the company's stance in response to the mix-up.

"Look," he said, "I just want to tell you straight out, this mistake is very disappointing; unacceptable to us. It's not how we do business here at Vertex. We're a high science company, and we pride ourselves on getting the science right and the numbers right.

"We assumed it was absolute, obviously presented it that way, and it was relative. As soon as we realized that, we corrected it. We're coming back to you not only with the corrected data but with more data to share with you why we're confident about this result. And obviously the final data available in midyear will trump all of this, and we will be able to tell you much more definitively about where we are. But I think what you're hearing from all of us today is that even with the corrected data, we have many reasons to be confident that the effect we're seeing here is real, at least from this interim analysis."

How often any company can play the "trust me card," as Smith called it, depends on the company. Within an hour, VRTX suffered its worst intraday drop since the crash of pralnacasan eight and a half years earlier, then recovered to $52.85 as the analysts repriced their forecasts. "This is a downward revision, but it is still good data in our opinion," ISI Group's Schoenebaum wrote in a bellwether note to clients.

By the next night, after a previously scheduled Bernstein investor conference in New York, Porges hosted a dinner for thirty analysts and fund managers so that Leiden could make the case personally. "People were upset—'How on earth could you do this?' 'How could you screw this up?' 'How could you not know?' " Partridge recalled. The takeaway for a few of them reflected long-term suspicions about the company that no amount of high-minded behavior or hand-holding would soon overcome. "From a management credibility perspective," Marshall Gordon, health care analyst for Clearbridge Advisors, told Barrons.com, "this looks bad. This has been known as a promotional management team in the past, and this smells like they hyped their data, prematurely. But I'm not convinced that this diminishes the clinical or commercial potential of the drug . . . it remains obvious that the drug is still active and should be able to get to the market."

Still upbeat, the analysts reproved Leiden and Vertex but didn't punish VRTX; on Wednesday shares rallied, climbing back over $60. Then, at once, another minor but untimely setback arose, further dinging the company's reputation. The FDA's Office of Prescription Drug Promotion routinely criticized a proposed piece of promotional literature for Incivek that Vertex had submitted for review, a story from a patient who received the drug and who was quoted as saying, "Six months after the treatment ended, I found out I'd cleared the virus. That made me feel so good. I was happy to know I'd be around a little longer to see my son grow up." Expressing concern that "this branded story misleadingly implies that most or all [patients] infected with hepatitis C will successfully achieve sustained virologic response" on Incivek and objecting to the use of "cleared," FDA officials asked Vertex to revise the draft materials, which hadn't been distributed publicly. The company agreed to cooperate.

The next day, the *Boston Globe* ran a business article conflating the overstated CF data with the FDA letter, calling it "the second setback for the Cambridge biotechnology company in the past week" and reporting that the earlier misstatement boosted Vertex shares by more than 55 percent on May 7, "enabling five senior executives and two directors to exercise stock options and sell shares worth millions of dollars." No matter that the stock was higher now than then, post-correction, or that the questionable Incivek ad was a draft, the paper abruptly planted and fed a new, circumstantial counter-narrative, the alluring and familiar story of a biomedical company systematically misrepresenting data and overstating claims to enrich insiders. Shades of Martha Stewart.

That narrative instantly assumed a life of its own. The next Monday, San Diego–based Shareholders Foundation, a self-described "professional loss prevention, settlement recovery, portfolio monitoring service"—in other words, a law firm—put out an alert trolling for plaintiffs, announcing it would probe whether Vertex had broken federal securities laws. Later that day, Senator Charles Grassley of Iowa, the ranking Republican on the Judiciary Committee, contacted SEC Chairman Mary Schapiro. "I write to you today to apprise you about a potentially troubling issue for investors in the pharmaceutical industry and for the federal government. I am disturbed by reports concerning the release of clinical trial data by Vertex Pharmaceuticals Inc. and stock sold by Vertex executives."

Citing the *Globe*'s reporting, Grassley urged the agency to investigate trades by Boger, Wysenski, Mueller, Sachdev, Kelly, and corporate controller Paul Silva. The largest number of shares—more than 365,300 in total; more than the others combined—were sold on May 7, 8, and 14 by Wysenski at prices ranging from $59 to $64, netting her more than $13 million. Like the others, her stock sales resulted from preexisting plans used by many companies to allow executives to automatically sell shares at regular intervals or set prices. "Despite Vertex's explanaton," Grassley concluded, "it could appear that these Vertex executives potentially took advantage of the spike in the stock knowing the news of the clinical data being overstated would be made public eventually, which in turn would negatively affect the stock value."

Hijacked by the developing story, Vertex's public posture suddenly resembled damage control, even when it wasn't. On Friday Vertex announced two executive changes. Both had been months in coming. Howton was replaced as chief legal officer, though he would stay on to smooth the transition. Wysenski, who in the wake of Emmens's retirement was left without her friend and mentor at the helm, retired. Company spokesman Zach Barber told the *Globe* that the moves were "completely unrelated" to the jump in the company's stock and subsequent events. Ten days later, the *Globe* published a glowing story about a novel $1.45 million partnership between Vertex and Boston to put advanced labs in one of the city's public high schools and the company's plans to build a 98,000-square-foot manufacturing plant on the waterfront. Though it had been in the works for months, the timing appeared more than coincidental—an attempt to change the subject.

In late June, Vertex released more comprehensive data from the CF combination study that, although similar to the revised findings, weren't quite as robust. Some investors again became upset. "I don't think we lost anybody along the way but all the investors hated the volatility," Partridge says.

The SEC, apparently finding nothing of interest, ignored Grassley's letter.

❖

Keith Johnson listened to the CF conference call. He believed on the basis of the company's explanation that his FEV1 numbers during the 770 trial qualified as "statistically significant and clinically meaningful"—Wright's, and the FDA's, classic definition of relative effectiveness. He couldn't understand why he was being blocked from getting Kalydeco when he had already shown that the drug had changed his life.

He imagined becoming a kind of experimental tool, a guinea pig, not just to aid Vertex's and the CF community's understanding of the drug, but also to dramatize an evolving model of personalized medicine, which the FDA was now trying to help accelerate by streamlining the process of clinical testing to allow, effectively, cohorts of one. He considered offering himself to the company, his doctors, and the foundation as a critical addition to their database—a patient with a CFTR folding mutation who nonetheless could be helped by a drug designed to improve gating activity. Johnson realized that by now he was consumed, but in his mental dialogue with the medical community, he believed his militancy was totally reasonable.

He honed his pitch: "This is what's in it for you. If I'm wrong you won't hear a peep from me again. If I did two percent, I did two percent. I'll move on with life. I can't go to sleep at night because I'm convinced that I'm way better than two percent. I need to take Valium-strength meds to go to sleep because it's maddening to me. So somebody has to show me: either I'm only two percent—then I'll leave you alone—or these numbers are meaningful and significant, and let's agree that they are. I need to know."

Van Goor had long ago anticipated that when personalized medicines became available, diseases would become more personal too. Since your genes would qualify you as a promising patient, more and more you would need to build a case that a treatment might work for you, then convince the medical world to let you have it, even if a year's supply cost more than your house. No one knew how this new system would work. Advocating for himself, trying to keep up his health until 809 or

661 reached approval, Johnson was becoming an inadvertent, crusading forerunner on a new medical frontier: the informed genotypic consumer.

On June 27 the Supreme Court upheld Obamacare, removing the issue of its constitutionality from the wider political debate, and giving Obama an unanticipated boost toward reelection. A few days later Glaxo agreed to plead guilty to criminal charges and pay $3 billion in fines for promoting its bestselling antidepressants for unapproved uses and failing to report safety data about a top diabetes drug. The largest settlement ever involving a pharmaceutical company, the agreement came on the heels of a $1.6 billion settlement against Abbott over its marketing of its antipsychotic drug Depakote to nursing homes and as much as $2 billion in fines against J&J for the same offense.

❖

During the next quarterly call in late July, Vertex and VRTX reached an uncommon equilibrium, one of the rare junctures in Vertex's recent history when Wall Street's on-again, off-again faith in its talents and its future trumped pessimism and uncertainty. Though Incivek sales dropped, auguring a lower, truncated revenue curve as doctors and patients awaited the arrival of all-oral regimens, shares surged 5 percent based on promising viral kinetic data from a small human study of Alios's lead nuc, ALS 2200. The molecule was as potent, and apparently as clean, as Gilead's lead compound, yet unlike Gilead and BMS the company hadn't spent billions of dollars to acquire it. "It puts them back in the hep C race," Garret and Co. analyst Brian Skorney told the daily industry news service FierceBiotech. "It makes them a real player again."

Not that this was news inside the company. Vertex had never stopped advancing. If the drubbing of VRTX during the previous year's hep C mania had a major internal effect, it was to incite the ET to step up its offense while forcing its members to impose stricter austerity on their organizations. Lowering the company's sales forecasts while acknowledging that Gilead was the new front runner meant everyone would have to do with less money than they thought they would have, while accepting the likelihood that things might get worse for a couple of years before they got better. For the sales force, it meant selling harder to fewer doctors, trying to reach those patients who couldn't wait two years for the next generation of drugs. Near-term expectations dimmed even as the long-term outlook improved. The launch had been a blur. People throughout the company were tired but acted resolved, as usual, not to hold anything back.

"If you go away for a couple of hours," an industry wag noted, "things change in HCV." Now that a herd of companies was racing hard to deliver the next wave of treatments against the virus and dominate a possible $20 billion market, this advice became literally true. The shape of the race shifted again in just the next few days. BMS announced it was shutting down the combo trial with Inhibitex's nuc after a patient who took the drug died of heart failure. By the time the company killed the development program three weeks later a total of nine patients had been hospitalized, raising long-standing safety concerns about nucs and leaving executives to explain the crash and burn in less than eight months of a ballyhooed $2.5 billion investment. The FDA, concerned about similar modes of action, called for a hold on development of the two nucs from Idenix that Adam Koppel believed would have solved all Vertex's problems.

Vertex, playing catch-up, suddenly caught up, not to Gilead or Abbott in the next phase of what FierceBiotech called "the hep C pill race"—coming to market with the first interferon-free regimen—but the phase after that, when people with encroaching liver disease would likely choose among several treatment options based on a combination of viral genotype and their own genetic profile. Even if, as Wall Street assumed, Gilead introduced an all-oral cocktail in 2014, how many patients could the company treat? Seventy-thousand a year would be a stellar number. Yet in the United States alone, more than 3.2 million people were infected, 75 percent of whom didn't know it. With hepatitis C, as always, the larger prize was the demographic bulge of millions of people who weren't sick now but whose livers would become scarred and cancerous as they aged.

In late August, six years after Boger and Sachdev sat down to figure out how to leverage Washington to address the public-health dimensions of hepatitis C, the Centers for Disease Control urged all baby boomers to get a onetime blood test for HCV. CDC officials said they decided to issue the recommendation after they calculated that the number of Americans dying of hepatitis C–related disease had nearly doubled between 1999 and 2007 and because two drugs had hit the market in 2011 that promised to cure many more people than was possible before. "Unless we take action," CDC director Thomas Frieden said in a call with reporters, "we project deaths will increase substantially."

Boger's plan had come together: get to market with a breakthrough drug in time to wake up the government and the world to the fact that unless infected people get treated, they will develop serious liver disease, costing everyone dearly. In fact, it may have worked too well for long-term investors and some inside the company. By leading the industry across the threshhold to effective, direct-acting antiviral drugs against HCV and promoting the expansiveness of the opportunity beyond, Vertex emboldened competitors who, for the moment, seemed to be racing ahead, costing the company its dominance and forcing it to scale back its plans while exposing it to takeover threats for longer than it preferred or was healthy. Still, there seemed to be no question that Vertex's aggressive, all-in scientific program and astute government lobbying had led the way in sparking a remarkable boon for patients, taking the world from one in which you had to suffer miserably for a year to have a 4-in-10 chance of being cured, to the likelihood in a couple of years of only having to take a few pills for twelve weeks for a 9-in-10 chance. Few areas of medicine have improved so dramatically in so short a time.

One problem, perhaps, of being a visionary leader is that once you've pried back the future, your vistas don't just end. As much as you've advanced things, what stirs you is the next challenge, and you're prone to be impatient and disappointed with a world that doesn't keep up. So long as Vertex didn't get taken out, Boger could content himself with its strategy in HCV. What worried him was its decision to be less brash in its public ambitions, to yield to gravity. Internally, the company still expected itself to do what others wouldn't, or couldn't, do, but as it presented itself to the Street, its arrogant insistence on being both exceptional and right seemed to be lost.

That's always what we did. I was always accused of giving guarantees that I never uttered. People read their own narrative onto those: "Telaprevir is gonna be monotherapy, absolutely." I was just painting you a vision of what the data said at the time. This is a possibility, and until it's not a possibility, it's a possibility. Why is that not the thing we still should be doing? It worked pretty damn well, and works pretty damn well for people that have high p/e ratios in other industries. Apple gets its high p/e ratio not because people have analytically projected out that they will control cell phones through 2020. That may be true, but why does Apple get that presumption? They get it because they take that ground. Apple doesn't promise that ground. They just *act like that's true.*

We don't externally act like it's true anymore. We don't act like we have the best ideas in, say, JAK3 anymore. We get caught up in "Well, I can't prove it. No, we have to get data." That's not inspiring. I think we're in a perfect position. Don't overpromise at all about HCV sales. Don't even get into that conversation. Just keep blowing the numbers out. But absolutely be incredibly excited about all the things that are coming along. Why can't those two live in the same world? I don't know why we have to be similarly buttoned down on our research. Research is all about hope. It's about possibility. It's not about certainty.

As summer slouched into fall, Vertex approached the run-up to the Liver Meeting, back in Boston this year at the Hynes, as the next engagement in a long war of attrition. The memory of last year's humiliation no longer stung. If Gilead or Abbott took away the company's lead in hepatitis C in a couple of years, Vertex would find ways to cope. Cumbo stoked the morale of the sales team, which had the most to lose from an onrushing, interferon-free world. To motivate them to promote Incivek while the drug remained the standard of care—and to keep them from defecting—he negotiated a generous retention package, winning support from Smith and Leiden. Meanwhile, the company announced partnerships with GlaxoSmithKline and Johnson & Johnson to test VX-135—formerly Alios's ALS-2200—with their antivirals. A disappointing quarterly earnings report in late October dragged the stock price back into the low 40s. In the airwalk between JB-I and JB-II the company added a colorful new banner alongside the three Vertex values: *Patients First.*

The more things sped up and changed, the more they remained the same. Kauffman, after a brief respite from investor relations, plunged back in during the weeks leading up to AASLD, becoming once again the company's incorruptible avatar, sanctifier of all data. Unlike Boger, he was immune to grandiosity. Asked by FierceBiotech's Ryan McBride about the breakneck, roller-derby-like "rush toward pharma gold" in hepatitis C, Kauffman commented: "It's not just crowded, but the speed of change I think is very unique. It's really quite remarkable and we are very happy to have been at the forefront of this. Incivek paved the way for direct-acting antivirals."

❖

Cumbo felt upended as Vertex evolved quickly from a twenty-year-old development-stage company staking its survival on a quick rout in hepatitis C to a cash-rich profit maker with a coming bonanza in cystic fibrosis and a smorgasbord of pills in mid- and late-stage trials for other diseases. He had built the Incivek sales team to last a decade, marketing multiple drugs. But the onrush of the first all-oral regimens meant that for an indefinite period, starting as early as 2014, they would have no competitive product to sell. Meanwhile, Incivek sales had peaked, down 40 percent year-to-year and projected now to slump steadily until they cliffed. "The tide is going out," Cumbo said, "and I got caught in it."

He had been all-in for two and a half years, coming to understand as well as anyone else at Vertex the thankless challenges and inevitable crises of operating a business driven, at its core, by research. Neither Wall Street nor the marketplace had shoved Vertex off its stance that R&D was paramount. Mueller's clinical strategy maintaining telaprevir as its cornerstone and his aversion to nucs had momentarily beggared Smith's determination to finance as broad a pipeline as possible, leaving Vertex to face a period of unexpected austerity without a midterm play in HCV and with little clarity beyond that. The field force would not have a chance, as Cumbo put it, to "kick Gilead's ass"—not for several years at least. CF was expected to pick up the slack.

A builder builds. In August Cumbo flew to San Diego to dive with great white sharks, descending in a battered steel cage off Guadalupe Island, Mexico, to come face-to-face with their prehistoric snouts and razor teeth. What did he learn? "To stay in the cage." Fending off feelers from Abbott and other large players in hepatitis C, he quietly edged throughout the fall toward joining a seventy-person company, Sarepta Therapeutics, as head of business development as it prepared to launch an orphan drug for muscular dystrophy. In early December, Cumbo turned in his notice, delaying his announcement to his team until after the first of the year.

The so-called "fiscal cliff" loomed in Washington. Obama's reelection clarified the near-term prospects for the $2.7 trillion US health care economy, meaning Obamacare would soon take effect, and the drug industry, for the first time in a decade, introduced a bumper crop of new medicines. Nearly half the year's thirty-nine new drugs approved by the FDA, like Kalydeco, treated rare diseases, signaling the coming-of-age of the new post-genomic paradigm in pharmaceuticals: the prevalence of eye-wateringly high-priced medicines that transform the lives of a handful of patients. As Vertex had learned, the vision of finding such drugs, the reduced obstacles to market, and the incomparable value to be derived from them were irresistible to nearly all companies now, smaller ones especially.

On December 19 Vertex issued two product updates, each in its way reflecting the company's maturation under Leiden. The National Health Service in England, after pushing back on price, agreed to pay for Kalydeco for the 270 patients in the United Kingdom with the G551D mutation. The decision, which was expected to consume half its budget for *all* cystic fibrosis patients, was cleared only after Vertex agreed to an undisclosed discount. It was one thing to convince American insurers and managed care companies to pay $300,000 for a drug; convincing austerity-hit foreign governments was another. With health care systems around the world under increasing financial stress, it wasn't too early to think that the super-orphan commercial strategy might become unsustainable, sooner rather than later.

The company also announced that the FDA had placed a black box warning label on Incivek, cautioning prescribers that the drug had caused fatal skin rashes in at least two patients. More than fifty thousand patients had started on the medicine, but despite the company's extensive rash management protocols, two patients in Japan had developed a potentially fatal skin condition, Lyell's syndrome, one dying of multi-organ failure, the other surviving after she discontinued treatment. Another woman died after being hospitalized. Vertex redoubled its instructions to doctors to halt therapy as soon as a serious skin reaction was identified, as the fatal cases resulted when patients stayed on therapy even after severe rash was diagnosed.

A black box warning at launch, or for a mass-market pharmaceutical, could break a product, but in specialty diseases where you're dealing with very sick patients, such advisories are both common and expected—"a hiccup," Cumbo called it. Every drug he'd ever launched had eventually earned one. As anxieties flared among the field force, Cumbo and Vertex's marketing leaders fanned out to steady the organization. "That was torture," he says. "I did all the conference calls. Area directors. Regional marketing managers. Reps. Treatment educators." Where a year earlier some voices inside the company might have been apoplectic, now there was no crisis. Wall Street shrugged off the news, preferring instead to celebrate the Kalydeco ruling. VRTX finished the day up 1.5 percent, closing at $45.85.

The area under the Incivek curve no longer ruled the Street's calculus in pricing the company. Selling nearly $2.5 billion of the drug in the twenty months since launch, Vertex turned a corner. Of seventeen firms providing ratings, according to a year-end summary, thirteen were positive, four neutral, and—notably—none negative. The analysts all now loved the company's CF franchise, viewing it as a major growth driver. The bulls were especially optimistic about its agreements with Glaxo and J&J to develop cocktails around the Alios nucs. No short players appeared on the horizon.

Vertex showed Cumbo its appreciation by not walking him out the door the day he said he was leaving and by letting him keep his computer through the holidays. Gearing to move up to business development at Sarepta, a stratum leap that would raise him into the realm of M&A trades and global licensing deals, he prepared for his debut: the Morgan conference. Sarepta had already arranged his San Francisco meeting schedule. Vertex was his first appointment of the week.

❖

Within thirty-six hours of announcing his departure on January 2, 2013, Cumbo answered 130 texts and 200 people had visited his Linked-in page. What his leaving signaled for Vertex wasn't yet fully clear, but it traumatized an already disheartened sales force. "They're shocked and they're scared," he said. Cumbo was a long way from hiring at Serepta, just entering late-stage trials with its lead drug and with early-stage and midstage development programs in hemmorhagic viruses and flu, but he relished the chance to help build a company from scratch, and he could imagine working with the best of them again in the future. By the end of the week three of them turned in their notice. Whether an exodus would follow depended on other considerations, but Vertex now had to face that grim prospect too.

Of all the costs of yielding its lead in hepatitis C, holding on to an anxious, dispirited field force was among the most pressing but far from the most crucial. In the longer term, the company's years of delay in bringing Incivek to market and the shrunken window of opportunity brought on by the looming arrival of nuc-based, all-oral regimens had forced a reckoning. Incivek may have been a booster rocket, as Emmens said, but what if it hadn't boosted the business high enough, fast enough to reach escape velocity, to soar? Vertex was an operating company at long last. Now what?

Could it continue to innovate and forge ahead when it failed to keep people who "bled purple," not just Cumbo and Murcko, who had landed on his feet—consulting, teaching at MIT and Northeastern, and mentoring young scientists pro bono—but also fellow positive deviants Ann Kwong, who left to start a drug-innovation consultancy, and medical affairs director Cami Graham, who returned full-time to clinical practice? What had senior management learned that would keep it from losing its lead in CF and other disease areas? Big Pharma may not know how to innovate but it remains effective at copying and incorporating ideas that work. It's not good enough to be "first and worst," as one Boston fund manager liked to describe Vertex's dilemma as a pacesetter. Mueller's mystique aside, it had been five years since the company had brought a breakthrough molecule out of its own labs and into development. A leader leads. Had Vertex become, in the end, as Vertexian operationally as it was scientifically? Was it truly the twenty-first-century New Pharma prototype it claimed to be?

At the Morgan confab, the perennially dismal pharmaceutical world bristled with new excitement, buoyed by the surge in FDA approvals. "Rational exuberance," someone called it. Leiden used his twenty-minute slide show to refocus the company's strategy and prioritize key business goals for the year. Here was a clear statement of bittersweet success. Vertex, he said, was redirecting development toward specialty diseases—hepatitis C, CF, Huntington's disease, multiple sclerosis, and cancer—and it would seek partners and outside funding to advance its JAK3 inhibitor and flu drug. He announced that the FDA had granted Kalydeco and 809 its first two "breakthrough drug" status labels, as part of a larger effort to speed important new medicines to patients. "Innovation is our lifeblood," Leiden told his audience, "our special sauce."

Leiden told Bloomberg News in an interview that the company would consider collaborations enabling it to get VX-509 to patients with RA and other autoimmunities without having to fund large-scale trials and a big commercial team, a deal that "would maintain value long term." He also said Vertex wouldn't fund trials of the flu molecule VX-787 with its own money, setting back the program indefinitely and demoralizing those who thought it represented the best chance of becoming a lifesaving medicine from among those in Vertex's current pipeline. "There are some things we are not going to do," he said. "That's new for Vertex."

Elsewhere, in another suite, Cumbo sat down for his first meeting as a strategist and deal maker. Across the table was Christiana Stamoulis, Vertex's senior vice president for business development. Stamoulis had the usual pedigree: two degrees from MIT, a consulting stint, leadership positions at Goldman Sachs and Citigroup, a veteran of some of the biggest deals in industy history. Cumbo was reassured to discover that he was treated well, respected.

Vertex was something else to him now, something like what Merck had been to Boger: an education, a crucible, a competitor, a potential partner; maybe, in some future, a takeover threat, or in some alternative one a target. "Strange world this bd thing . . ." he texted later in the morning, "but I think I like it!!! It will be challenging again . . . It will be fun to watch and see how they (bd pll) discount me because I don't have an MBA from Harvard and come from 'bama . . . made a living off people underestimating me."

❖

In February 2013 Vertex announced that it would start a pair of late-stage studies—1,000 patients at 200 sites—to test its first combination therapy in CF. Pressed on by the FDA's determination to show that it could speed up approvals, the company decided to evaluate two doses of VX-809 in combination with Kalydeco for twenty-four weeks. With patients in urgent need of new treatments, Olson and the CF franchise steering group spurred on the timetable for FDA review. "If the Phase III is ultimately successful (late 2013/early 2014), upside for this stock could exceed anything in our coverage universe," ISI's Schoenebaum speculated. "But with that upside risk comes downside risk as well. We remain 'Buy' rated on Vertex, but admit it's quite binary and not for everyone."

A few weeks later, Keith Johnson went on intravenous antibiotics for the first time in three and a half years—since December 2009, a few months before he started on VX-770. He hoped to qualify for the combination trial but his FEV1 percentages were declining and, after his experience in the last trial with VX-661, he wasn't sure that enrolling in clinical studies was his most effective route. He had concluded that his best hope of getting access to Kalydeco was to wait until the combo was approved for people with his genotype. "I'd rather get what I know works," he said, "and leave the science to somebody else at this point."

On a Monday in mid-April, Patriots' Day, two bombs exploded at the finish line of the Boston Marathon, killing three people, injuring hundreds of others, and plunging Boston and Cambridge into a cycle of fear, apprehension, grief, and resilience, the reeling epicenter of the most intensive antiterrorist manhunt and media storm since 9/11. Vertex had received data from the midstage study combining VX-661 and Kalydeco—Johnson's trial. Patients taking a 100-milligram dose of 661 had a 9 percent relative improvement in lung function while those taking 150 milligrams improved 7.5 percent. The drug was well tolerated. The company, weighing internal and external events, announced the findings as soon as the markets closed on Thursday. During the conference call shortly after four thirty, Mueller disclosed that it had advanced a third corrector, VX-983, into development.

Investors drove VRTX shares up $26.88—a 51 percent gain—to $79.75 in a little over three hours in after-hours trading. Their enthusiasm sprang less from the prospects represented by the molecule itself than from the heady sense that Vertex's strategy in CF would not fall prey to the same mistakes it had made with hepatitis C, that it had positioned itself aggressively to protect a lucrative franchise against any and all comers. The take-home: Vertex was well on its way to delivering combinations of multiple drugs, experimenting with a range of cocktails to help even the sickest patients, and with no real competitors in sight. Its market capitalization vaulted to more than $16 billion, more than triple what it was eighteen months earlier, landing improbably near the bottom of the S&P 500.

Shortly after the last of the ET left for home, two immigrant brothers from Cambridge murdered an MIT campus policeman about a mile away down Vassar Street, then carjacked a Mercedes SUV and told the driver that they were the Marathon bombers. They led police on a high-speed chase into neighboring Watertown, where one of them was killed in a ferocious gun battle. The younger assailant climbed back in the car, floored it, ran over his brother, and got away. By morning, Boston, Cambridge, Watertown, and several surrounding cities were locked down. Amtrak service from Boston to New York was cut.

VRTX traded up throughout the day, cresting above $86, as sirens wailed throughout the paralyzed region and saturation media coverage enthralled the nation. A nineteen-year-old suspect with still-to-be-known affiliations, capabilities, and plans had forced a major American city to "shelter in place," as Governor Patrick said. With Vertex locked down, Smith and Partridge took investors' calls from home. After the younger brother was captured that night and in the weeks ahead as his double life as a college student and lone-wolf terrorist was revealed, VRTX drifted back into the high 70s, remaining there through the Q1 conference call at the end of the month, when despite a continuing drop-off in Incivek revenues and a decision to write down the non-nuc VX-222, the company beat analysts' expectations. Hope trumped concern. For now.

The future is repriced every day.

# AFTERWORD

Soon after I returned to Vertex, John Thomson gave me a copy of *Pharmaplasia*, a critical examination of the drug business written by Michael Wokasch, a onetime Merck sales rep and former general manager of a piece of the Aurora acquisition, now a consultant. Wokasch diagnoses the underlying cause of what's wrong with the drug giants: "rapid uncontrolled growth in a pharmaceutical company that exceeds its capacity to be managed effectively, resulting in a series of unintended consequences."

Pharmaplasia usually "presents," as doctors put it, as a symptom of grotesque size, and there's no doubt that much of what ails Big Pharma is managerial dysfunction and a loss of philosophical grounding brought on by mergers, buyouts, and the ever-escalating arms race to dominate worldwide markets with marginally improved products, primarily to satisfy Wall Street and investors. But it's the "rapid uncontrolled growth"—the breakneck, hypermetabolic expansion itself—not the resulting giantism that Wokasch has identified as the culprit, and I think he's right.

In business it's grow or die. You go through stages. Vertex's expansion as a commercial company has been a noteworthy success. Its current concentration on specialty diseases positions it to help lead the way as the industry pivots away from mass-market products for widespread chronic illnesses and it saves the company from having to run massive clinical trials and build up and maintain expensive field operations. Since it has no plans in the midterm to grow to more than a few thousand people, its leanness should keep it nimble. But the question now is whether it has retained enough creative DNA from its early days, particularly in research, to adapt as well during its second quarter-century as it did during its first.

With Boger and his torchbearers lighting the way, Vertex tried to build itself to last, but the future is impossible to predict. What a company becomes depends on its management. It's too early to know if Leiden, Mueller, and the current leadership will inspire first-in-kind, first-in-class breakthroughs to equal Incivek and Kalydeco, but they will surely be measured on whether they do.

Anticipating a long sales cycle for Kalydeco and the company's CF franchise, Wall Street has again made VRTX a vogue investment. The anxieties about when Incivek will cliff and the "2015 problem" are dim memories. But at what cost? The question arose again in May when Eric Olson announced he was leaving to join a small development-stage company as its new chief scientific officer. Inside Vertex it was well understood that Olson's passion and leadership had been invaluable in guiding the CF franchise and its celebrated collaboration with the foundation. How many positive deviants must a company lose before it's clear that the culture isn't working as intended and that management has a problem far deeper than a few disappointing quarterly reports?

Vertex has set itself a high standard. For insurance, the company let me in—twice now—to record it all in detail. If Vertex is condemned to repeat the mistakes of the "bigs" as it grows, at least no one can say that it didn't once know how to do things right. Its struggles to keep to and build on Boger's corporate vision and sense of possibility have been amply reported for everyone to see. There's proof-of-concept data.

My access to Vertex, as when I wrote *The Billion-Dollar Molecule*, depended on my allowing two people from the company—Ken Boger and Megan Pace—to review the book for proprietary disclosures. Before publication I gave Pace the manuscript. With her perspective as spokeswoman, I imagined she would have her defenses up, and she did—in a hopeful way, I think, since what she seemed to want to protect was the actual Vertex and its sense of itself as a place with an important obligation to live up to its potential, not business decisions or executive performance or the feelings of individuals about how they're portrayed. "Well, what'd you think?" I asked.

"I thought, 'We better not screw this up.' "

Boger has considered the impact he thinks the company can have as it matures and expands around the world. Pharmaplasia doesn't intimidate him, though it wouldn't be unlike him to expect coming generations of right-thinking, R&D-based scientist-executives, armed with the Vertex values, to invent a cure for it. Either way, his sights are set. He knows what the top of the mountain looks like. Midway through the Vertex Vision Process, during an off-site retreat in 2006, Bink Garrison asked all those on the executive team to develop a vivid description of where they saw the company headed. Boger went off and came back an hour later with an impromptu two-page, single-spaced memo dated 12 April 2039. It was a note to shareholders from the 2038 annual report of Vertex Health Inc., celebrating the accomplishments of the company's fiftieth year in business.

"Driven, as always, by our passion for innovation," it began, "we introduced three new therapies to the worldwide health market, in Alzheimer's disease, in the remaining drug resistant cancers, and against the so called 'Mars virus'. This marked our tenth straight year of three major product introductions." He went on:

> Most of us cannot remember when cancer was untreatable or, worse, when the treatments rivaled the disease for harm to the patient. Now dealing with cancer is no more trouble than a tune-up of your home fusion reactor; plug in and diagnose the imbalance and then apply the right combination of available treatments. How could we imagine the world without the wisdom and creativity of the emerging "second-lifers," "generation triple-X," that increasing demographic of ninety- to one-hundred-twenty-year-olds blessed with their full mental capacities, enhanced by the wisdom and perspective of a century of adult experience. We are only beginning to understand what transformations in art, in science, and in literature they will bring us.

The looming social and economic burdens of coping with the ever-extending health needs and other impacts of tens of millions of empowered, productive centenarians didn't faze him. Indeed, he had long believed that by becoming dominant in its markets Vertex would eventually begin to be able to bend the cost curve at the root of the crisis in medicine, at which point active, contributing old people could add more value to society than they consume. He concluded:

> The last year marked the fifth successful year of our new global pricing model, now the industry standard. Introduced in 2030 to a skeptical equity market at the time, we pioneered, as you recall, getting paid for our products by a share of the value of the increase in health we brought to each market. We were gratified, and our investors were rewarded, when the model exceeded even our own expectations, allowing more rapid adoption of our newest products on a worldwide basis. Indeed, as our latest earnings indicate, Vertex products are now the single largest driver of economic growth in the shrinking "third world," contributing almost a third of our annual profits. We are especially pleased to receive, as the first companywide recipient ever, the 2038 Nobel Prize in Economics, adding to the five Nobel Prizes in Medicine received by our scientists over the last decade and the Peace Prize awarded to the company in 2037.
>
> Our new economic model has been widely successful. Based on our "benefit sharing" pricing model, Vertex now controls 16 percent of the world economy (30 percent outside America and the EU), up from 14 percent in 2037. Cash and investments outside our core businesses increased by 15 percent to $12.5 trillion. "With great power comes great responsibility," as the old saying goes, and we are using our newfound economic power to drive the Vertex values of innovation and public benefit in other industries as well. We believe that is good for those companies as it has been good for Vertex and for the world. We are all stewards of a legacy from our predecessors, and it is our duty to leave the world—perhaps older and healthier than our parents—in a better state than we came into it.

Boger would be eighty-eight when the letter was written, not quite a "second-lifer." He signed it XXXXXXX, Chief Executive Officer. He wasn't thinking of himself. But his ease and quickness in writing it suggested that his thoughts were not new. "Until it's not a possibility, it's a possibility." In his mind, he was already there.

NORTHAMPTON, MASSACHUSETTS
OCTOBER 9, 2013

# APPENDIX 1

## MOLECULES

| NUMBER | TARGET | DISEASE | PARTNERS | NAME |
|---|---|---|---|---|
| **VX-330** | HIV protease | AIDS | | |
| **VX-478** | HIV protease | AIDS | Burroughs/Wellcome | amprenavir/Agenerase® |
| **VX-740** | ICE | RA/psoriasis | Roussel Uclaf | pralnacasan |
| **VX-745** | p38 MAP kinase | RA/psoriasis | Kissei | |
| **VX-175** | HIV protease | AIDS | Glaxo/Kissei | fosamprenavir/Lexiva® |
| **VX-497** | IMPDH | hepatitis C | | merimepodib |
| **VX-950** | HCV protease | hepatitis C | Mitsubishi/J&J | telaprevir/Incivek® |
| **VX-770** | CFTR (gating) | cystic fibrosis | Cystic Fibrosis Foundation (CFF) | ivacaftor/Kalydeco® |
| **VX-809** | CFTR (folding) | cystic fibrosis | CFF | lumacaftor |
| **VX-661** | CFTR (folding) | cystic fibrosis | CFF | |
| **VX-222** | HCV polymerase | hepatitis C | | |
| **VX-765** | caspase-1 | epilepsy | | |
| **VX-509** | JAK 3 | RA | | |
| **VX-787** | | influenza | | |
| **ALS-2200** | HCV polymerase | hepatitis C | Alios | |

# APPENDIX 2

## ABBREVIATIONS and ACRONYMS

**AASLD**—American Association for the Study of Liver Diseases

**AD COMM**—Food and Drug Administration advisory committee

**AZT**—azidothymidine: early AIDS drug

**BHAG**—Big Hairy Audacious Goal

**BIO**—Biotechnology Industry Organization

**BMS**—Bristol-Myers Squibb

**CEO**—chief executive officer

**CF**—cystic fibrosis

**CFF**—Cystic Fibrosis Foundation

**CFO**—chief financial officer

**CFTR**—cystic fibrosis transmembrane conductance regulator: mutated protein that causes CF

**CMC**—chemistry, manufacturing, and controls

**CRO**—contract research organization

**DCF**—discounted cash flow

**DDI**—drug-drug interaction

**delta-F508**—CFTR gene mutation, folding

**DNA**—deoxyribonucleic acid

**EASL**—European Association for the Study of the Liver

**EPO**—erythropoietin

**ET**—executive team

**EU**—European Union

**E&Y**—Ernst and Young

**FDA**—Food and Drug Administration

**FEV1**—forced expiratory volume in one second

**FTE**—full-time employee

**G551d**—CFTR gene mutation, gating

**GSK**—GlaxoSmithKline

**HBE**—human bronchial epithelial cell

**HBS**—Harvard Business School

**HCV**—hepatitis C virus

**HIV**—human immunodeficiency virus

**HMR**—Hoechst Marion Roussel

**H&Q**—Humbert and Quist health care conference (now Morgan Healthcare)

**ICE**—interleukin-1 beta converting enzyme

**IMPDH**—inosine-5'-monophosphate dehydrogenase

**IMS**—IMS Health; leading provider of medical market data

**IPO**—initial public offering

**JAK**—Janus kinase

**JB-II**—Joshua Boger Innovation Center II

**J&J**—Johnson & Johnson

**KOL**—key opinion leader

**M&A**—mergers and acquisitions

**NDA**—new drug application

**NIH**—National Institutes of Health

**p38 MAP kinase**—mitogen activated protein

**peg-riba**—pegylated interferon and ribavirin

**PhRMA**—Pharmaceutical Research and Manufacturers of America

**PI**—product insert, prescription information: FDA-approved label

**PK**—pharmacokinetics

**POC**—proof of concept

**PZA**—pyrazinamide

**RA**—rheumatoid arthritis

**R&D**—research and development

**Reg FD**—Regulation Fair Disclosure

**RGT**—response-guided therapy

**RNA**—ribonucleic acid

**ROV**—real options valuation

**SEC**—Securities and Exchange Commission
**SJS**—Stevens-Johnson syndrome
**SVR**—sustained viral response; cure
**VIP**—Vision into Practice
**VRTX**—Vertex NASDAQ symbol

# ACKNOWLEDGMENTS

Embedded reporting earned itself a black eye during the wars in Iraq and Afghanistan, but often it's the only route a journalist can take to find a story. Having embedded with Vertex once before, I've now finished my second deployment, so to speak, and what I've found again—happily—is an amped-up, open culture where people are deeply passionate and proud about the work they do and eager to have others understand it. Whatever the hundreds of people inside the company with whom I spoke did not, or could not, tell me pales, I think, next to their willingness, patience, and good humor in having me along through an intensely productive and stressful period in their lives. I am especially grateful to Josh Boger, Ken Boger, and Matt Emmens for thinking it might be valuable for Vertex, and readers, to have me around.

It was not a condition of employment at Vertex, past or present, that you had to talk with me, but perhaps because people were as curious about me as I was about them I felt greeted warmly wherever I went, the San Diego site in particular. For their time and assistance, many kindnesses, and tolerance for my bumbling questions as I tried to find my way I also want to thank Bambang Adiwijawa, John Alam, Richard Aldrich, Valerie Andrews, Mike Badia, Zach Barber, Virginia Carnahan, Paul Caron, Karolyn Cheng, Heather Clark, John Condon, Pat Connelly, Peter Connolly, Bo Cumbo, Paul Daruwala, Dave Deininger, Maria DeLucia, Diane Ferrucci, Matt Fitzgibbon, Ted Fox, Bink Garrison, Shelley George, Carol Gonsalves, Steve Goodstein, Cami Graham, Jim Griffith, Peter Grootenhuis, Sabine Hadida, Matt Harding, Beth Hoffman, Tom Hoock, Ty Howland, Trish Hurter, Marc Jacobs, Dawn Kalmar, Bob Kauffman, Lisa Kelly-Crosswell, Tara Kieffer, Liz Kula, Ann

Kwong, Jeff Leiden, Chris Lepre, Judy Lippke, David Livingston, Jon Moore, Peter Mueller, Mark Murcko, Dallan Murray, Mark Namchuk, Victoria Narausky, Paul Negulescu, Tim Neuberger, Eric Olson, Megan Pace, Michael Partridge, Debra Peattie, David Rodman, Amit Sachdev, Vicki Sato, Priya Singhal, Ian Smith, Russ Smith, Craig Sorensen, Cynthia Spencer, Christiana Stamoulis, Megan Steel, Pam Stephenson, Ernst ter Haar, John Thomson, Roger Tung, Fred Van Goor, Alissa Van Zee, Jack Weet, Chris Wright, and Nancy Wysenski.

Many people who have observed Vertex at close quarters helped inform my perspective. Bob Beall, Bob Brown, Doug Deiterich, Clint Gartin, Adam Koppel, John McHutchison, Geoff Porges, Charles Rice, Mark Robinson, and David Stein were insightful and generous. Erica Jefferson helped at the FDA. John Hallinan, Mark Jones, Arnold Thackray, and Jamie Cohen-Cole brought timely attention to my work among their communities and colleagues. Keith Johnson volunteered to share his experience, opening an unexpected vista on the conundrum of personalized medicine.

I owe a special debt to all those who agreed to help me through the early drafts of this project and who, as it sprawled and slogged and lumbered finally toward coherence, probably wished they hadn't: Hilda Werth, Kathy Goos, Alex Werth, Alan Sosne and Fred Eisenstein. Cathy Bouffides Walsh, Richard Levine, Jackie Austin, David Weintraub, Emily Filloy, Susan Eisenberg, and Susan Werth put me up during my travels. Jamie Moore helped with a fine transcription that unfortunately ended up on the cutting room floor. Steven Shapin, Sam Freedman, and Tony Giardina lent timely encouragement. Clear-eyed professional and trusted friend Chris Jerome again saved me from my worst tendencies as a writer.

At Simon & Schuster, Alice Mayhew, Jonathan Karp, and Jonathan Cox saw—and brought out—the best in the story and, I hope, in me. Elisa Rivlin, Phil Bashe, and Mara Lurie helped immeasurably with turning the manuscript into a book. I want to thank George Turianski, Kyoko Watanabe, and Jackie Seow for packaging it and pushing it out the door. I'm grateful to Julia Prosser, Stephen Bedford, and Kate Gales for marketing and publicity.

As an unaffiliated journalist-historian, I am grateful to the Smith College American Studies Department for providing me with an operating base, especially Rick Millington, Michael Thurston, and Dan Horowitz. As always, I am indebted to my unfailing agent, Amanda Urban, and her assistant, Margaret Southard. No words can do justice to the sustenance I get from my family—Kathy Goos, Emily Werth, and Alex Werth—who know that every book I write is like a long illness, and that eventually it ends.

# NOTES

## CHAPTER 1: April 28, 1993

I was in attendance at both scenes described here. (For more on the origins of Vertex's HIV program, see *The Billion-Dollar Molecule*.) Additional sources:

(pp. 11–16) Interviews with Josh Boger, Rich Aldrich, and Mark Murcko. Rupert Cornwell, "Clinton Lambasts Greedy Drug Firms," *Independent*, February 13, 1993; Elizabeth Rosenthal, "Research, Promotion and Profits: Spotlight Is on the Drug Industry," *New York Times*, February 21, 1993; Philip J. Hilts, "U.S. Study of Drug Makers Criticizes 'Excess Profits,' " *New York Times*, February 26, 1993; Tom Petruno, "Penny Pinching Squeezes Growth Stocks," *Los Angeles Times*, June 21, 1993; Thomas Stossel, "The Discovery of Statins," *Cell*, September 19, 2008.

❖

(pp. 16–20) Interviews with John Thomson, Josh Boger, Vicki Sato, Mark Murcko, and Roger Tung. Lawrence K. Altman, "Conference Ends with Little Hope for AIDS Cure," *New York Times*, June 15, 1993; Editorial, "The Unyielding AIDS Epidemic," *New York Times*, June 17, 1993.

## CHAPTER 2: August 22, 1993

Most material for this chapter was contemporaneously reported. Other sources:

(pp. 21–24) Interviews with Debra Peattie, Charles Rice, John Thomson, and Vicki Sato. Gina Kolata, "Mysterious Epidemic of Furtive Liver Disease," *New York Times*, January 19, 1993.

❖

(pp. 24–27) Interviews with Josh Boger, Vicki Sato, Richard Aldrich, Roger Tung, and David Deininger. David Gold, "Highlights from the First Conference on Human Retroviruses," *Gay Men's Health Crisis: Treatment Issues*, March 1994; ACT-UP Capsule History, 1989 (www.actupny.org/documents/cron-89.html); Huntley Collins and Shankar Vedantam, "8 Years and $700 Million Later, How a Better Drug Was Found," *Philadelphia Inquirer*, March 17, 1996.

❖

(pp. 27–30) Interviews with Rich Aldrich, Josh Boger, Vicki Sato, Mark Murcko, John Thomson, and Roger Tung.

❖

(pp. 30–32) Interviews with Roger Tung, Josh Boger, Vicki Sato, and Carl Dieffenbach. Collins and Vedantam, "8 Years and $700 Million Later . . ."; John James, "Searle Abandons Its Protease Inhibitor," *AIDS Treatment News*, November 4, 1994.

❖

(pp. 32–35) Interviews with Mark Murcko, Paul Caron, John Thomson, Ted Fox, and Matt Fittzgibbon.

❖

(pp. 35–36) Interviews with Rich Aldrich, Josh Boger, and Ken Boger. David Dunlap, "From AIDS Conference, Talk of Life, Not Death," *New York Times*, July 15, 1996; *Time*'s Man of the Year: Cristine Gorman, Alice Park, and Dick Thompson, "Dr. David Ho: The Disease Detective," December 30, 1996. (Remarkably, the story credits Ho for coming up with the idea of combination therapy, while giving one company, Abbott, a single mention in the thirty-fourth paragraph.)

❖

(pp. 36–38) Interviews with Mark Murcko, Paul Caron, Ted Fox, John Thomson, and Josh Boger. Lisa Benavides, "Hepatitis C Discovery Could Be Boon for Vertex," *Boston Business Journal*, October 18, 1996; Lawrence Fisher, "Schering-Plough and Lilly Sign Liver Drug Deals," *New York Times*, June 13, 1997.

## CHAPTER 3: April 11, 1997

(pp. 39–42) Interviews with Ann Kwong, Vicki Sato, John Thomson, and Roger Tung.

❖

(pp. 42–44) Interviews with Rich Aldrich, Josh Boger, and Vicki Sato.

❖

(pp. 44–49) Interviews with Josh Boger, Vicki Sato, and Ann Kwong. Walter Isaacson, *Steve Jobs*, Simon & Schuster, 2011. Veronica Hope Hailey and Julia Balogun, "Devising Context Sensitive Approaches to Change: The Example of GlaxoWellcome," *Pergamon*, 2002; Janet Kelly, "GlaxoWellcome Cultural Change," *Management Development Review*, 1996; GlaxoWellcome: Fighting Disease and Improving Health (http://folk.uio.no/ivai/ESST/GlaxoSmithKline_Case_Study.pdf); Wendy Orent, "Out of the Shadows (On the Long Road to Fighting Hepatitis C)," *Proto*, Summer 2007; Jon Cohen, "Chiron Stakes Out Its Territory," *Science*, July 2, 1999; "Chiron's Hepatitis C Patents," *Hepatitis C Harm Reduction Project*, June 22, 2004.

❖

(pp. 49–53) Interviews with Josh Boger, Rich Aldrich, and Michael Partridge. Gabi Horn, "Vertex Vortex," *POZ*, September 1998; Alana Kumbier, PopPolitics.com, "Despite Ad Images, HIV Still Not Carefree," *AlterNet*, posted June 12, 2001; Tom Abate, "Passing the 'BioBucks'—Small Investors Aren't In on the Joke/Inflated Deal Values Sometimes Only Way Firms Can Raise Funds," *SFGate.com*, June 14, 1999.

❖

(pp. 53–60) Interviews with Mark Murcko, Josh Boger, Ken Boger, Rich Aldrich, and Vicki Sato. Andrew Pollack, "Finding Gold in Scientific Pay Dirt," *New York Times*, June 28, 2000; Philip Ball, "Bursting the Genomic Bubble," *Nature*, published online March 31, 2010; Siddhartha Mukherjee, *The Emperor of All Maladies*, Scribner, 2010; Daniel Vasella, *Magic Cancer Bullet*, HarperCollins, 2003.

## CHAPTER 4: January 22, 2001

(pp. 61–64) Interviews with Josh Boger, Ann Kwong, Roger Tung, Dave Deininger, Vicki Sato, and John Thomson.

❖

(pp. 64–67) Interviews with Josh Boger, Vicki Sato, and Mark Murcko. Amy Tsao, "The Vertex Vortex: Drug Development at Hyperspeed?" *BusinessWeek*, March 15, 2001; Andrew Pollack, "Vertex Buys Biotechnology Rival for $592 Million," *New York Times*, May 1, 2001; Kevin Fogarty, "Speed Is the Vertex Creed," *BioIT World*, April 7, 2002.

❖

(pp. 67–70) Interviews with Josh Boger, Vicki Sato, Bob Beall, and Rich Aldrich. Vertex's decision to partner with the nonprofit Cystic Fibrosis Foundation is the subject of an excellent academic case study, which I relied on heavily in reporting this section: Robert F. Higgins, Sophie Lamontagne, and Brent Kazan, "Vertex Pharmaceuticals and the Cystic Fibrosis Foundation: Venture Philanthropy Funding for Biotech," *Harvard Business School*, October 2007; revised July 2010.

❖

(pp. 70–73) Interviews with John Alam, Josh Boger, and Vicki Sato.

❖

(pp. 73–76) Interviews with Josh Boger, Ken Boger, Vicki Sato, and Mark Murcko. United States District Court, District of Massachusetts, In Re VERTEX PHARMACEUTICALS, INC. SECURITIES LITIGATION, Master File No. O3 11852 PBS.

❖

(pp. 76–78) Interviews with Josh Boger and Vicki Sato. Andrew Pollack, "Announcement on a Hepatitis C Drug Is Expected Today," *New York Times*, January 7, 2002; Tom Abate, "H&Q Conference Has Matured Along with the Biotech Industry," *SFGate.com*, January 7, 2002.

❖

(pp. 78–81) Interviews with John Thomson, Paul Negulescu, Mark Murcko, Roger Tung, and Eric Olson.

❖

(pp. 81–83) Interviews with Vicki Sato, Ann Kwong, Josh Boger, Roger Tung, and John Thomson. Ryan McBride, "How Eli Lilly Let a Billion-Dollar Molecule Slip Away, and Make a Fortune for Vertex," Xconomy, August 4, 2010.

## CHAPTER 5: January 6, 2003

(pp. 84–88) Interviews with Josh Boger, Ian Smith, Ken Boger, and Vicki Sato.

❖

(pp. 88–90) Interviews with Ian Smith, John Thomson, Jon Moore, Vicki Sato, and Chris Lepre. Andrew Pollack, "Despite Billions for Discoveries, Pipeline of Drugs Is Far from Full," *New York Times*, April 19 2002; Andrew Pollack, "Awaiting the Genome Payoff," *New York Times*, June 14, 2010; Allison Connolly, "Bio-Layoffs Cool Once Booming Job Market," *Boston Business Journal*, June 30, 2003.

❖

(pp. 90–94) Interviews with Josh Boger, Vicki Sato, and Peter Mueller. Again, deeply reported academic case studies were invaluable here for piecing together critical developments during a formative period: Gary Pisano, Lee Fleming, and Eli Peter Strick, "Vertex Pharmaceuticals: R&D Portfolio Management (A)," *Harvard Business School*, June 20, 2006; Francesca Gino and Gary Pisano, "Vertex Pharmaceuticals: R&D Portfolio Management (B)&(C)," *HBS*, April 25, 2006.

❖

(pp. 94–96) Interviews with Ann Kwong and Josh Boger.

❖

(pp. 96–99) Interviews with Geoffrey Porges and Josh Boger. *HBS* case studies; *Time* cover story, "Medicine: What the Doctor Ordered," August 18, 1952; Jim Collins and Jerry Porras, *Built to Last*, HarperBusiness, 1994; "Vertex Slips 37% After Arthritis Test Halted," *Boston Business Journal*, November 11, 2003; Geoffrey Porges, "Vertex Pharmaceuticals: Still Floundering," *Bernstein Research Call*, October 17, 2003.

❖

(pp. 99–102) Interviews with John Alam, Josh Boger, and Geoff Porges. Charles Pierce, "Boston's Biotech Moment," *Boston Globe Magazine*, December 14, 2003; David Hamilton, "The FDA's Approval of Drugs Doesn't Ensure Biotech Riches," *Wall Street Journal*, October 29, 2003; Geoff Porges and Marshall Gordon, "VRTX: Investor Day Yields More Positives Than Negatives; Uncertainty Remains, Upgrade to Marketperform," *Bernstein Research Call*, December 4, 2003.

## CHAPTER 6: February 14, 2004

(pp. 105–9) Interviews with Ian Smith, Josh Boger, Ken Boger, John Alam, and Ann Kwong. David Margolis, "11th Annual Retrovirus Conference," *Conference Reports for NATAP* (National Aids Treatment Advocacy Project), February 8–11, 2004; Ann Kwong, Robert Kauffman, Patricia Hurter, and Peter Mueller, "Discovery and Development of Telaprevir," *Nature Biotechnology*, November 2011.

❖

(pp. 109–11) Interviews with Peter Mueller, Mark Murcko, Josh Boger, John Alam, and Vicki Sato.

❖

(pp. 111–14) Interviews with Ian Smith, Vicki Sato, Trish Hurter, and Pat Connelly.

❖

(pp. 114–17) Interviews with Josh Boger, Bink Garrison, and Vicki Sato. Gardiner Harris, "Drug Makers Seek to Mend Their Fractured Image," *New York Times*, July 8, 2004; Jim Collins and Jerry Porras, *Built to Last*, HarperBusiness, 1994.

❖

(pp. 117–21) Interviews with Vicki Sato, Roger Tung, Paul Negulescu, Eric Olson, Peter Grootenhuis, Fred Van Goor, and Sabine Hadida. Jerome Groopman, "Open Channels," *New Yorker*, May 4, 2009 (Groopman was a member of Vertex's original scientific advisory board); Penni Crabtree, "Poised to Be a Star: Cystic Fibrosis Project Has San Diego Unit of Vertex on Verge of Treatment," *San Diego Union Tribune*, October 21, 2005; Matthew Herper, "A Drug of Your Own," *Forbes*, July 20, 2011.

❖

(pp. 121–24) Interviews with Josh Boger, Bink Garrison, Matt Emmens, and Trish Hurter. Anna Wilde Matthews and Barbara Martinez, "E-Mails Suggest Merck Knew Vioxx's Dangers at Early Stage," *Wall Street Journal*, November 1, 2004; Alex Berenson, Gardiner Harris, Barry Meier, and Andrew Pollack, "Despite Warnings, Drug Giant Took Long Path to Vioxx Recall," *New York Times*, November 14, 2004.

❖

(pp. 125–28) Interviews with Tim Neuberger, Paul Negulescu, and Vicki Sato.

❖

(pp. 128–32) Interviews with Robert Kauffman, John Alam, Josh Boger, and Ian Smith. Scott Gottlieb, "Magic Bullet for Hepatitis C," *Forbes*, January 24, 2005; Geoff Porges and Neil Agran, "VRTX: VX-950—'Billion Dollar Molecule' in New SCB Market Model," *Bernstein Research Call*, June 8, 2005.

❖

(pp. 132–35) Interviews with Trish Hurter, Bink Garrison, and Josh Boger.

❖

(p. 135) Interview with Bink Garrison.

## CHAPTER 7: January 9, 2006

(pp. 136–40) Interviews with Josh Boger, John Alam, John McHutchison, and Ian Smith. Andrew Pollack, "Hoping a Small Sample May Signal a Cure," *New York Times*, February 7, 2006; Scott Kirsner, "Why Biotech CEOs Need to Think Like Steve Jobs," *Boston Globe*, August 26, 2007.

❖

(pp. 140–43) Interviews with Eric Olson, Bob Beall, and Ken Boger.

❖

(pp. 143–46) Interviews with Josh Boger, John Thomson, Mark Murcko, and Roger Tung. Manuel A. Tipgos and Thomas J. Keefe, "A Comprehensive Structure of Corporate Governance in Post-Enron Corporate America," *CPA Journal*, 2004; Bruce Morton, "Two NC Democrats Vie for a Shot at Helms," *All Politics/CNN TIME*, May 6, 1996.

❖

(pp. 146–50) Interviews with Josh Boger, Ann Kwong, Tara Kieffer, and Ian Smith. Andrew Pollack, "New Medicine for AIDS Is One Pill, Once a Day," *New York Times*, July 9, 2006; "Vertex: A Promising Hep-C Play," *BusinessWeek*, October 9, 2006; Peter Kang, "Vertex's J&J Deal Key to Future Success: Analyst," *Forbes.com*, June 30, 2006; Brian Lawler, "Vertex's Billion-Dollar Drug," *The Motley Fool*, April 18, 2007; Andrew Pollack, "2 Winning Drug Tests, One Expected and One a Surprise," *New York Times*, November 2, 2007.

❖

(pp. 150–51) Interview with Bob Kauffman.

❖

(pp. 151–57) Interview with Josh Boger. Boger Blog, with permission from the author; "Case Study: The Brain Power," *Boston* magazine ranking of the city's most powerful people, May 2008; "Hawking Takes Zero Gravity Flight," *BBC News*, April 27, 2007; Luke Timmerman, "Gov. Patrick Travels West to Plug Massachusetts' Life Sciences Initiative at BIO," Xconomy, June 16, 2008.

❖

(pp. 157–62) Interviews with Josh Boger, Bink Garrison, and Trish Hurter.

❖

(p. 162) Interviews with Josh Boger, Amit Sachdev, and Lisa Kelly-Crosswell.

❖

(pp. 162–65) Interviews with Josh Boger, Amit Sachdev, and John Alam. Jacalyn Duffin, *Lovers and Livers: Disease Concepts in History*, University of Toronto Press, 2005; Gardiner Harris, "Medical Marketing—Treatment by Incentive: As Doctor Writes Prescription, Drug Company Writes a Check," *New York Times*, June 27, 2004; Douglas T. Dieterich, MD, "IDEAL Study COMMENTARY: A Healthy Dose of Curiosity: Clinical Trial Results Require Careful Interpretation," *Liver Health Today*, January–March 2008); Andrew Pollack, "2 Winning Drug Tests, One Expected and One a Surprise," *New York Times*, November 2, 2007; Robert Langreth, "Viral Vertigo," *Forbes.com*, November 26, 2007.

❖

(pp. 165–67) Interviews with John Condon, John Thomson, John Alam, and Josh Boger.

❖

(pp. 167–74) Interview with Josh Boger. *Boger Blog©*, with permission.

**CHAPTER 8: February 11, 2008**

(pp. 175–78) Interview with Michael Partridge. Vertex's Q4 2007 earning call transcript, available online: http://seekingalpha.com/article/64143-vertex-pharmaceuticals -inc-q4-2007-earnings-call-transcript; Vikas Bajaj and Louise Story, "Mortgage Crisis Spreads Past Subprime Loans," *New York Times*, February 12, 2008.

❖

(pp. 178–82) Interviews with Eric Olson, Virginia Carnahan, Paul Negulescu, Peter Mueller, Sabine Hadida, and Ken Boger. Kate Kelly, "Inside the Fall of Bear Stearns,"

*Wall Street Journal,* May 9, 2009; "Vertex Achieves Breakthrough in Treating Basic CF Defect," *Commitment* (news publication of the Cystic Fibrosis Foundation), Spring 2008.

❖

(pp. 182–84) Interviews with Josh Boger, Peter Mueller, Ann Kwong, John Alam, Bob Kauffman, Jack Weet, and Ken Boger. "Pharmasset Presents Results of 4-Week Combination Study of R7128 for the Treatment of Chronic Hepatitis C," *Drugs.com Mednews,* April 25, 2008; "Hepatitis C Drug Development Projects That Have Been Terminated, Transferred to Other Companies or for Which Information Is No Longer Available," http://www.hcvdrugs.com, August 3, 2009.

❖

(pp. 184–87) Interviews with Matt Emmens, Josh Boger, and Ian Smith. Andrew Pollack, "Genentech Rejects Takeover Bid from Roche," *New York Times,* August 14, 2008; " 'Standstill' Agreements Limit Potential Buyout Deals," *IN VIVO,* September 11, 2008; Andrew Ross Sorkin, "Lehman Files for Bankruptcy; Merrill Is Sold," *New York Times,* September 15, 2008; Luke Timmerman, "Vertex Sells Royalty Rights to HIV Drugs, Bets on Hepatitis C," Xconomy, June 3, 2008; Elizabeth Bumiller and Jeff Zeleny, "First Debate Up in Air as McCain Steps Off the Trail," *New York Times,* September 24, 2008.

❖

(pp. 187–91) Interviews with Josh Boger and Matt Emmens. Matt Emmens and Beth Kephart, *Zenobia: The Curious Book of Business,* Berrett-Koehler Publishers, 2008.

## CHAPTER 9: January 12, 2009

(pp. 192–95) Interview with Josh Boger. Ron Winslow, "Investor Prospects Look Grim," *Wall Street Journal,* January 12, 2009; Luke Timmerman, "Vertex CEO Josh Boger Retiring in May; Matthew Emmens to Fill Role," Xconomy, February 5, 2009. The Boger-Huckman interview can be viewed online, http://video.cnbc.com/gallery/?video=996324419.

❖

(pp. 195–98) Interviews with Josh Boger, Judy Lippke, Matt Fitzgibbon, Mark Murcko, Bink Garrison, Geoff Porges, and Ian Smith. Andrew Ross Sorkin and Duff Wilson, "Pfizer Agrees to Pay $68 Billion for Rival Drug Maker Wyeth," *New York Times,* January 26, 2009; Catherine Arnst, "Pfizer-Wyeth Merger Isn't the Cure-All," *BusinessWeek,* January 24, 2009; Mark Murcko, "This is a test," email to his colleagues, February 6, 2009, by author's permission.

❖

(pp. 198–200) Interviews with Matt Emmens, Ian Smith, Michael Partridge, and Josh Boger. Todd Wallack, "Vertex Feeling Growing Pains: Firm Scrambles for Workers, Space as Drug Shows Promise," *Boston Globe,* November 2, 2007; Luke Timmerman, "Vertex Raises $320 Million in Secondary Stock Offering," Xconomy, February 19, 2009; Luke Timmerman, "Out with Hedge Funds, In with Blue Bloods," Xconomy, February 20, 2009; Natasha Singer, "Merck to Buy Schering-Plough for $41 Billion,"

*New York Times*, March 10, 2009; Toni Clarke, "Vertex Out-Foxes Big Pharma to Buy ViroChem," *Forbes.com*, March 4, 2009; Andrew Pollack, "Roche Agrees to Buy Genentech for $46.8 Billion," *New York Times*, March 13, 2009; "Biotech Could Follow Pharma's M&A Lead," *Investor's Business Daily*, March 20, 2009.

❖

(pp. 200–5) Interviews with Jack Weet, Matt Emmens, Peter Mueller, Josh Boger, Bink Garrison, and Amit Sachdev. The BIO podcast of Boger and Greenwood discussing the industry perspective can be heard online at http://www.bio.org/articles/ bio-leaders-joshua-boger-and-jim-greenwood-discuss-challenges-and-opportunities -biotech-ind; "Consequences of Hepatitis C Virus: Costs of a Baby-Boomer Epidemic of Liver Disease," Milliman, Inc., May 2009; Peter Baker, "Obama Was Pushed by Drug Industry, E-Mails Suggest," *New York Times*, June 8, 2012.

❖

(pp. 205–7) Interviews with Bink Garrison, Mark Murcko, Ann Kwong, and Matt Emmens.

❖

(pp. 207–9) Interviews with Jack Weet, Ann Kwong, and Tara Kieffer. Andrew Pollack, "FDA Warning Is Issued on Anemia Drug's Overuse," *New York Times*, March 10, 2007.

❖

(pp. 209–14) Interviews with Matt Emmens, Nancy Wysenski, Bink Garrison, and Adam Koppel. Robert Langreth, "Hard to Swallow," *Forbes.com*, May 13, 2002; SEC Form 8-K for Endo Pharmaceuticals Holdings Inc., August 31, 2009 (Changes in Directors or Principal Officer, Financial Statements); "2009 Exits/Financings Deal of the Year Nominee: Vertex's Milestone Sale," *IN VIVO*, December 16, 2009; Adam Feuerstein, "Vertex Raising Money—Again! Biobuzz," TheStreet, December 2, 2009.

❖

(pp. 214–17) Interviews with Matt Emmens, Christopher Wright, Mark Namchuk, and Josh Boger. The Emmens/Huckman interview can be viewed online at www.cnbc .com/id/34845387. David Kirkpatrick, "White House Affirms Deal on Drug Cost," *New York Times*, August 6, 2009; Aelok Mehta, "Seizures, Epilepsy Linked to Immune Reaction," *The Dana Foundation*, April 2009.

❖

(pp. 217–20) Interviews with Nancy Wysenski and Matt Emmens. Michael Cooper, "GOP Senate Victory Stuns Democrats," *New York Times*, January 21, 2010; David D. Kirkpatrick, "White House Affirms Deal on Drug Cost," *New York Times*, August 6, 2009; Sheryl Gay Stolberg and Robert Pear, "Obama Signs Health Care Overhaul Bill, with a Flourish," *New York Times*, March 23, 2010.

❖

(pp. 220–24) Interviews with Bob Kauffman and Jack Weet. Luke Timmerman, "Vertex Maps Out Combo Drug Game Plan for Treating Hepatitis C," Xconomy, March 8, 2010; Andrew Pollack, "Hepatitis C Drug Raises Cure Rate in Late Trial," *New York Times*, May 25, 2010.

❖

(pp. 224–29) Interviews with Bo Cumbo, Jack Weet, Bob Kauffman, and Matt Emmens.

❖

(pp. 229–33) Interviews with Bo Cumbo, Jack Weet, Peter Mueller, Bob Kauffman, and Matt Emmens. Robert Weisman, "Vertex Seeks Fast Approval for Drug," *Boston Globe*, November 24, 2010; "Vertex Seeks FDA Green Light for Hepatitis C Drug: Chomps at the Bit for Fast Review," Xconomy, November 23, 2010.

## CHAPTER 10: January 9, 2011

(pp. 237–39) Interviews with Matt Emmens and Jack Weet. Ed Silverman, "JPMorgan Event: Narrow Hallways and Velvet Ropes," *Pharmalot.com*, January 14, 2011; Mike Huckman, "Vertex Adds Color to the JPMorgan Healthcare Conference," *CNBC .com*, January 13, 2011; Adam Feuerstein, "Merck Beats Vertex to FDA Hep C Filing," TheStreet, January 6, 2011; Thomas Gryta, "Vertex CEO Unfazed by Matchup Against Merck," *Wall Street Journal*, January 10, 2011; Julie M. Donnelly,"Vertex Gears Up for Its Big Year," *Boston Business Journal*, December 31, 2010.

❖

(pp. 239–41) Interviews with John Condon and Jack Weet. Casey Ross, "City Draws Cambridge Drug Firm to Fan Pier," *Boston Globe*, January 25, 2011; Jerry Kronenberg, "Cambridge Prepares Counter-Bid to Keep Vertex Away from Hub," *Boston Herald*, February 14, 2011.

❖

(pp. 241–43) Interviews with Bo Cumbo and Josh Boger.

❖

(pp. 243–44) Interviews with Megan Pace, Peter Mueller, Jack Weet, Bob Kauffman, Ken Boger, and Josh Boger.

❖

(pp. 244–47) Christopher K. Hepp, "New Merck CEO Kenneth C. Frazier Has Philadelphia Roots," *Philly.com*, December 1, 2010; Linda A. Johnson, "Earnings Preview: Merck to Tout Pipeline in Report," *BloombergBusinessweek*, February 2, 2011; Tom Randall, "Merck's Risky Bet on Research," *BloombergBusinessweek*, April 23, 2011; Tom Randall, "Merck, Pfizer Research Strategies Diverge on Spending," *Bloomberg*, February 3, 2011; "Pfizer vs. Merck and the Future of R&D: Déjà Vu All Over Again, *INVIVO*, February 10, 2011; "Goldman Sachs Is Bullish on Vertex," *Bloomberg News*, February 9, 2011.

❖

(pp. 247–51) Interviews with Peter Mueller, Ian Smith, Ken Boger, Bob Kauffman, Matt Emmens, Geoff Porges, and Eric Olson. Andrew Pollack, "Trial Shows Cystic Fibrosis Drug Helped Ease Breathing," *New York Times*, February 23, 2011; Luke Timmerman, "Vertex Nails Pivotal Study for Cystic Fibrosis, Racing Toward Market with Second Drug," Xconomy, February 23, 2011; Matthew Herper, "Vertex May Make History with Cystic Fibrosis Drug," *Forbes*, February 23, 2011; Matthew Herper, "A Big and Dangerous Day for Personalized Medicine," *Forbes*, February 23, 2011.

❖

(pp. 252–55) Interview with Keith Johnson.

❖

(pp. 255–58) Interviews with Jack Weet, Bob Kauffman, and Matt Emmens.

❖

(pp. 258–61) Interview with Matt Emmens.

❖

(pp. 261–63) Interviews with Matt Emmens, Ann Kwong, Ian Smith, and Michael Partridge. Naomi Kresge, "Pharmasset to Challenge Vertex Hepatitis C Treatment, BMO Says," *Bloomberg*, March 8, 2011; Adam Feuerstein, "Pharmasset Hep C Data Wows Investors," TheStreet, March 8, 2011; Robert Weinstein, "Pharmasset: The Real Numbers Behind the Hype," *Seeking Alpha*, March 11, 2011; Katan Desai, "Who Will Win the Hepatitis C Market?" *Seeking Alpha*, March 14, 2011; Adam Feuerstein, "Hep C Drug Stocks in the Spotlight," TheStreet, March 28, 2011; NBC Evening News, March 30, 2011.

## CHAPTER 11: April 27–28, 2011

I was in attendance at all scenes described in chapters 11 to 14 unless otherwise specified. Additional sources:

(pp. 264–68) Interviews with Bob Kauffman, Josh Boger, Camilla Graham, Amit Sachdev, Nancy Wysenski, and Jack Weet. Emily P. Walker, "FDA Panel Endorses Boceprevir for Hepatitis C," medpagetoday.com, April 27, 2011; Heidi Ledford, "Regulatory Advisors Recommend New Hepatitis C Drug," nature.com, April 28, 2011.

❖

(pp. 268–73) Interviews with Nancy Wysenski, Jack Weet, Megan Pace, Bob Kauffman, and Matt Emmens. "Noteworthy Pharmacist, Patrick Clay, Pharm.D.," *TheBody .com*, HIV Leadership Awards 2005; on Clay's research funding from Merck, *American Journal of Pharmaceutical Education*, 65 (Winter 2001), p. 426.

❖

(pp. 273–77) Interviews with Peter Mueller, Jack Weet, and Bob Kauffman. Jason Brudereck, "New Drug Has City Woman Free of Hepatitis C," *Reading (Pa.) Eagle*, June 1, 2011; Luke Timmerman, "Vertex Wins FDA Panel's Recommendation for New Hepatitis C Drug," Xconomy, April 28, 2011; Robert Weissman, "Hepatitis C Drug from Vertex Sails Through Test," *Boston Globe*, April 29, 2011; Richard Knox, "New Drugs for Hepatitis C Called Game Changers," *NPR.org*, April 28, 2011; "Vertex Hepatitis Drug Still Holds Edge on Merck After FDA Panels," *Wall Street Journal*, April 29, 2011; Brian Orelli, "Coronations for New Drug Royalty," *The Motley Fool/Fool.com*, April 2011; "Goldman Sachs Raises Price Target on Vertex (VRTX), Sees 100% Chance of Approval Now," *StreetInsider.com*, April 29, 2011.

❖

(pp. 277–80) Interviews with Bo Cumbo and Ken Boger.

❖

(pp. 280–83) Interviews with Ian Smith and Matt Emmens.

❖

(pp. 283–85) Interviews with Bo Cumbo, Matt Emmens, and Nancy Wysenski. Luke Timmerman, "Merck, Genentech Team Up on Hepatitis C Drugs, Raising Ante in Vertex Rivalry," Xconomy, May 17, 2011; Brian Orelli, "If You Can't Beat 'Em, Use 'Em to Beat Your New Rival," *The Motley Fool/Fool.com*, May 18, 2011; Tracy Staton, "Can Merck/Roche Hep C Deal Put Victrelis on Top?" FiercePharma, May 18, 2011; "Merck and Co.: HCV Marketing Juggernaut," *UBS Investment Research*, May 17, 2011; Linda A. Johnson, "Merck, Roche Expand Hepatitis C Drug Promo Deal," *Bloomberg-Businessweek*, July 20, 2011.

❖

(pp. 285–88) Interviews with Nancy Wysenski, John Condon, Paul Daruwala, Josh Boger, and Adam Koppel. Matthew Herper, "Vertex's Biggest Advantage," *Forbes*, May 24, 2011; "Vertex's Sales Force Gears Up for 'David vs. Goliath' Marketing Push," *Wall Street Journal*, May 24, 2011; Bill Berkrot and Lewis Krauskopf, "Vertex CEO Unfazed by Competition, Future Rivals," Reuters, May 25, 2011.

❖

(pp. 288–91) Interview with John Thomson.

❖

(pp. 291–94) Interviews with Matt Emmens and Josh Boger.

**CHAPTER 12: June 6, 2011**

(pp. 295–99) Interviews with Ken Boger, Ian Smith, Peter Mueller, and Geoff Porges. Thomas Gryta, "Vertex Reports Positive Test of Cystic Fibrosis Combo," *Wall Street Journal*, June 9, 2011; "Vertex Cystic Fibrosis Combo Shows Promise," Reuters, June 9, 2011; Adam Feuerstein, "Vertex Cystic Fibrosis Drug Combo Hits Bump," TheStreet June 9, 2011; Marley Seaman, "Vertex Falls on Cystic Fibrosis Study Data," *Forbes*, June 9, 2011; Geoff Meacham, "Vertex Pharmaceuticals: Our Thoughts Ahead of VX-809/VX-770 Combo Data in Cystic Fibrosis," *JP Morgan North America Equity Research*, May 2, 2011.

❖

(pp. 299–303) Interviews with Bo Cumbo, Matt Emmens, Amit Sachdev, and Michael Partridge. Julie M. Donnelly, "Vertex CEO Emmens Keeps Emphasis on Science," *Boston Business Journal*, June 9, 2011; "Vertex Bolsters HCV Position with Potential $1.5B+ Alios Deal," *BioWorld*, June 14, 2011; Brian Orelli, "Going for Seconds in the Hepatitis C Space," *The Motley Fool/Fool.com*, June 14, 2011; Gardiner Harris, "Federal Research Center Will Help Develop Medicines," *New York Times*, January 22, 2011; Geoffrey C. Porges, Amrita Rahmani, and Aleksander Rabodzey, "VRTX: More on Cracking the Code in CF: Incivek Launch Early Signals Positive," *BernsteinResearch*, June 15, 2011.

❖

(pp. 303–7) Interviews with Keith Johnson and Ken Boger.

❖

(pp. 307–12) Interviews with Michael Partridge, Matt Emmens, Ken Boger, and Josh Boger. Peter Loftus, "Vertex Hepatitis Drug Takes Early Lead over Rival from

Merck," Dow Jones Newswires, June 29, 2011; "Vertex Reports Strong Initial Incivek Sales, Sees 2012 Profit," *Wall Street Journal*, July 28, 2011; Adam Feuerstein, "Vertex Earnings: Incivek's Boffo Launch," TheStreet, July 28, 2011; Brian Orelli, "Vroom! There Goes Vertex," *The Motley Fool/Fool.com*, July 28, 2011; Bill Berkrot, "New Vertex Hepatitis Drug Shines Out of Gate," Reuters, July 28, 1011; Geoffrey Porges, Amrita Rahmani, Aleksander Rabodzey, "Vertex: Q2 Strong Early Incivek Result Suggests Consensus Could Be Crushed; CF Gathering Steam, New T/P $82," *Bernstein-Research*, July 29, 2011; Adam Feuerstein, "Dendreon: Parsing Provenge's Problems," TheStreet, August 4, 2011; Adam Feuerstein, "Biotech Stock Mailbag: Dendreon's Aftermath," TheStreet, August 5, 2011; Eric Rosenbaum, "Merck Hep C Drug Draws More Attention Than Job Cuts," TheStreet, July 29, 2011.

❖

(pp. 312–15) Interviews with Michael Partridge, Ian Smith, and Matt Emmens. For a survey of the federal debt ceiling "crisis," see "Times Topics," *New York Times*, http://topics.nytimes.com/topics/reference/timestopics/subjects/n/national_debt_us/index.html; Damian Paletta and Matt Phillips, "S&P Strips US of Top Credit Rating," *Wall Street Journal*, August 6, 2011; Partridge's note to employees, courtesy of the author; Luke Timmerman, "Dendreon Wounds Are Self-Inflicted, Not the Start of a Biotech Industry Virus," Xconomy, August 8, 2011; Val Brickates Kennedy, "3 Biotech Stocks Battle the 'Dendreon Effect,' " *MarketWatch*, August 18, 2011; Matthew Herper, "Biotech, Where Winners Lose," *Forbes*, December 21, 2010; Steve Worland, "Dramatic Changes in Hepatitis C Treatment Expected to Continue," Xconomy, September 6, 2011.

❖

(pp. 315–19) Interviews with Matt Emmens and Geoff Porges. Geoffrey Porges, Amrita Rahmani, and Aleksander Rabodzey, "Vertex—SCB HCV Focus Groups Point to Solid Launch, Strong Preference for Incivek," *Bernstein Research*, June 23, 2011; Matthew Herper, "Could Vertex Sell $1 Billion of Its Hepatitis C Drug This Year?" *Forbes*, August 8, 2011; Christine Levoti, "Vertex, Merck Face Little Payer Pushback with Newly Marketed HCV Drugs," *FT.com*, September 2, 2011; Adam Feuerstein, "Vertex Arthritis Pill Shines in Mid-Stage Study," TheStreet, September 6, 2011; Ed Silverman, "Vertex CEO Chides Analyst in Front of Investors," *Pharmalot.com*, September 14, 2011; Luke Timmerman, "Stirring the Pot Once in a While Doesn't Hurt, and It Could Help Biotech Break Its Malaise," Xconomy, September 19, 2011; Alex Philippidis, "Pfizer Edges Toward Lipitor Patent Cliff as Exclusivity Extensions Near End," *GEN*, October 18. 2011.

## CHAPTER 13: September 23, 2011

(pp. 320–24) Interviews with Matt Emmens, Josh Boger, and Mark Murcko. "Innovation and Research: The Human Factor," *Science Careers, Science*, September 16, 2011; Bill Berkrot, "Analysis–Vertex Takes Early Rounds of Hep C Bout with Merck," Reuters, September 29, 2011.

❖

(pp. 324–27) Interviews with Ian Smith, Michael Partridge, Josh Boger, and John Thomson.

❖

(pp. 327–33) Interviews with Michael Partridge, Ian Smith, Matt Emmens, and Geoff Porges. Adam Feuerstein, "Vertex's Hep C Drug Needs a Growth Injection," TheStreet, October 11, 2011; "Pharmasset Expands Hepatitis C Trial; Shares Rise," Reuters, October 10, 2011; "Vertex (VRTX) Shares Sink in Late-Day Trade, Volume Picks Up," StreetInsider.com, October 11, 2011; Alex Nussbaum, "Vertex Shares Rise on Optimism for Higher Incivek Sales," Bloomberg, October 13, 2011; Toni Clarke, "IMS Revises Incivek Drug Data; Vertex Shares Jump," Reuters, October 13, 2011; Julie M. Donnelly," Vertex Shares Rise on Hep C Drug Sales Tracking SNAFU," Boston Business Journal, October 13, 2011; Adam Feuerstein, "Hep C Drug Updates: Vertex and Anadys," TheStreet, October 13, 2011; Emmens's cover letter to Vertex directors, courtesy of Emmens.

❖

(pp. 333–38) Obviously, I wasn't at the meeting between Mark Murcko and Peter Mueller, but I was party to numerous conversations about it over the following days and weeks, and I was present throughout the remainder of this section. Other sources include interviews with Mark Murcko, John Thomson, Jon Moore, Paul Negulescu, Eric Olson, and Chris Wright; Ransdell Pierson and Bill Berkrot, "Abbott Says Hepatitis C Combo May Be a Blockbuster," Reuters, October 21, 2011; Peter Mueller's R&D reorganization email, with permission from Mueller.

❖

(pp. 338–42) Interviews with Matt Emmens, Ian Smith, Michael Partridge, and Nancy Wysenski. Andrew Pollack, "Vertex Bests Merck in New Hepatitis C Drug Sales," New York Times, October 28, 2011; Adam Feuerstein, "Vertex Earns First Profit, Backed by Blockbuster Pace of Hep C Drug," TheStreet, October 28, 2011; Luke Timmerman, "Vertex Flips into the Black for First Time, as Hepatitis C Drug Beats Expectations Again," Xconomy, October 27, 2011.

## CHAPTER 14: November 2, 2011

(pp. 343–46) Interviews with Adam Koppel, Michael Partridge, Geoff Porges, Ian Smith, and Bob Beall. Adam Feuerstein, "11 Biotech Stocks Hedge Funds Love and Hate," TheStreet, August 17, 2011; Marshall Hargrave, "Bain Capital's Bet Against Romney," insidermonkey.com, October 5, 2012.

❖

(pp. 346–49) Interviews with Karolyn Cheng and Bo Cumbo. Liz Highleyman, "AASLD: PSI-7977 Plus Ribavirin Can Cure Hepatitis C in 12 Weeks Without Interferon," www.hivandhepatitis.com, November 8, 2011.

❖

(pp. 349–51) Interviews with Ian Smith, Michael Partridge, Nancy Wysenski, Amit Sachdev, and Bob Kauffman. Adam Feuerstein, "Pharmasset Takes Lead in Race to Develop Hep C Therapy by Pill," TheStreet, November 1, 2011; Luke Timmerman,

"Vertex Stock Drops 17 Percent over Two Days, as Potent Hep C Rivals Emerge," Xconomy, November 8, 2011; Brett Chase, "Pharmasset Winning Hepatitis C Drug Race," *Minyanville*, November 7, 2011; Marley Seaman, "Vertex Continues to Slump on Threats to Incivek," *BloombergBusinessweek*, November 8, 2011.

❖

(pp. 351–56) Interviews with Adam Koppel, Geoff Porges, and Matt Emmens.

❖

(pp. 356–59) Interviews with Michael Partridge, Bo Cumbo, Ann Kwong, Nancy Wysenski, Ian Smith, Matt Emmens, Josh Boger, and Peter Mueller. Andrew Pollack, "Gilead to Buy Pharmasset for $11 Billion," *New York Times*, November 21, 2011; Kimberly Ha, Claudia Montato, Yana Morris, and Ashley Armstrong, "Gilead's 'Big Bet' on Pharmasset Hinges on Future Results," *Financial Times*, November 22, 2011; Bill Berkrot, "Gilead Could Have Had Pharmasset Cheap: Founder," Reuters, November 22, 2011; Bert Wilkison, "Vertex Trading Near 52-Week Lows After Gilead Acquired Pharmasset," *Seeking Alpha*, November 22, 2011; Todd Campbell, "Gilead's Pharmasset Acquisition Makes Vertex Look Cheap," *Seeking Alpha*, November 22, 2011; Luke Timmerman, "The Hepatitis C Market: Biotech's Version of the Daytona 500," Xconomy, December 12, 2011; Robert Weisman, "Jeffrey Leiden Will Head Vertex, Which Gets Priority Review for Cystic Fibrosis Drug Candidate," *Boston Globe*, December 15, 2011; Ryan McBride, "Interview: Vertex CEO Concerned About Investors' 'Hyper-Focus' on Hep C," *FierceBiotech*, December 21, 2011; Adam Feuerstein, "The Best Biotech CEO of 2011 Is . . . ," TheStreet, December 14, 2011.

## CHAPTER 15: January 10, 2012

(pp. 360–63) Interviews with Michael Partridge and Eric Olson. Robert Weisman, "In Hepatitis C Market, Vertex Gets a Big New Rival," *Boston Globe*, January 10, 2012; Luke Timmerman, "Vertex Vows to Fight On with Alios Drugs in High-Stakes Hepatitis C Race," Xconomy, January 24, 2012; Drew Armstrong, "Vertex Falls as Analyst Cuts Sales Estimates on Hepatitis C Pill," *Bloomberg*, January 30, 2012; Adam Feuerstein, "Vertex Hep C Sales Growth Nears End," TheStreet, January 31, 2012; Anna Yukhananov and Bill Berkrot, "FDA Approves Vertex Cystic Fibrosis Drug," Reuters, January 31, 2012; Robert Weisman, "Vertex Gets Early OK for New Drug," *Boston Globe*, February 1, 2012; Andrew Pollack, "FDA Approves New Cystic Fibrosis Drug," *New York Times*, February 1, 2012; Tracy Staton, "Vertex Backs Up Pricey New CF Drug with Co-Pay Help," FiercePharma, February 1, 2012; Tracy Staton, "How Do the 12 Priciest Drugs in the US Stack Up," FiercePharma, February 7, 2012; Luke Timmerman, "Vertex's Big Day Felt Like Moon Landing, Seattle Researcher Says," Xconomy, February 1, 2012; "In Trial, Hep C Patients Saw Viral Relapse: Gilead," Reuters, February 17, 2012; Luke Timmerman, "Vertex Stays in HepC Game, as All-Oral Combo Passes Small Study," Xconomy, February 23, 2012.

❖

(pp. 363–64) Interview with Keith Johnson.

❖

(pp. 364–65) Interview with Michael Partridge. "A Cystic Fibrosis Treatment Is Called 'Game-Changing,' " Reuters, May 7, 2012; Robert Weisman, "New Data on Cystic Fibrosis Drug Lifts Vertex Stock," *Boston Globe*, May 8, 2012; Matthew Herper, "A One-Two Punch Against Cystic Fibrosis, and Maybe Someday Other Diseases Too," *Forbes*, May 7, 2012; "Vertex Pharma Continues to Rise on Upgrade," Associated Press, May 14, 2012.

❖

(pp. 365–67) Interviews and correspondence with Josh Boger. Steven Syre, "Tiny Start-Up Lands a Former Vertex CEO," *Boston Globe*, May 23, 2012.

❖

(pp. 367–71) Interview with Michael Partridge. Meg Tirrell, "Vertex Revises CF Combo Data Showing Less Benefit," Bloomberg, May 29, 2012; Val Brickates Kennedy, "Analysts Still Upbeat on Vertex," *Marketwatch.com*, May 29, 2012; Teresa Rivas, "What Next for Vertex?" *Barrons*, May 29, 2012; Robert Weisman, "FDA Says Vertex Promotional Material Overstates Benefits of Hepatitis C Drug," *Boston Globe*, May 31, 2012; Matthew Herper, "Clearing Up Vertex's Data Bungle," *Forbes*, June 1, 2012; Beth Healey, "Two Vertex Executives Are Stepping Down," *Boston Globe*, June 8, 2012; Ed Silverman, " The Curious Timing of Those Vertex Stock Sales," *Forbes*, June 11, 2012; Casey Ross, "Vertex to Fund Partnership with Boston Schools," *Boston Globe*, June 18, 2012; Luke Timmerman, "After Big Oops, Vertex Plows Ahead with Cystic Fibrosis Drug Combo," Xconomy, June 28, 2012.

❖

(pp. 371–72) Interview with Keith Johnson. Katie Thomas, and Michael S. Schmidt, "Glaxo Agrees to Pay $3 Billion in Fraud Settlement," *New York Times*, July 2, 2012.

❖

(pp. 372–76) Interviews with Josh Boger and Bo Cumbo. John Carroll, "Vertex Surges as Rival Hep C Contender Plays Catch-Up in Clinic," *FierceBiotech*, July 31, 2012; Adam Feuerstein, "Bristol's Hep C Drug Blow Up May Benefit Gilead, Idenix, Vertex Pharma," TheStreet, August 2, 2012; "Idenix Shares Plunge on Hepatitis C Treatment Fears," Bloomberg, August 16, 2012; "CDC Recommends One-Time Test for Hepatitis C for All Baby Boomers to Check for Infection," Associated Press, August 16, 2012; Robert Weisman, "Hepatitis C Testing May Lift Vertex's Market," *Boston Globe*, August 18, 2012; Adam Feuerstein, "Vertex Advances One of Two Hep C Drugs," TheStreet, September 25, 2012; Meg Tirrell, "Vertex Joins Glaxo, J&J in Testing Hepatitis C Combos," *BloombergBusinessweek*, November 1, 2012; Ryan McBride, "Hep C Pill Race Report 2012," *FierceBiotech*, November 14, 2012.

❖

(pp. 376–79) Interview with Bo Cumbo. Susan Fernando, "Vertex' Kalydeco Faces UK Price Pushback Though Solid Cystic Fibrosis Data Warrants Funding Settlement," *Financial Times*, November 29, 2012; Martin Barrow, "Cystic Fibrosis Drug Kalydeco Gets NHS Funding Go-Ahead," *London Times*, December 20, 2012; Ben Hirschler, "Analysis: Entering the Age of the $1 Million Medicine," *Chicago Tribune*, January 3, 3013; Matthew Herper, "Inside the Pricing of a $300,000-a-Year Drug," *Forbes*, January 3, 2013; Ben Hirschler, "Cashing in on Rare Diseases," *Times Colonist*, January 6, 2012.

❖

(pp. 379–81) Interviews with Bo Cumbo, Mark Murcko, and Peter Kolchinsky. Robert Weisman, "Surge in Federal Approvals Buoys Drug Makers," *Boston Globe*, January 8, 2012; Meg Tirrell, "Vertex Refocuses Drug Development to Specialty Diseases," Bloomberg, January 9, 2013; Julie M. Donnelly, "Vertex Hepatitis C Drug Revenues Plummet," *Boston Business Journal*, January 29, 2012; Luke Timmerman, "If You've Got a Real Breakthrough, the FDA Wants to Talk," Xconomy, January 14, 2012.

❖

(pp. 381–83) Interview with Keith Johnson. John Carroll, "Vertex Plots a Race Through Phase III for 'Breakthrough' Combo CF Therapy," *FierceBiotech*, February 26, 2013.

## AFTERWORD

(pp. 385–89) Interviews with Josh Boger, Ken Boger, and Bink Garrison. Boger's 2039 Annual Report, courtesy of the author.

# INDEX

# ABOUT THE AUTHOR

Barry Werth is an award-winning journalist and the acclaimed author of six books. His landmark first book, *The Billion-Dollar Molecule,* recounts the founding and early struggles of Vertex. Werth's articles have appeared in *The New Yorker, The New York Times Magazine,* and *GQ,* among others. He has taught journalism and nonfiction writing at Smith, Mount Holyoke, and Boston University.